THE BUTTERFLIES AND LARGER MOTHS

LINCOLNSHIRE
NATURAL HISTORY BROCHURE
No. 10

THE BUTTERFLIES AND LARGER MOTHS

Of Lincolnshire and South Humberside

J. DUDDINGTON
President Lincolnshire Naturalists' Union 1977-8

and

R. JOHNSON M.Ed., F.R.E.S.

Published by
LINCOLNSHIRE NATURALISTS' UNION

LINCOLN
1983

c Lincolnshire Naturalists' Union

ISBN : 0 9500353 8 6

Editor
JOHN EAST, F.I.M.L.S.

The Lincolnshire Naturalists' Union gratefully
acknowledges financial help given by the
Willoughby Memorial Trust towards the printing
of this publication

Cover photograph
Peacock (Nymphalis Io)
by Rex Johnson

Printed in Great Britain by
Lincolnshire Standard Group

CONTENTS

Chapters   *Page*
1. LINCOLNSHIRE AS A HABITAT   13
   Introduction: Mr. R. Johnson, M.Ed., F.R.E.S.
   A. The Geology and Scenery of Lincolnshire – Mr. D. N. Robinson, M.Sc.   22
   B. The Flora of Lincolnshire and South Humberside – Mrs. I. Weston, B.Sc.   26
   C. The Pattern of Weather in Lincolnshire – Mr. B. Fox-Holmes   36
   D. Wildlife and Conservation in Lincolnshire – Mrs. A. Goodall.   51
   E. The Ecology of Certain Lincolnshire Butterflies and Moths – the Authors   58
2. SYSTEMATIC CHECK LIST OF ALL RECORDED SPECIES   69
3. RECORDS OF LINCOLNSHIRE SPECIES. .   89
4. CONTRIBUTORS TO THE AUTHORS AND TO THE LINCOLNSHIRE NATURALISTS' UNION RECORDS.   251
5. LIST OF L.N.U. LEPIDOPTERA RECORDING SECRETARIES SINCE 1893.   255
6. REFERENCES CONSULTED.   257
7. SILHOUETTES TO ASSIST IN THE IDENTIFICATION OF A NUMBER OF THE LEPIDOPTERA FAMILIES – Drawn by J. G. A. Johnson.   261
8. SKETCH MAPS OF THE COUNTY.   267
   Fig.1. Lincolnshire Topography   268
   Fig.2. Important Habitats Frequently mentioned in Lepidoptera Records   269
   Drawn by Mr. E. J. Redshaw.
   Fig.3. Rocks/Soils and the distribution of Lincolnshire Nature Reserves.   270
9. INDEX OF SCIENTIFIC NAMES OF RECORDED SPECIES.   273
10. INDEX OF ENGLISH NAMES OF RECORDED SPECIES.   283
11. A CODE FOR INSECT COLLECTING. . . .   297

PREFACE:

In recent years a considerable number of books concerned with butterflies and moths have flooded on to the market. There has been a demand for them – as there has also been a demand for titles concerned with birds and flowers by a reading public wanting to become increasingly aware of the natural order and variety of the wildlife in their environment. A wide choice of lepidoptera books is available. A number have been written by noted entomologists. Some have excellent descriptive content, some are superbly illustrated, and a few are very, very expensive.

In one respect, however, these books fail. Almost without exception they are too universal. To be commercially viable they have to provide coverage of "The British Isles", of "Palaearctic Europe", or even "The World". They can do little to answer the questions a Lincolnshire naturalist may have about the distribution of wildlife within a few miles of his own doorstep. It is to provide this local information, that we, local practising naturalists, have written this book.

Perhaps it is relevant to state here, that when Lincolnshire is referred to in this book, we mean the County as it was prior to Government re-organisation and the partitioning of it to create "South Humberside".

Where does one normally turn to discover in detail what exists in a particular area? The answer is that it is necessary to consult the appropriate County Faunal Lists. In one recent publication, a major new series entitled "The Moths and Butterflies of Great Britain and Ireland", Heath, J. (1976), Vol.1., any budding lepidoperist wanting to discover where he could look up the most up to date County list for Lincolnshire would read (p.137):
LINCOLNSHIRE, Vice-counties 53 and 54,
Mason, G. W., 1905. A List of Lincolnshire Butterflies
Trans.Lincs.Nat.Un.l; 76-85
1905-12, The Lepidoptera of Lincolnshire.
Ibid.2: 75-103, 176-219, 3:59

Our most recent published records are, then, some seventy or more years old, and substantially in need of revision. Additionally, these old Lincolnshire Naturalists' Union Transactions are not easily accessible.

Occasionally it is possible to pick up local information from a "neighbourhood" resident experienced naturalist. In this respect I was most fortunate since I came into the study of butterflies and moths in the early 1960s as a result of a chance meeting with a certain Mr. Joe Duddington. This meeting changed the pattern of my life so far as my leisure activities are concerned. Through Mr. Duddington I developed a working knowledge of butterflies and moths in this part of the county, and became interested in the range and ecology of these insects across Europe. I became a Fellow of the Royal Entomological Society, and joined a number of natural history societies – enjoying sharing the interests of other naturalists in, for example, Lincolnshire Naturalists' Union field meetings. In Mr. Duddington I could not have had a more experienced field naturalist as a teacher and friend, and the fact that we have spent so many hours together, drawing on his expertise in the writing of this book, is a tribute to him, and to the regard I have for him.

Joe Duddington, co-author, past President of the Lincolnshire Naturalists' Union, past Chairman of the Natural History section of the Scunthorpe Museum Society, etc., is well known in Lincolnshire's natural history circles for his dedication to the study of Lincolnshire's Lepidoptera, and for the depth of his knowledge. He is currently the L.N.U. lepidoptera records secretary, and holds all records submitted to the Union since the 1800s. These volumes of records, written by successive lepidoptera secretaries are invaluable to anyone with an interest in this field.

Out of our realisation that an up to date county record was long overdue, we set about writing this book in 1978. Had we known what would be involved in presenting our findings in a format suitable for publication – how many hundreds of man hours would be expended in reading, writing, re-writing, checking, typing, etc., we may have delayed the project indefinitely.

Neither of us is a professional entomologist. Joe is a retired dental technician, and all my qualifications are in education. Joe's contribution has been his expert knowledge gained through a lifelong interest in entomology, mine has been with a pen – setting up an indexed card system of all recorded species, writing longhand, rewriting and co-ordinating what has gone into the book.

Joe's personal records are concentrated in the main on the Northern half of our County. A tribute must be paid immediately to another expert lepidopterist whose own indexed record system covers a large territory more to the South. I refer to Dr. R. E. M. Pilcher. Joe and Dr. Pilcher are the two major contributors to the records. Both retired from work (Dr. Pilcher a surgeon), but both working hard at their hobby, these two enthusiasts have between them well over a hundred years of studying butterflies and moths. It is their knowledge, their records, and the records of other dedicated people which has made this work a practical proposition at this moment. It has been my concern that I should manage to get the information in the minds and the records of our more "senior experts" logged while it was possible, for the benefit of a wider public.

Countless hours of observation by local naturalists have gone into notebooks to make this volume possible. We owe a great debt of gratitude to all our recorders, and to the L.N.U. lepidoptera secretaries who have collected this information over the years. Here, special mention must be made of Mr. G. A. T. Jeffs. Some years ago he undertook a revision of the recording system, and made it the informed reference work it is today. His work has been immensely useful to us.

Without the offer of publication by the Lincolnshire Naturalists' Union, and the support of Union officers, nothing could have been achieved. It is for the Union that we have willingly given our time and endeavoured to complete this book.

REX JOHNSON, November 1981.

ACKNOWLEDGEMENTS

We are particularly indebted to the following people for their assistance, contributions or encouragement. The listing is not in any significant order.

Mr. J. East, L.N.U. Hon. Editor.
Miss N. Goom, L.N.U. Hon. General Secretary.
Mrs. A. Goodall, one-time L.N.U. Publications Officer, and guest author, Wildlife and Conservation.
Mr. D. Robinson, guest author, Lincolnshire Geology.
Mrs. I. Weston, guest author, Lincolnshire Flora.
Mr. B. Fox-Holmes, guest author, Lincolnshire Climate.
Mr. E. J. Redshaw, contributor of maps/diagrams.
Mr. P. Walter, for checking indices.

Other members of the Johnson family – James for drawing the Silhouettes, Andrew for his valued companionship in the field, and Wendy, for her considerable patience, and involvement in our hobby.

All contributors to the records.

Lastly, but by no means least, we give special thanks to Mrs. Patricia Hesslewood. She has typed for us, and then re-typed after changes in format, all copies of this work prior to publication. Overcoming the problems posed by the typing of all the Latin names in the indices required for this book has alone been a considerable achievement.

Joe Duddington
Rex Johnson

CHAPTER 1

LINCOLNSHIRE AS A HABITAT

INTRODUCTION   R. JOHNSON, M.Ed., F.R.E.S.

Why should a particular insect be found in a particular place at a particular time? Why should a colony of butterflies thrive in one locality, and not in another nearby, when seemingly identical conditions exist, and everything points to the second area being totally favourable? It will soon become clear that there is no simple answer.

Ford (in "Moths", 1967, p.118) stated "There is always a reason why any species of moth lives in a particular locality. It may be simple or very recondite, but, simple or complex it is rarely known". He later wrote "In general terms, organisms solve the problem of their distribution in one of two ways. They may fit themselves to tolerate a wide range of environment or they may become adjusted with increasing perfection to one particular type of habitat". On the subject of "Distribution, Types of Habitat and Geographical Races", in his book "Butterflies" (1967, p.122) he said much the same thing – "many of our butterflies are only to be found in a particular type of habitat – some are catholic in their tastes, others have requirements much more strictly defined" – in a reference to ecological preferences.

To begin to understand some of the factors which make a habitat suitable to a particular insect, one has to branch out into the study of Ecology. This was defined by L. H. Newman (in "Man and Insects", 1965, p.92) as "the complex system of relationships that exist between insects and their environment".

First and foremost this environment must be able to provide food, and all creatures obtain food directly or indirectly from plants. Plants make their own food – using the light from the sun to join together chemicals obtained from the minerals of the soil where they are growing, and utilising the gases of the air. Clearly, therefore, whatever controls the distribution of plants must to a large extent affect the distribution of animals also.

Two of the most important limiting factors are the restrictions imposed by the surface geology (the raw materials for the plants'

14 THE BUTTERFLIES AND LARGER MOTHS

food-making within the soil), and the restrictions imposed by climate (since weather will directly influence the growth of these food plants, as well as influence the chances of survival of the insects feeding on them). More will be said of these two factors later in this chapter.

Other lesser, but very important considerations are also capable of having a critical impact on insects breeding in a habitat. These include the incidence of predators, parasites and diseases which attack the various stages of the life cycle; the presence of physical features which may set up ecological barriers, or alter the microclimate, and check the free movement of the members of a population (e.g. hedges, rock outcrops, wide expanses of water, sand, cultivated crops etc.) and not least in importance, the activities of man.

Human activities can be particularly diverse and far reaching. Consider –

1. Widespread drainage activities which lead to the lowering of the humidity level or water table of an area, with the subsequent modification of the local fauna, and the change in the animal/insect life this supports.
2. The move to more intensive agricultural practice, including the widespread use of insecticides, the ploughing up of natural grassland to put it to crop, the ripping out of hedgerows, and the burning off of "rough" land.
3. The felling of forests – and the policy of replacing native deciduous woodland with "foreign" conifers.
4. The building of dams, artificial lakes, reservoirs.
5. The pollution of air, land and water by fumes and a variety of domestic and industrial effluent.
6. The spread of the human population – urban sprawl, and the movement of the population from the towns to the villages.
7. Expanding industrialisation – open cast mining, extension of factory space, industrial estates etc.

The list is far from complete.

So, what happens to an insect population in circumstances such as these? As Newman (1965) put it, "The more sensitive species disappear, the more adaptable survive". He pointed out also, that the survivors may even increase, as they may, in the changed circumstances, suffer less from the attack of their predators and parasites (which may have suffered more in the upheaval than the insects themselves), and there may be less competition for food. To be fair, a number of the previous human activities have a plus factor, which may lead to the increase of the number of species being found in the particular area. For example, the formation of new habitats like lakes or reservoirs can establish a new flora, and tempt the spread of insects from a similar habitat not too far distant. Many of the previous activities of man, in fact, led to change in species populations, rather than having too negative an effect on the whole natural history of the area.

The effect of climate on a habitat was mentioned in passing early in this chapter. It warrants further coverage, as its influence can be critical. We live in a region where our weather is "temperate", and can vary considerably from year to year. Rapid changes, or extremes of weather can obviously contribute to insect mortality. All that is needed is a bad "spell" at a critical time, for far reaching effects to be noticeable. Some insects have only a short time as adults, and for example, if bad weather, with heavy rain or strong winds make it impossible for them to find mates, pair, and lay eggs, then the numbers must be reduced in the next generation, and a small colony may be put in a precarious position, or even be wiped out.

Sometimes the effect of the weather isn't quite what one would expect. People associate butterflies with sunny weather, and generally seem to think that the hotter the weather, the better our butterflies will thrive. This is a misconception as Newman (1965, p.93) pointed out when referring to the hot dry summer of 1959. He wrote "Instead of increasing as many people had hoped, many species of butterfly almost vanished from their old haunts because the herbage dried up and the caterpillars could find no food. Sudden heatwaves, when the temperature soars high above the normal for even a short time, can prove fatal. Heat can also have an adverse influence on fertility".

Robert Gooden (1977, "Butterflies and Moths" p.22) added to this argument by writing "One of the most important requirements of butterflies and moths is moisture rather than food; they dehydrate very quickly. In dry weather butterflies live much shorter lives".

A further effect of heat and drought is the hardening of the ground, so that certain species which pupate underground may find emergence impossible.

At the other end of the scale, are the possibilities of extremely low temperatures and prolonged wet weather. The former of these need not have too detrimental an effect. R.L.H. Dennis in "The British Butterflies" (1977, Adaptations to the Environment, p.109) tabulated the January and July "temperature means noted on the periphery of each species range in oceanic Britain and continental Europe" (see his Appendix I.), and drew attention to the fact that all species can withstand "far more severe winter temperatures than are ever experienced in these islands". He continued "winter cold is an essential requirement for diapause stages; certainly a comment often made is that warm winters adversely effect the preponderance of butterfly fauna in the following summer". Continual wet weather and dampness can have a particularly bad influence causing water-logging, death by drowning, or the spread of the fungal diseases to which the insects are susceptible.

In the book "Notes and Views of the Purple Emperor" (Heslop, I. R. P., Hyde, G. E., Stockley, R. E., 1964) Heslop included in Paper 3, an interesting comment on the effect of the weather on this frequently described and documented species.

He wrote: "The Purple Emperor is essentially a species of the Temperate Region. It does not occur either in Arctic regions, or in regions in which the winter is nominal only. In the latter connection it is perhaps worth noting that in England it appears to thrive better in hard winters than in mild ones. The caterpillar is well adapted to damp. And while genial summers are the exception rather than the rule in England, it is clear that the purple emperor does not like great heat. A breeze suits it better than dead calm; and it will fly in quite a stiff wind. It prefers temperatures below 70deg.F., and the male will not fly at all when the thermometer is above 85deg.F. On the other hand I have several times seen it flying under absolute overcast conditions; and even two or three times in light rain".

An interesting article entitled "The Seasonal Abundance of the British Lepidoptera" appeared in "The Entomologist" in March 1947 (Vol. LXXX, No. 1006, article by Bryan P. Beirne, Ph.D., M.R.I.A., F.R.E.S.).

The following are just a few of the considerations under discussion in the article.

1. . . . severe and prolonged frost in winter is favourable to the majority of Lepidoptera. This probably is due to large numbers of insectivorous birds being killed by the cold, as birds are probably the most important of the natural enemies of most lepidoptera.

    . . . severe frosts in the winters of 1878-9, 1880-1, 1894-5, 1916-17, 1928-29, 1939-40 caused heavy mortality amongst bird life, many species taking years to recover.

    . . . in most cases these winters were followed by marked increases in the abundance of the Lepidoptera.

2. A protracted frost in winter . . . is said to be detrimental to those lepidoptera which hibernate as larvae, but frost causes the greatest damage when it occurs in late spring or early autumn.

    There is some evidence that mild winters are unfavourable to most lepidoptera . . . because they are favourable to the natural enemies.

3. . . . a succession of seasons of gradually increasing rainfall is detrimental to Lepidoptera.

    . . . periods of scarcity of Lepidoptera coincided with the latter parts of the periods of gradually increasing rainfall.

4. Regrettably little is known of the diseases of Lepidoptera. There is a possibility that the scarcities during the successions of relatively wet years were due to the increases in the incidence of disease as, . . . diseases of Lepidoptera are caused

LINCOLNSHIRE AS A HABITAT 17

by Protozoa, fungi, bacteria and viruses . . . moisture plays an important part in the spread of an infection by entomogenous fungi.
5. A summer drought is very unfavourable to most species. The larvae and pupae may die from lack of moisture, the hard condition of the ground may prevent the adults from emerging from subterranean pupae, and the dry and sunny weather favours the activities of parasitic insects and other natural enemies.
6. High humidities favour the activities of nocturnal moths, which is a reason why drought is unfavourable to them.
7. Wind deters Lepidoptera from flight and thus a season or longer period of continuous wind may result in the decreased abundance of many species.
. . . Cold winds in spring have similar effects to spring frosts, especially when coming after a mild period.

etc.,

More recently Michael J. Ford of the Climatic Research Unit, University of East Anglia, Norwich wrote on "The Nature of Climatic Change in Britain during the Period under Consideration – 1886-1975". (in the British Butterfly Conservation Society News No.19 October 1977). He reported "it can be seen that there was a marked rise in temperature from the end of the 19th century until about 1950. The biological significance of this trend will be obvious when it is realised that the magnitude of the temperature rise . . . would have resulted in the extension of the growing season of plants by several weeks.
. . . In response to the warming . . . a number of European animals (especially insects) whose distribution in Britain was formerly restricted to the South for thermal reasons, . . . were able to extend their range northwards".

Also in various recent B.B.C.S. News, there has been considerable comment on the unusually hot summer of 1976. For instance, David Robinson, (News No.20 – Editorial) wrote "After the memorable summer of 1976, when even uncommon species came into our gardens, 1977 was a great disappointment. All reports seem to indicate that most species emerged a fortnight later than is normal, which in some cases was three weeks later than 1976".

Frank Masters (News 21 – Letter) noted of 1976: "it has been a most interesting season, as I have seen species not recorded in the last ten years". He wrote about "parched conditions" when "the grasses and flowers withered and the butterfly population in 1977 has been sadly reduced".

Andrew Seats (News 22 – Asst.Ed.Jnr.Section) talked of "Larval foodplants desiccated" and "nectar plants badly affected" in 1976.

18　THE BUTTERFLIES AND LARGER MOTHS

David Robinson (News 22 - Editor, p.18) wrote "it can now be seen that the combination of hot weather and very few days of rainfall in 1976 brought a spectacular increase in the number of butterflies on the wing and incidentially attracted scores of migrants". Subsequently "the result was a loss of food plant and a lowering of population in 1977".

Let us revert now to the subject of surface geology, and its associated flora. Clearly the more the variety of plants within a county, the more diverse will be the types of habitat opportunity available to insects, and the more likely it will be that the recorded County specimens will be extensive in number.

Another intricate web of relationships and reactions determines what plants can survive in a particular community. The plant's genetic system must be able to tolerate the area's mineral conditions. Anne Bülow-Olsen (1978: Plant Communities, p.110) talking of mineral balance, wrote "If there is too small a quantity of a single nutrient, the plant's growing strength is greatly reduced". Conversely "An overdose of nutrient can also have violent effects on vegetation". She illustrates this later (p.118) by writing "move a heather plant to a place with nutrient-rich clay soil, (and) it soon dies. Other plants cannot survive an opposite move".

Plants equally cannot survive extremes of temperature. If it is too hot, there may well be too much water evaporated from the leaves, and the plant dies. If it is too cold, the liquid in the plant cells may freeze and rupture the cell walls - which is equally fatal.

Wind may scorch plants in an exposed situation, and dry off protruding shoots.

Water shortage or excess can affect plants seriously, and all plants need a suitable level of light to thrive.

Within an area the size of one of the English counties, a rich variety of types of habitat may exist side by side - each region differing from and blending into neighbouring areas, due to change in geological, geographical, climatic etc. factors. We may have dunes along a coastal strip, cliffs, sandy heaths inland chalkland areas, limestone crags, acid moorland, clay deposits, fens, areas of mountainous granite uplands, extensive forests etc.

Perhaps, before reading about Lincolnshire specifically, it would help non-specialist readers to have some very generalised information about these widely differing habitat types.

Dunes are often thought of as being "wet" areas. In fact, since sand drains very rapidly, plants on dunes are almost always faced with a shortage of water, and they often need to be adapted to life in dry places. To conserve water, or reach it, dune plants often have specialised (e.g. fleshy, spiny or hairy etc.) leaf systems, and deep root systems. As the dune is built from sandy drift containing shells etc. its soil is at first alkaline, but due to rapid leaching and washing out with water, the level of·alkalinity changes with age.

LINCOLNSHIRE AS A HABITAT

Plants need, however, a limestone tolerance, as well as an ability to cope with salt spray in the drying wind, and possibly to keep salt out of the root system.

Sandy Heaths have an acid soil – either from underlying rocks or from the peat layer near the surface. Generally the soil is deficient in nutrient, badly aerated, of poor quality and plants often have a shallow root system. Heaths are dryish places, populated by birch, pine, bracken and plants like heather and gorse which can tolerate these living conditions.

Acid Moorland (found from 750ft upwards to 3,000ft or more) has a much deeper peat layer (perhaps several yards thick). As a result of the altitude, rainfall may be high, it may often be cloudy, or foggy, and moorland peat tends to be wet, and to retain moisture. Sometimes lowland moors are formed in estuaries or silted up river valleys. The soil is acid, typically covered with heathers, sedges, cotton grass, low shrubby bushes, etc. with boggy areas having sphagnum moss and rushes.

Chalklands contain chalk or limestone deposits (the former being the softer of the two). Chalk and limestone both come to the surface in the east of England, though large expanses of chalk can be covered by sand or clay deposits. The soil is obviously alkaline, it is porous, drains rapidly, and is often shallow. A very wide variety of plant types are supported. There can be a mass of colourful flowers on grassy areas – like vetches and clovers.

There may be beech or ash woodlands, scrub with plantains, thistles, knapweeds etc., possibly bushy places with buckthorn, hedges and fields, pools, marshes – all rich in animal and insect life. There will be no heathers or bracken.

Limestone areas are similar to those on chalk – alkaline, with similar plant and animal life. Again there may be ash or birch woods, scrubby trees, willows, sallows in wet areas, pastures, hedges, meadows, grasses, colourful flowers.

Clay covered regions can be acid or alkaline depending on the content of humus in the soil, or calcium carbonate respectively. Oak woodlands form well on heavy, clay soils which are moderately alkaline. Shrubs thrive – e.g. hawthorn, blackthorn, bramble, honeysuckle, hazel, willows. Clay supports many species of flowers (less colourful supposedly than those on chalk). There can be cultivated fields, hedges, open meadows etc.

Forests are probably the richest of habitats for animal/insect communities. Coming down through the forest storeys, there is the tall tree layer, the layer of shrubs and bushes underneath, the herbs and flowers under the shrubs, and algae, liverworts and mosses below that. The soil is rich in nutrient from decaying shed leaves and the remains of the animals which live, excrete and die in the forest.

Enough of the generalities. An interested reader can obtain more accurate information from other books – this preamble is

hopefully included to introduce the bewildering different gradations in the living conditions available to our flora and fauna.
The following list is a synopsis of the main habitat types to be found in the British Isles: (based on J. Heath (Ed.) 1976).

A. COASTLANDS
1. **Saltmarshes** – found in river estuaries and bays, and more rarely along open coasts. Also, these occur in association with shingle beaches (as in Lincolnshire and Norfolk).
2. **Sand-dunes** – often popular as holiday resorts, containing hills and hollows which may contain water, and be very rich in vegetation.
3. **Shingle-beaches** – some devoid of vegetation (and therefore of insect populations), some with thin vegetation and some associated with sand-dunes.
4. **Cliffs** – (not found between the Thames and Humber).

B. WOODLANDS
1. **Broadleaved woods** – providing a very wide range of habitats, and therefore rich in lepidoptera. Trees are very important larval food plants. In "The Caterpillars of British Moths", W. J. Stokoe, (1958) nearly 120 species are listed as feeding on Oak. Additionally, many species are associated with the herbs and shrubs found growing in forests.
2. **Coniferous woods** – mainly non-native species of trees and not, therefore supporting many lepidoptera, unless rides or belts of broad leaved trees have been interspersed with the conifers.

C. GRASSLAND, HEATHS AND SCRUB
1. **Calcareous grassland** – found on a range from soft cretaceous chalk, through to hard carboniferous limestone. These areas carry a rich flora, and rich insect fauna.
2. **Acid Heath** – acid soil with heather, together with gorse and heath. Often has birch and pine trees.
3. **Scrub** – develops on grassland when conditions support such a change. Can be a stage in the development to woodland, or can be a stable habitat in its own right.
4. **Breckland and heath** – sandy soils over chalk.

D. PEATLANDS
1. **Acid bog** – sphagnum, with heather or heath. Typical example – Thorne Moor.
2. **Fens** – with reed, bulrush, alder, sallow, buckthorn.

E. UPLANDS
– high regions North and West of a line drawn from the Bristol Channel to Flamborough Head – harder non-calcareous rocks like granite, shales etc.,

F. SPECIAL HABITATS

Ponds, lakes, rivers, birds' nests, decayed wood, bees' or wasps' nests, lichen/fungi, man-made habitats with introduced plants etc.,

People living outside the County of Lincolnshire are often unaware of its true nature. Their general impression can be that the land is all flat and put to such agricultural use as bulb growing or to potato or sugar beet cultivation. Indeed, many Lincolnshire inhabitants are equally unaware that this large county is one of the most varied in the country, and from the natural history point of view it is particularly rich and diverse in the living conditions available to its wild life. A study of the above range of habitat types will confirm this, as almost all the possibilities can be found somewhere in the county. We have no cliffs, no breckland or mountains, but apart from these every other type of habitat is there for the naturalist willing to seek it out. For the entomologist, the county certainly provides a wonderful range of insect fauna.

Surface geology, flora, climate, man-made habitats, etc. are all, then, intimately enmeshed in making Lincolnshire what it is as a habitat. Some of these aspects will now be expanded, and described as they are found within the county, by people who are expert in the particular field. The final section of the chapter will then correlate this information and relate it to the pre-requisites of certain of the butterflies and moths recorded in the county.

THE GEOLOGY AND SCENERY OF LINCOLNSHIRE
by D. N. Robinson MSc

The shape of the traditional county of Lincolnshire is that of a blunted peninsular with a long maritime border stretching from the Wash into the Humber estuary. Geologically it forms part of the eastern lowlands of England with alternating bands of hard and soft rocks, although much of the pattern is masked by superficial deposits. Three-quarters of the county is under 30m in altitude and much of this is only a little above sea level, although the central Wolds near Normanby-le-Wold contains the highest ground (168m) in eastern England between Kent and Yorkshire.

There are two main lines of hills – the limestone Heath trending north-south and the mainly chalk Wolds trending north-west to south-east, between which is a clay vale broadening south into the Fens. To the west is the Vale of Trent and the Isle of Axholme, and to the east is the Lincolnshire Marsh fringed by a sand-dune and saltmarsh coastline. Drainage is north to the Humber (Trent and Ancholme) and east to the Wash (Welland, Witham and Lymn), with a series of small streams from the Wolds to the North Sea. The basically simple pattern of a belted lowland has been significantly modified by ice during two periods of glaciation, and the coastline altered by rising sea level in post-glacial times.

It is convenient to consider the regions of Lincolnshire from west to east. The **Isle of Axholme** – really an island within an island – is composed of red Keuper Marl, the hard skerry bands within which give a series of low flat-topped hills at 15-45m. Covering the Marl and banked against the hills are blown sands with thick peat extending over the county boundary; to the east are the alluvial warp lands of the Trent valley.

The meandering River Trent flows through a broad vale of alluvium. Overlooking Trent Falls, where the Trent joins the Humber estuary, is the impressive 60m Burton Cliff of Rhaetic and tough Hydraulic Limestone. These rocks also form a series of low hills at Laughton and Scotton, again with blown sand, and a narrow clay vale between the Burton Cliff and the limestone hills is drained north by the Winterton Beck. At Gainsborough and south is a series of low 15-30m hills with a bluff of Keuper Marl overlooking the Vale of Trent. The wide Till valley, on Lias Clays drains south-east to the Witham.

South of the Roman Fossdyke and west of Lincoln are expanses of sands and gravels, deposited by former glacial meltwater drainage through the Lincoln Gap. Southwards again is the upper course of the Witham with the vale of its tributary the Brant parallel to the Heath escarpment, both flowing on Lias Clays.

The distinctive cuesta of the **Lincoln Heath** is a north-south line of hills capped by the oolitic Lincolnshire Limestone. Some three miles wide and up to 80m in height north of Lincoln (known as the 'Cliff'), it rises to over 115m between Lincoln and Ancaster (the 'Heath') and broadens to 15 miles wide at 130-145m on the south Kesteven Plateau. The hills are cut by the well-defined Lincoln Gap and, at a higher level, the Ancaster Gap with its infill of river gravels.

Underlying the oolitic limestone is a series of sandy and clay strata which form the lower part of the west-facing scarp slope. In the north these are in turn underlain by the Frodingham Ironstone. This is also an area of extensive Cover Sands – blown sand which gave rise to sandy warrens with miniature dune systems and waterlogged depressions; little remains, because of quarrying and agriculture, except at Risby and Twigmoor.

The hills between Kirton-in-Lindsey and Ancaster have an extensive pattern of dry valleys in the eastern dip slope. South of the Ancaster Gap the limestone ridge broadens rapidly into the Kesteven Plateau with an extensive covering of glacial Boulder Clay and is well wooded. The ridge here is dissected by the north-flowing Witham and south-flowing Glen and Eden. The latters' distinctive meandering courses cut through the Cornbrash, and Great Oolite clay and limestone into the Estuarine series. The Marlstone Ironstone forms a secondary escarpment to the west of Grantham. The Northants Ironstone of the Middle Jurassic has been extensively quarried round Colsterworth, and to the east sink holes occur in the limestone.

Between the Cliff and the Wolds is the broad **Clay Vale** of mid-Lincolnshire. Although it is floored by thick deposits of Oxford and Kimmeridge Clay these seldom appear at the surface because of the extensive covering of glacial boulder clay, gravel and blown sand. Where the vale narrows markedly to the north it is occupied by the River Ancholme and peaty Carr lands. The low 15-45m hills at Howsham and Kelsey are capped by Boulder Clay. Cover Sands are banked against the Wolds from just north of Caistor to Market Rasen. South of the headwaters of the Ancholme and south-east from the valley of the Langworth Beck is an undulating, well-wooded Boulder Clay cover. Most of this is derived from the underlying Upper Jurassic clays, but nearer the Wolds there is a greater chalky content. The woods and heathy moors of the Woodhall-Coningsby area are on a massive delta fan of sands and flinty gravels washed down from the Wolds during the Last Glaciation.

The extension of the **Fens** towards Lincoln, bounded by the Witham and by the Roman Carr Dyke at the foot of the boulder clay covered dip slope of the Heath, is predominantly of peat. Another major area of peat occurs in Deeping Fen and continues over the county boundary; a third area of peat is in the East Fen. There is a line of Fen edge gravels from Potterhanworth to

Timberland and south from Swaton to Tallington, Langtoft and Deeping St James; the other area of gravel is in the northern part of Wildmore Fen. The rest of the Fens is divided into alluvial and finer silt soils, the latter occupying the eastern part nearer the Wash, particularly the higher silt bank zone of the 'Townlands'. The most extensive reclamations of saltmarsh in the last three centuries have been between Skegness and Wrangle, and in the former Bicker Haven and the Welland and Nene estuaries.

The **Wolds,** capped by chalk, rise to a plateau of about 130m and converge towards the Cliff in the northern part of the county. The simple west-facing escarpment is only apparent north of Caistor; the clearly defined eastern margin was a series of chalk sea cliffs before the Last Glaciation which partially masked them with boulder clay. This glaciation also produced modified stream valley systems such as the in-and-out meltwater spillways at Ravendale, Hubbard's Hills and Swaby.

Underlying the White Chalk, which does not exceed 30m in thickness in the Wolds, is the Red Chalk producing a vivid splash of colour on some steep hillsides. Beneath this is the Carstone, a friable khaki-coloured sandstone and the Tealby series of clays, limestones and ironstones. The upper sandy ironstone – the Roach Stone – forms a noticeable ledge feature in the south. Also in the south the Spilsby Sandstone with its sour soil occurs at the surface over a wide area, particularly round the Lymn valley.

The northern Wolds have dry valleys in the dip slope, but in the centre (north of Binbrook) the headwaters of the Waithe Beck have cut through the chalk to the lower Cretaceous rocks, and the western escarpment has been deeply dissected by the Nettleton and Ussleby Becks and the River Rase.

The southern Wolds is a complex area. The true escarpment (Donington – Tetford – Langton) faces the hills of the Lower Cretaceous ridge with cappings of glacial deposits to the south-west, or it overlooks the Lymn valley. The chalk becomes an increasingly narrow tongue towards Candlesby. The River Bain rises near Ludford and flows south along the western edge of the hills towards the Witham. The River Lymn rises in a combe near Belchford and tumbles through the spectacular New England gorge into the main valley which is surrounded by a wide ledge of Spilsby Sandstone and floored by Kimmeridge Clay. A number of its tributaries notably Snipe Dales (and Sow Dale occupied by a stream flowing south into the Fens) have steep, dry valley sides of sandstone and a flatter floor of clay. The valley emerging at Keal Carr is of similar form. The headwaters of the Calceby Beck have, like the Waithe Beck, cut through the chalk to the Lower Cretaceous rocks creating deep combes and leaving only a very narrow crest to the Wolds, particularly at Oxcombe.

East of the Wolds is the **Lincolnshire Marsh.** Its main divisions are the Middle Marsh – of hummocky, undulating Boulder Clay, the result of the Last Glaciation – with outwash gravels in the Thornton Abbey-Keelby and Alford-Bonthorpe areas; and the Outmarsh. The latter is of flat marine silts about 3m OD. Occasional hummocks of boulder clay project through the silt cover near Hogsthorpe and the Middle Marsh reaches the coast at Cleethorpes. In the northern part of the Middle Marsh 'blow wells' occur where sand and gravel lenses allow water from the chalk to reach the surface.

Before the stormy 13th century the Lincolnshire **Coast** was protected by a series of off-shore boulder clay islands and sand bars from Holderness to Norfolk. The final destruction of the barrier gave rise to a storm beach which became the basis for the present sand-dune system, and is best seen at North Somercotes Warren. Between Mablethorpe and Skegness the coastline has retreated. At Saltfleetby-Theddlethorpe a strip of fresh-water marsh occurs in the dunes; north-west of Saltfleet accretion has been considerable and the dunes and reclamation banks are fronted by a wide beach and extensive 'fitties' or saltmarshes. South of Skegness accretion by longshire drift and by material derived from offshore has extended the dune line to Gibraltar Point and built the coast outwards in a complex of dune ridges and strip saltings. The Wash coast is fringed by saltings some half a mile wide which are subject to reclamation from time to time. The oldest remaining saltmarshes are between the estuaries of the Witham and Welland, and at Holbeach and Gedney.

THE FLORA OF LINCOLNSHIRE AND SOUTH HUMBERSIDE
by Mrs. I. Weston, B.Sc.

The very large number of species of vascular plants listed in VC53 and VC54 the Watsonian Vice Counties of South and North Lincolnshire, is compliant both with the wide span of habitat types found within the historical County of Lincolnshire and with the fact that a number of these species, commoner in other parts of Britain are found at the extreme edges of their geographical distribution in Britain, within this county.

The habitat range from the Trent eastwards to the coast and from the Humber Bank southwards to Stamford and the Wash incorporates a great variety of soil types and drainage levels. The Vale of Trent provides areas of sands and gravels and supports a typically calcifuge flora, the Isle of Axholme an area of peatlands, sandy moorlands and warplands. The two lines of hills – the limestone cliff and heath and the Chalk wolds give calcareous habitats both of limestone and chalk outcrops with quarries and of calcareous boulder clay, where calcicole plants flourish. In the North the Cover Sands provide areas where an admixture of species is found – the sand over the limestone being of varying depth and many grassland, heathland, woodland, warren and dune systems occur, with rich wetter areas interspersed between. These sandy heathlands show succession to scrub and woodland. The Southern Wold valleys are distinctive, being cut through the chalk and sandstone to the Kimmeridge clay and wet woodland carrs rich in species occur. The Eastern Wold woodlands are rich in species, many occurring only in this part of Lincolnshire.

Between the Limestone and Chalk ridges the mid clay Vale has both grassland and woodland habitats. Neutral to slightly calcareous grassland species are found in old meadows and the deciduous small leaved lime woodlands of Central Lincolnshire are particularly significant in Britain. The fen edge deposits of sand and gravel in the Woodhall area produce a contrasting acid heathland and woodland type.

The peats and silts of the Fens in the southern part of the County, once rich habitats for water plants are now mainly under cultivation, but the aquatic range of habitats here and in the rest of the County support a range of such plants. The eastern marshes have rich neutral grassland areas and the coastal region provides salt marsh, sand dune and mudflat systems. The Humber and Wash saltmarshes are nationally important. The National Nature Reserve at Saltfleetby contains a fresh water dune slack system of national significance. Both this reserve and the County Reserves at Gibraltar Point and Donna Nook are of high ecological value in

LINCOLNSHIRE AS A HABITAT

providing a large variation of ecosystems covering much of the coastal range.

The distribution of British plants is such that Lincolnshire has many northern limit plants – on the coast these include marsh mallow, rock sea lavender, amongst others. In calcareous grassland man orchid, pasque flower, and bastard-toadflax occur. In woodland wild service tree and midland hawthorn are present with wood spurge only in the South. The two fluellens once much more common are arable weeds also at their Northern limit in the county.

The wide range of species thus produced for the Lincolnshire list does not mean that vast numbers of wild flowers occur in the county. The quantity has been much diminished by losses of habitat due to many factors – agricultural improvement, use of herbicides, drainage, reclamation in the fens and on the coast, sand and gravel workings on the heathland and moorland, ironstone, chalk and limestone quarrying and also urban development. However the range of species has been preserved in the nature reserves, in small remnants of habitats under sympathetic private ownership and by agreement with the Forestry Commission. Food and shelter plants for dependent insect species are thus available, and the list of plants is increased also by the wide range of crops grown in the county.

The flora of the County by E. J. Gibbons was published in 1975 by the L.N.U. and incorporates the records of past and present botanists of the County. This first published Flora for Lincs is worked on the 18 Natural History Divisions of the County devised by the Rev. E. A. Woodruffe-Peacock. It also indicates the 10Km sq. National Grid system recording which is now used and relates this to the Division System. The 10Km. sq. records are already published in the Atlas of the British Flora and new square records are kept by the Recorders. A detailed chapter on the Flora of the Major Habitats is given in the Flora and also a description of the early Recording Divisions.

Some of the major plant habitats and plant communities in Lincolnshire are further discussed below.

A. Coastal
1. Saltmarsh communities

The saltmarsh communities stretch from the Wash, up through Gibraltar Point, Saltfleetby – Theddlethorpe, Donna Nook, Cleethorpes, and the South Bank of the Humber to the Trent. These mud flats and saltings are rich in species. Stages in colonisation can be seen throughout the area. Different species dominate in different areas. A fine area of mature saltmarsh at Gibraltar Point shows an expanse of common sea-lavender. At Saltfleetby the thrift occurs in sheets of pink whilst at Goxhill and Barrow Haven the area is dominated by stands of sea aster, the plants here being very tall.

Saltmarsh plants include samphire, sampher or glasswort. This has a traditional use in Lincolnshire as a pickle and vegetable. Seablite is found extensively as a coloniser and in the older marshes and is found in morphologically diverse forms. Shrubby sea-blite is very scarce and at its N. limit at Gibraltar point. The common sea-lavender is extensive and rock sea lavender plentiful at Gibraltar Point. Sea aster is present overall, with the rayless form in some areas. Sea-purslane is present densely in the higher marsh communities and flowers well, and sea wormwood is found but not in quantity. The scurvy grasses, sea plantain, buck's-horn plantain and sea-arrowgrass are all common. Sea-milkwort forms meadows on the drier parts of the saltmarsh but sea heath is rare. The sea-spurreys are frequent along the coast. Thrift is found more plentifully in the North.

The salt marsh grasses and rushes are well represented and include *Juncus gerardii, Festuca rubra, Agropyron pungens, Puccinellia maritima* and *P. distans, Parapholis strigosa* and *P. incurva* and the native cord grass *Spartina maritima* which is scarce. *Spartina anglica* has spread throughout the range and is a common coloniser on new mudflats, invasive on the older marshes where it has to be controlled and introduced on the Humber Bank to prevent erosion.

2. **Sand-dune communities**

The sand dune systems are extensive. Large dunes are found on the National Nature Reserve at Saltfleetby, at Gibraltar Point and Donna Nook and a smaller system at Cleethorpes. The major dune plants are sea buckthorn (a feature of the Lincolnshire coast), elderberry, common ragwort, dewberry, vipers bugloss, hound's-tongue, centaury. Where the shell content is high the calcicole plants are abundant – yellow wort, pyramid and bee orchids, fairy-flax, carline thistle, bird's-foot-trefoil. Field mouse-ear is found-commonly and also inland on the blown sand and other sandy habitats. The very good area of dune grassland at Cleethorpes has the specialities *Poa bulbosa* and *Apera interrupta* together with strawberry clover and the slender Bird's-foot-trefoil. Bird's-foot-trefoil is much more common and yellow rattle very plentiful. The sand dune grasses include *Agropyron junceiforme, Psamma arenaria, Elymus arenaria and Festuca rubra* with var. *arenaria*. Sand sedge is frequent here and occurs inland on blown sand. The strandline and the younger dunes have a number of typical plants including sea rocket, sea sandwort, and prickly saltwort. Henbane appears occasionally in quantity at Gibraltar Point and hemlock is also found. The duen ephemerals – storksbill, soft crane's-bill, wall speedwell, cornsalad and small forget-me-nots are regularly recorded.

There is very little shingle and plants such as sea-holly, yellow horned-poppy and sea campion are scarce. Sea bindweed is rare and other sand dune plants found commonly on West Coast dunes – creeping willow, burnet rose, squinancywort, restharrow and to a lesser extent kidney vetch are not seen on the dune system, although some are recorded inland.

3. **Fresh water slacks**

A very important slack system at Saltfleetby includes spectacular populations of the marsh orchids – southern and early marsh and their hybrids. Also common spotted orchids. The slack notably contains the great water dock, greater pond sedge, yellow flag iris, water plantain, gipsywort, water mint, marsh marigold, skullcap, lesser spearwort, as well as the sea rush, and spike rush. The fenland marsh pea is present in quantity and this slack is a unique feature in the County.

B. **Woodland Communities**

1. Deciduous woodland

Both primary and secondary deciduous woodlands occur. Old woodlands are present at Broughton, Skellingthorpe, Wragby, Revesby, Kesteven, in the S.Wold valleys and on the E.Wolds. New areas have been planted up extensively at Brocklesby and scattered elsewhere. The composition of the woodland flora varies with the soil types and the species list is rich. Dominants vary – oak, birch, elm, ash, oak-ash etc. Birch is extensive on the lighter soils, oak and ash on the clay and more calcareous soils. Birch woodland is also a succession type from heathland and is commonly mixed with pine and oak. Willow scrub is found in the wetter areas. Species include the two birches, common oak, ash and elm species. These also occur as hedgerow trees. Alder is present in the wet alder carrs. Willows and aspen are frequent. Small leaved lime occurs both as standards in high forest and as coppice particularly in the clay woodlands of the Wragby/Bardney area and notably with wild service tree. Scattered trees of small leaved lime occur elsewhere.

The scrub layer also varies and significant species include hazel, (as coppice, particularly in old woods), common and midland hawthorn (– the latter in the old woodlands, the former also in hedgerows and scrub from ungrazed grassland), guelder rose, dogwood (in old woodland and hedges), bramble, dewberry, spindleberry, spurge-laurel and purging buckthorn. Maple is common on woodland edges, but there are some fine old trees in the county. Wild cherry is found over a wide area but sparingly. Crab apple is common. The dog rose and field rose are found as woodland and hedgerow shrubs.

On the more acid soils rowan is widespread and particularly

abundant at Woodhall Spa. Alder buckthorn occurs sparingly and sweet gale is found in a few scattered localities.

The ground flora is very varied. Primary woods on clay soils have primroses, bugle, water avens, wood anemone, greater butterfly-orchid, early-purple orchid, twayblade, dog mercury, spotted-orchid, yellow archangel, wood forget-me-not, wood violets, greater stitchwort woodruff and wood sorrel. Lily-of-the-valley and bluebells are found on lighter soils and herb paris in more calcareous woods. Sanicle and common cow-wheat are not common but occur in some of the woodlands.

The flora of the alder carrs can be very spectacular with red campion, wood forget-me-not, marsh marigold and sometimes the two golden saxifrages. Climbing corydalis is often a ground cover plant on the lighter soils of the fen edge and on the silts. Some of the woodland plants are very restricted in the county – wood speedwell is found mainly on the Eastern Wold woods, the large bellflower chiefly in the North and the nettle leaved bellflower in the south, wood spurge only in the extreme south. Ransoms is found in many old woods but more common in the North.

Woodland grasses include *Festuca gigantea, Bromus ramosus, Milium effusum, Melica uniflora, Poa nemoralis, Calamagrostis epigejos* and *C. canescens. Hordelymus europaeus* is very rare. *Deschampsia caespitosa* is abundant in wet woods on clay soils and *Deschampsia flexuosa* is widespread in acid woodland. *Molinia caerulea* is widespread in wet acid woodland. Wood sedge and remote sedge are common as are the smaller woodrushes, but greater woodrush is rare.

Secondary woodlands have a poorer species list, but characteristically contain nettles, brambles, wood avens, sweet violet, lords and ladies, ground ivy, enchanter's nightshade, celandine, bugle, *rosebay willowherb and honeysuckle together with some ferns*. Some acid woodlands e.g. Woodhall, Tumby and Skellingthorpe are rich in ferns. Ferns are not common overall, but found profusely in certain areas. Both buckler ferns are found, narrow-buckler less commonly overall, but in some woods on the lighter soils it can be abundant. Male fern and lady fern are widespread. Hard fern is uncommon and the shield ferns very restricted in distribution and mainly east of the wolds. Bracken occurs abundantly on the light acid soils.

Sycamore, common in the county, is an invasive species particularly in secondary woodlands. Poplar plantations occur in several areas but the native black poplar trees are very rare.

2. Coniferous and mixed woodlands

The Forestry Commission still holds a large acreage of primary woodlands, but some areas have been felled and replanted, either with mixed woodland species or entirely with conifers. The latter lack the rich ground flora, but the rides still have interesting relic plants of the old woodland. The dense conifer stands of older plantations have little or no ground flora and needle litter covers the ground. Where conifers are planted exclusively on non-woodland sites the same lack of species occurs. Mixed plantations retain good ride flora. Some species – rosebay willowherb, bramble and nettle are extensive.

On the acid heathlands extensive coniferous plantings have occurred and small clear areas within the forests are rich in ground flora. Mixed woodlands in these areas retain ling and bell heather and other acid plants. The mixed birch and pine woodland, which develops as a natural succession contains relics of the heathland wavy hair-grass and heath bedstraw on the drier areas and purple moor-grass in the wetter parts. Willows are common in the damper areas and some sites have aspen.

Hedgerow plants include elms, sycamore, oak, ash, blackthorn, dogrose and hawthorn commonly whilst the older hedges, particularly where the hedge was part of an old woodland edge are much richer. Field maple, hazel, guelder rose, dogwood and field rose occur in some of these. Honeysuckle, bramble, white bryony and more rarely black bryony occur as climbers and in the hedge bottoms the umbels – cow parsley, hedge parsley, chervil and hogweed are frequent as are hedge woundwort, nettle, greater stitchwort and garlic mustard. White deadnettle is a common weed of hedge bottoms.

C. Grassland Communities

1. Calcareous grasslands

Large stretches of grassland are very few and both on the limestone and chalk the road verges and quarries are important habitats for the grassland flowers.

The grasses include the the calcareous indicators *Bromus erectus Brachypodium pinnatum,* seen very well at Red Hill and at Ancaster and also usually associated with them quaking-grass, crested hair-grass and sheep's-fescue. Another calcareous indicator species, hoary plantain, is widespread. The calcicoles include felwort, rockrose, pyramidal orchid, bee orchid, dwarf thistle, carline thistle, thyme, salad burnet, yellow-wort, small scabious, greater knapweed, burnet-saxifrage, fairy flax, marjoram and kidney vetch. Bird's-foot-trefoil is very common but horseshoe vetch is rare. Man orchid, fragrant orchid, bastard-toadflax and pasque flower are all found in the County with interesting distribution ranges. Wild calamint and basil are uncommon. Basil thyme and ploughman's-spikenard are not common but found plentifully in one quarry which also has in some years very large numbers of bee orchids and a large population of tall broomrape parasitic on knapweed. This quarry also has a few plants of hartstongue fern which is rare. Glaucus sedge is a common plant of the grassland areas. Nodding thistle is fairly widespread but woolly thistle is rare. Good areas where these plants may be seen are at Red Hill, Tetford, Ancaster and Clipsham. The quarries, the mound at Holywell and the Valley at Ancaster are prime sites. Where calcareous grassland is not mown, hawthorn scrub develops.

2. Clay Grassland

The clay grasslands are most interesting communities. Cowslips are abundant in a few Lincolnshire neutral clay meadows and on roadside verges, sometimes associated with green-winged orchid, adder's-tongue fern, yellow hay-rattle, sneezewort, saw-wort, greater burnet, dyer's greenweed and dropwort. Red bartsia, betony and devil's-bit scabious are widespread but meadow saxifrage is decreasing. The improvement of the grass meadows has been widespread and the plants have diminished overall. Lincolnshire is particularly fortunate in having several meadows as Trust Reserves including those in the Isle of Axholme, Bratoft and at Little Scrubbs.

3. Acid Heathland Communities

These are found on acid soils and are quite a feature of the county being much more extensive in the past. They are found on the blown or cover sands and on river and glacial sands and gravels. Also on the fen edge. In the North they were commons and rabbit warrens. They support calcifuge species and show succession to birch and pine woodland with some oak. Heather, gorse and bracken present vivid colourful areas. Dry and wet areas present a mosaic of vegetation and interesting species. Wavy hair-grass forms dominant swards often associated with heath bedstraw, wood sage and devil's-bit scabious and ling heather is present on the drier and on more peaty areas. Bracken is invasive on some of the heathlands. A large number of small annuals occur on these sands as one the dunes – dove's-foot crane's-bill, storkbill, wall speedwell, shepherd's cress, birdsfoot, parsely piert and cornsalad. Sheep's sorell is widespread in certain areas. Heath speedwell, hare's-foot clover and centaury occur.

In the Woodhall, Market Rasen and Gainsborough areas there are extensive grass heathlands. In the drier parts the wavy hair-grass dominates. Ling heather and occasionally marsh gentian occur with *Molinia* in slightly damper areas and also bell heather and cross-leaved heath. In wetter places *Molinia* dominates. Cotton grass and bog asphodel occur locally, sundews rarely, but marsh pennywort is common. Bog pimpernel, bogbean and marsh cinquefoil are recorded from some of these damp patches and creeping willow, common in the west, is found in a few areas,

LINCOLNSHIRE AS A HABITAT

notably Linwood and Messingham amongst *Molinia*. Marsh violet has a limited distribution but a striking population occurs at Moor Farm on the wet heath. This site also has heath spotted-orchid, common spotted-orchid, lousewort, adder's-tongue, devil's-bit scabious, moor mat-grass and many sedges. At Kirkby Moor the damper grassland near the reservoir is rich with southern marsh-orchid, marsh pennywort, twayblade, lesser spearwort, cotton-grass, spikerush, common reed, and *Carex demissa*. Sallows are invasive.

Bilberry is singularly absent from Lincolnshire except for one extending population at Linwood Warren. Large areas of the heathlands are now afforested e.g. Linwood and Willingham, Woodhall, Scotton and Laughton.

Areas of sand over limestone as at Risby Warren in the North of the County show different plant associations. Purple milk-vetch and *Brachypodium pinnatum* occur in large separate stands adjacent to patches of ling heather, sheep's-fescue grassland and large areas of sheep's sorrel and sand sedge. Elsewhere the purple milk-vetch is mixed up with the ling heather. The vegetation indicates where the sand lies thin on the limestone. In places the cover sands are very deep and plants reminiscent of the coastal dunes such as hound's tongue occur. This type of coverage along parts of the Lincoln limestone cliff gives rise to interesting habitats. The wet and dry areas at Messingham and Twigmoor are ecologically interesting. Field mouse-ear forms sheets of white over the cover sands.

D. Peatlands and Wetlands

1. Acid bog

Areas of acid peatland are found in the North of the County. Shallow areas build up on acid heathland and patches of sphagnum bog develop. An extensive area of deep peat occurs at Crowle Waste adjacent to Thorne and Hatfield Moors. Cottongrasses – common and harestail, dominate the old peat cuttings with sphagnum moss in open water. *Carex curta* is also present. Marsh cinquefoil, bog rosemary, cranberry and *Cladium mariscus* are found. At Epworth and Haxey Turbaries there is a succession to sallow/birch woodland. Bracken and sheep's sorrel are found extensively on the drier areas. Buckler ferns are common.

2. Fens

These comprise the whole of S.E. Lincolnshire, south of the Wolds and east of the limestone plateau and extending along the Witham to Lincoln. Drainage of the fens has produced very rich agricultural land and very little original fenland remains. The areas of peat fen are now well inland, the area toward the coast being reclaimed salt marsh. Fen plants include *Phragmites arundinacea* –

reed grass as a common drain plant. Examples of fen vegetation can be seen at Baston Fen and plants include greater pond-sedge, yellow flag iris, great water dock, marsh marigold, with marsh arrowgrass on the grassland area. Fen edge drains at Tattershall are rich in species and include water violet, bladderwort, arrowhead, marsh woundwort, many pondweeds and stone-worts.

3. Drains

In Lincolnshire the drain flora includes many *Potamogeton* species. Drain-bank sides are notable in the North. Large meadow rue is found on some grassy banks together with agrimony, others are species rich with cowslips and spotted orchid.

Other Habitats

1. The aquatic habitats of the sand and gravel workings as in Burton and Tattershall show a succession of habitats from open water to swamp and carr vegetation. Greater spearwort is found both at Burton and Tattershall. Other notable plants include amphibious bistort, water violet, greater and lesser bur-reed, gipsywort, greater and lesser reedmace, bulrush, water and marsh horsetails, marestail, arrowhead, duckweeds, yellow flag iris, flowering rush, common reed-grass and reed canary-grass, the large water reed sweet-grass and many rushes.

2. Railway ballast pits and clay pits are important man-made habitats. A pit near Crowle has marsh helleborine and a very large pit at Killingholme now filled with rubbish had a magnificent flora which included pyramid, bee and twayblade orchids, sickle medick, melilot, lucerne, adder's-tongue, eyebright, carrot, rest-harrow, trefoils and many others. It was important for its wealth of flowers and yielded large numbers of lepidoptera. Also large populations of the rare Yarrow broomrape and the common broomrape, parasitic on the legumes.

3. Disused railway lines are becoming increasingly important as plant habitats. This has been markedly shown in the Isle of Axholme where green-winged orchids and many other flowers are now found on the banks.

Arable Weeds and Farming

Lincolnshire is primarily an arable county and has an extensive range of cropping. Barley and wheat cover large acreages, together with potatoes, sugar beet, peas, oilseed rape and a small quantity of fodder maize, some beans, brussels sprouts and carrots. In the south there are more vegetables, – cabbages, onions and beetroot. In the Holland division there is a large acreage of flower bulbs, but daffodils are included on a smaller scale as part of the farm croppings elsewhere.

The list of arable weeds in the county is extensive and includes corn marigold, redshank, cornspurrey, large and common hempnettle, poppy, fat hen, fumitory, pennycress, the three spurges – sun, common and dwarf, the corn, prickly and smooth sowthistles, speedwell, mayweeds, pineapple weed, wart-cress. Annual cornflower, corncockle and corn buttercup have virtually disappeared and the two fluellens and Venus'-looking-glass are decreasing. Grass weed species include twitch, tall oat, wild oat, blackgrass and sterile brome.

Perennial nettle, mugwort, knotgrass, yarrow and common fleabane are common along roadsides. The slender speedwell is spreading extensively in grassland over the county. Also Oxford ragwort in waste places. Rubbish dumps have yielded a large list of plants including a number of aliens. A recent comer – *Amsinkia*, seems to be spreading over the sandy areas. Some weed plants rapidly colonise waste ground and notable amongst these are the populations of coltsfoot over vast areas of ironstone workings at Scunthorpe and the sand and gravel workings at Lincoln and Woodhall Spa. Blue fleabane and canadian fleabane are also found in these situations.

WEATHER AND CLIMATE OF LINCOLNSHIRE AND SOUTH HUMBERSIDE
by Mr. B. Fox-Holmes

INTRODUCTION

Although readers will be very familiar with the two words, 'weather' and 'climate', I consider it important that they have a clear understanding of the meanings. **Weather** is the state of the atmosphere over a given region at a given time. **Climate** indicates the characteristic features of the weather in that region over a long period.

Weather has been with us always, but it is only in more recent times that comprehensive records have been maintained, eventually being based on readings of reliable, standardised instruments read under standard conditions, and thus permitting legitimate comparisons.

I have tried to compare the climate of Lincolnshire and South Humberside with that in other parts of the British Isles, and later considered climatic variations over the Lincolnshire and South Humberside area. Statistics are available from a large number of reporting stations within the area, but for varying periods, some broken, thus necessitating the exercise of caution when comparing values of a particular parameter at various stations. It is usual to determine average values over a thirty year period, but it is not always possible to obtain data for the same 30 year period when comparing the climate at a number of stations, and sometimes there are gaps in the period. Averages for the same place but for different periods will show differences, which may be quite large if the selected period is short. For example, the average July rainfall at Coningsby was:

1959-75	49.6mm
1969-79	39.9mm
1975-80	26.1mm

Finally, I have illustrated the variation of seasonal temperature and rainfall at a particular Lincolnshire station (Coningsby) over several years.

WEATHER AND CLIMATE OF LINCOLNSHIRE AND SOUTH HUMBERSIDE

1. The climate of the British Isles, largely dominated by Westerly winds arriving after a long sea track, is described as 'temperate', but this covers quite a wide range. For example it is markedly warmer in the south than in the north, whilst the east is much drier than the west.

LINCOLNSHIRE AS A HABITAT

2. The Climate of Lincolnshire and South Humberside in relation to that for the rest of the British Isles.

Charts of averages of mean daily temperature, annual rainfall, mean daily sunshine, wind speed and number of days per year with gales (Figs 1 to 5) enable Lincolnshire and South Humberside values of these elements to be compared with those for other parts of the British Isles.

The mean daily temperature is about average for the British Isles, mean daily sunshine is above average, annual rainfall is about the lowest, the number of mornings with snow lying is rather low especially in the east, and wind speeds below average.

3. Variations of Climate over Lincolnshire and South Humberside

Even when one relatively small area such as Lincolnshire and South Humberside is considered, significant variations are still evident.

Areal means, shown in table 1, for the areas indicated in Fig. 6. illustrate small differences in mean values of air temperature, rainfall and daily sunshine. Areas I & II are chiefly the old Kesteven and Lindsey areas of Lincolnshire divided by the western edge of the Wolds. Area III covers the low-lying districts around the Wash, including much of the old Holland division of Lincolnshire.

	AREA I			AREA II			AREA III		
LAT	53.2N			53.4N			52.8N		
AV HT	42m(138ft)			47m(154ft)			5m(16ft)		
HT RANGE	2-130m			0-126m			0-51m		
	AIR TEMP deg.°C	RAIN mm	SUN HR/DAY	AIR TEMP deg.°C	RAIN mm	SUN HR/DAY	AIR TEMP deg.°C	RAIN mm	SUN HR/DAY
JAN	3.0	53	1.6	3.1	57	1.7	3.1	48	1.7
FEB	3.3	41	2.5	3.4	44	2.4	3.4	38	2.4
MAR	5.5	42	3.6	5.4	45	3.7	5.6	38	3.8
APR	8.3	41	5.3	8.1	44	5.2	8.5	37	5.5
MAY	11.2	51	6.4	10.8	54	6.3	11.3	44	6.6
JUNE	14.3	48	6.7	14.0	49	6.7	14.4	47	7.0
JULY	16.1	55	5.8	15.7	57	5.8	16.1	56	6.1
AUG	15.7	64	5.5	15.5	67	5.4	16.0	60	5.6
SEPT	14.0	51	4.2	13.9	55	4.3	14.2	49	4.7
OCT	10.6	49	3.2	10.6	54	3.3	10.8	50	3.5
NOV	6.3	60	1.9	6.4	66	1.9	6.4	58	2.0
DEC	4.1	50	1.4	4.1	56	1.5	4.2	50	1.5
TOT		605			648			575	
MEAN LAST FROST	EARLY MAY			EARLY MAY			LATE APRIL		

TABLE 1. Areal Average Values of Daily Temperature, Monthly Rainfall and Daily Duration of Sunshine.

The topography of the areas and the distance from the North Sea will have influenced the climate of the areas, though the resulting differences will be rather obscured in the areal values quoted. None the less, it will be seen that Area III has higher mean daily temperatures, a lower monthly rainfall and greater daily sunshine duration than in Areas I & II. Although Areas I & II have similar mean daily sunshine, Area I is slightly warmer and drier than Area II.

The variation of rainfall over the area, especially in relation to the togography and distance from the sea, is well illustrated in Fig.7.

4. A Comparison of the Climate at Specified Stations in Lincolnshire & South Humberside

A number of individual stations (Cleethorpes, Lincoln, Cranwell, and Skegness) in various parts of Areas I, II & III (as defined in para.3) have been selected and comparisons made of long term (1941-70) values of average daily mean temperature, average monthly maximum temperature, average monthly minimum temperature, and average monthly total of sunshine. These are shown in Table II. (N.B. Several of these stations had some breaks in their 30 year record between 1941 and 1970). It will be noted that the average daily mean temperature shows little variation over the county, but inland the average maximum temperatures are significantly higher than at stations near the coast, whilst average minimum temperatures are lower inland than near the coast. This difference between places on the coast and inland may be illustrated by determining the average monthly range of temperature as shown in Table III. The range at the coastal stations (CL & SK) is less than at the inland stations (LI & CR), where the low-lying Lincoln station has the greater range. Sunshine hours show no marked pattern.

LINCOLNSHIRE AS A HABITAT

MONTH	AVERAGE DAILY MEAN TEMP (deg.°C) CL	LI	CR	SK	AVERAGE MON. MAXM TEMP (deg.°C) CL	LI	CR	SK	AVERAGE MON. MINM TEMP (deg.°C) CL	LI	CR	SK	AVERAGE MON. TOT. SUNSHINE HRS CL	CR	SK
JAN	3.5	2.7	2.7	3.2	11.0	11.8	11.4	10.9	-4.2	-7.9	-6.5	-4.4	48.5	54.8	52.3
FEB	3.7	3.3	3.1	3.5	11.9	12.2	12.1	11.6	-3.9	-7.5	-5.7	-4.1	65.3	71.6	68.6
MAR	5.7	5.5	5.3	5.3	16.0	16.6	16.3	15.1	-2.9	-6.2	-4.3	-3.0	109.9	113.1	115.4
APR	8.4	8.3	8.1	8.1	19.5	19.5	19.3	18.5	0.0	-3.4	-1.2	-0.3	157.4	157.7	163.4
MAY	11.0	11.1	11.1	10.8	22.0	23.4	22.7	21.5	2.5	-1.4	0.9	2.2	191.8	196.6	200.2
JUNE	14.2	14.2	14.3	14.0	25.4	26.1	25.9	24.8	5.7	2.1	4.6	5.7	200.6	201.0	203.6
JULY	16.0	15.9	15.9	15.9	26.4	27.1	27.2	25.5	7.9	4.4	7.0	7.5	184.0	180.0	187.3
AUG	15.9	15.7	15.7	15.7	25.1	26.3	26.1	24.8	8.0	3.7	6.7	7.4	167.7	165.9	174.3
SEPT	14.3	13.7	13.8	14.1	23.6	24.1	23.8	23.4	5.9	1.3	4.4	5.2	138.8	137.0	144.8
OCT	10.9	10.3	10.4	11.0	19.6	20.0	19.3	18.9	1.7	-2.1	0.3	1.6	103.5	103.2	104.5
NOV	6.7	6.1	6.1	6.7	14.4	14.7	14.1	14.3	-1.5	-5.0	-2.7	-1.8	55.9	60.2	61.0
DEC	4.5	3.7	3.9	4.4	11.9	12.3	12.2	11.8	-3.7	-6.5	-5.0	-3.7	46.1	52.4	47.5
Year	9.5	9.2	9.2	9.5	27.6 X	28.9 X	29.0 X	26.9 X	-6.1 X	-10.2 X	-8.3 X	-6.8 X	1469.5	1493.5	1522.9

TABLE II Average values of Mean Daily Temperature. Monthly max & min Temperatures. Monthly Sunshine, at four stations.

STATION DETAILS

STATION	(CL) CLEETHORPES	(LI) LINCOLN	(CR) CRANWELL	(SK) SKEGNESS
HEIGHT	7m	7m	62m	5m
NAT.GRID REF.	TA(54) 314079	SK(43)962719	TF(53)003494	TF(53)569631
PERIOD COVERED	1941-55 1958-70	1945-70	1941-70 with breaks	1941-70

X = Averages of Highest/Lowest Temperatures recorded in each year, regardless of the month of occurrence.

40 THE BUTTERFLIES AND LARGER MOTHS

MONTH	AVERAGE MONTHLY TEMPERATURE RANGE (deg.°C)			
	CL	LI	CR	SK
JAN	15.2	19.7	17.9	15.3
FEB	15.8	19.7	17.8	15.7
MAR	18.9	22.8	20.6	18.1
APR	19.5	22.9	20.5	18.8
MAY	19.5	24.8	21.8	19.3
JUNE	19.7	24.0	21.3	19.1
JULY	18.5	22.7	20.2	18.0
AUG	17.1	22.6	19.4	17.4
SEPT	17.7	22.8	19.4	18.2
OCT	17.9	22.1	19.0	17.3
NOV	15.9	19.7	16.8	16.1
DEC	15.6	18.8	17.2	15.5
YEAR	33.7	39.1	37.3	33.7

TABLE III Average values of Monthly Temperature Range at Four Stations.

The temperatures of coastal areas are influenced by the adjacent sea temperatures which vary much less than land temperatures, and this influence decreases with increasing distance from the sea. The sea temperature is higher than the mean land temperature in Winter and lower in Summer. Mean sea temperatures (1905-1954) for the sea near the coast from the Humber to the Wash are shown in Table IV.

MONTH	JAN	FEB	MAR	APR	MAY	JUNE
TEMP (deg.C)	5.5/6.0	4.5/5.0	5.0/5.5	6.5/7.0	8.5/9.0	11.5/12.0

MONTH	JULY	AUG	SEPT	OCT	NOV	DEC
TEMP (deg.°C)	13.5/14.0	14.5/15.5	14.0/14.5	12.5/13.0	9.0/9.5	6.5/7.0

TABLE IV Mean Monthly Sea Temperature of Adjacent North Sea

5. A comparison of Seasonal Values of Temperature and Rainfall each Season over a number of Years at a Particular Station (Coningsby) in Lincolnshire and South Humberside.

Values of the mean daily temperature and the rainfall recorded at Coningsby during Winter (Dec, Jan & Feb), Spring (Mar, Apr, May), Summer (June, Jul & Aug), and Autumn (Sep, Oct & Nov) will be compared for each season during the period from Autumn 1968 until Winter 1979/80.

LINCOLNSHIRE AS A HABITAT 41

YEAR	WINTER Temp	WINTER Rain	SPRING Temp	SPRING Rain	SUMMER Temp	SUMMER Rain	AUTUMN Temp	AUTUMN Rain	THROUGHOUT YEAR Temp	THROUGHOUT YEAR Rain
	°C	mm	°C	mm	°C	mm	°C	mm	°C	mm
1968							10.9			
1969	2.7	133.0	7.1	220.8	15.5	173.3	10.5	111.0	9.0	670.1
1970	3.0	145.6	7.6	126.2	16.0	115.7	10.9	173.3	9.5	535.2
1971	4.2	110.5	8.1	106.5	15.2	132.6	10.5	80.2	9.6	423.9
1972	4.5	109.2	8.6	147.2	14.2	127.5	9.5	112.7	9.1	523.8
1973	4.6	85.9	8.3	112.2	15.9	200.4	9.7	98.4	9.6	477.3
1974	5.1	117.6	7.9	49.5	14.9	165.3	8.7	216.6	9.4	552.8
1975	5.8	121.6	7.6	159.7	17.0	84.6	9.9	107.3	9.8	482.5
1976	4.9	97.8	8.3	83.6	17.6	21.9	10.1	245.3	10.0	456.6
1977	2.7	223.1	8.1	92.1	14.5	122.9	10.5	85.3	9.3	515.5
1978	3.5	181.7	7.9	123.7	14.5	227.5	11.4	55.6	9.2	669.1
1979	1.4	213.3	7.7	230.3	15.2	105.9	10.5	112.6	8.9	613.7
1980	4.2	210.9								

TABLE V Annual & Seasonal Values of Mean Daily Temperature and Rainfall at Coningsby.

From Table V it will be seen that the annual rainfall varies quite significantly from year to year and that the temperature also varies but to a lesser degree. However, it will be seen (Table V) that seasonal values exhibit even greater variations, and this may be of significance to the flora and fauna. The still greater monthly variations (not shown here) may need to be considered when studying the effect of weather on wild-life. The warm summers of 1975 and 1976, especially the latter, are shown clearly as is the drought of 1976.

6. **Surface Wind over Lincolnshire and South Humberside.**

In section 2, the averages of wind speed and the number of gales in a year over Lincolnshire and South Humberside were compared with the rest of the British Isles. Detailed wind statistics are available for some stations in this area, but are either in the form of complicated tables or of wind roses, and not suitable for a brief, – easily comprehendable presentation.

Fig.8, shows a typical wind rose, for Manby, and it shows clearly that the wind blows mainly from between South and Northwest. However, it does show that Northerly and Easterly winds are not infrequent, and Winds from these directions can have considerable significance, both to human beings and to the flora and fauna.

Sea breezes can occur from Spring to Autumn when suitable conditions prevail, carrying cooler and moister air, not merely to the holiday resorts, but for some considerable distance inland. Sometimes these may aid the development of afternoon thunderstorms, or nocturnal mists and fogs.

In Summer' Easterlies may bring inland warm dry continentall air masses.

In Winter, Northerlies and Easterlies bring our cold and snow spells.

Thus, although our 'temperate' climate, referred to in section 1, is maintained by the prevailing Westerlies for much of the time, the wind rose does confirm that we do have considerable spells of Northerlies and Easterlies when it may be far from 'Temperate'.

ACKNOWLEDGEMENTS

I wish to acknowledge my thanks to the Director-General of the Meteorological Office for allowing me access to Meteorological Office publications and in particular to use information on published diagrams (to draw up FIGS 1, 3, 4, 5) and tables (TABLES II, III, IV, & V) which appear in Meteorological Office publications or records, and for FIG 8 & TABLE I which were published in MAFF Technical Bulletin 35.

My thanks are also due to the Anglia Water Authority (Lincolnshire River Division) for allowing me to use the information on which FIG 7 is based.

Further, all the material is used with the permission of the Controller of Her Majesty's Stationery Office.

FIG.1. Average means of daily mean
temperature: 1901 – 1930

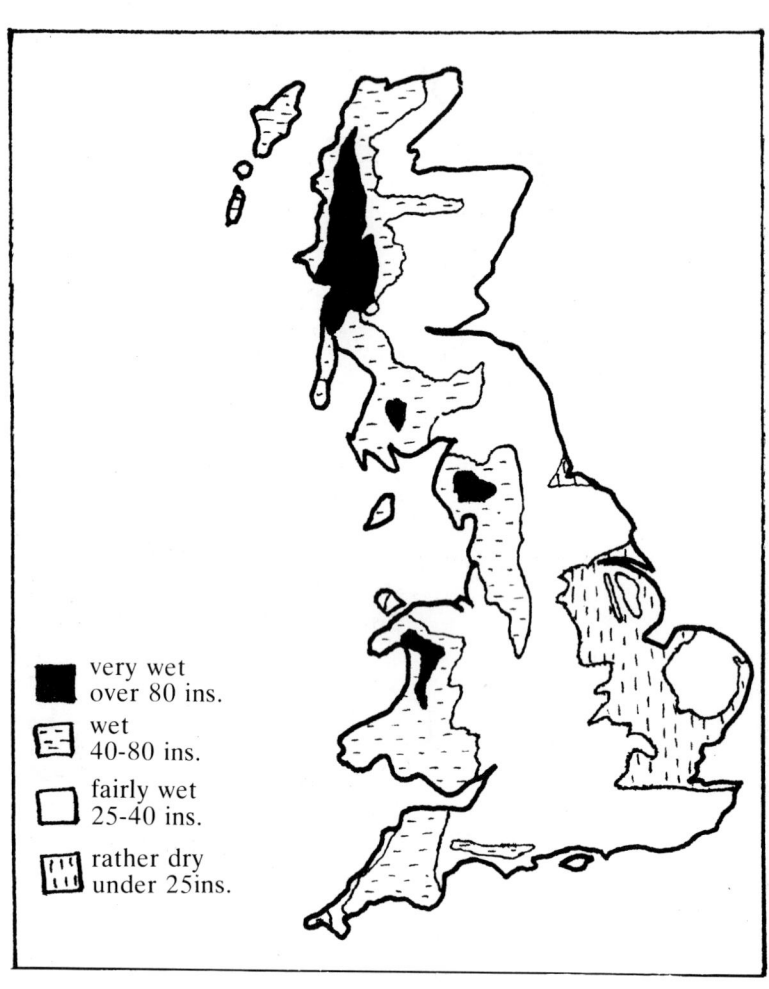

FIG.2. Average annual rainfall (simplified)

FIG.3. Average means of daily duration of bright sunshine: 1901 – 1930

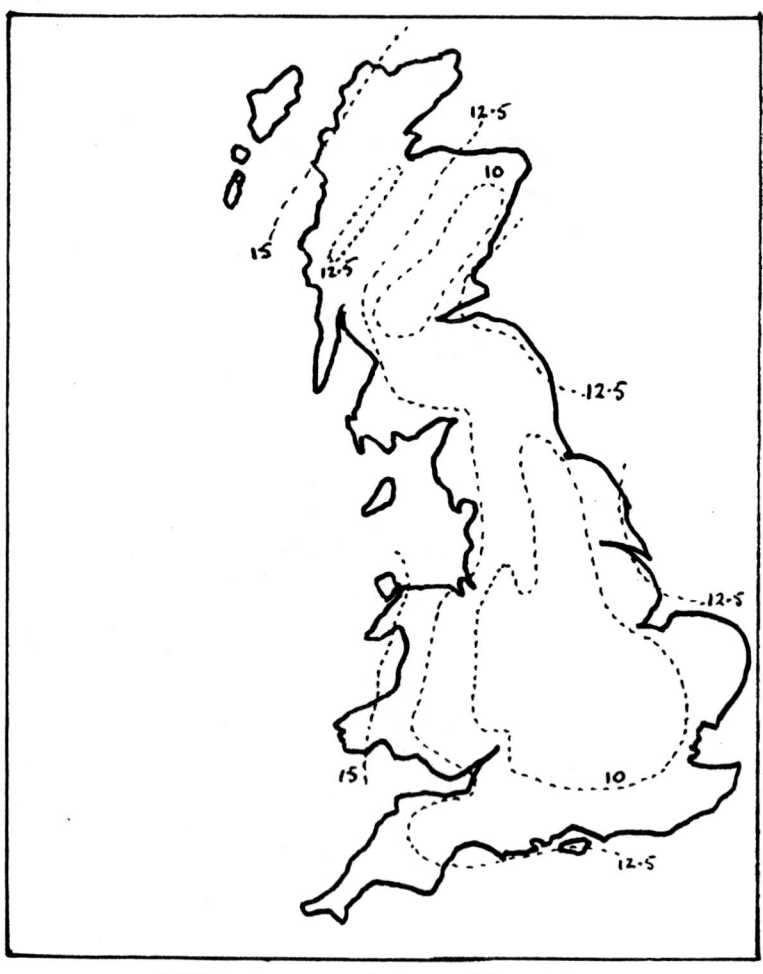

FIG.4. Average wind speed (m.p.h.):
1926 – 1940

LINCOLNSHIRE AS A HABITAT

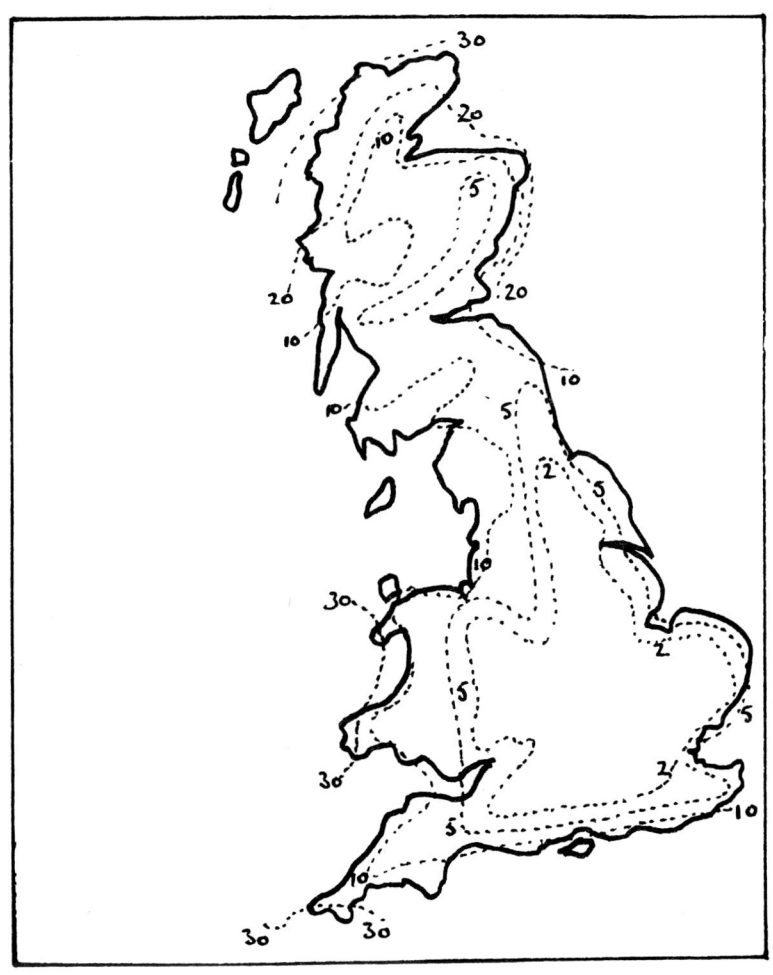

FIG.5. Approximate average annual number of days with gale: 1918 – 1937

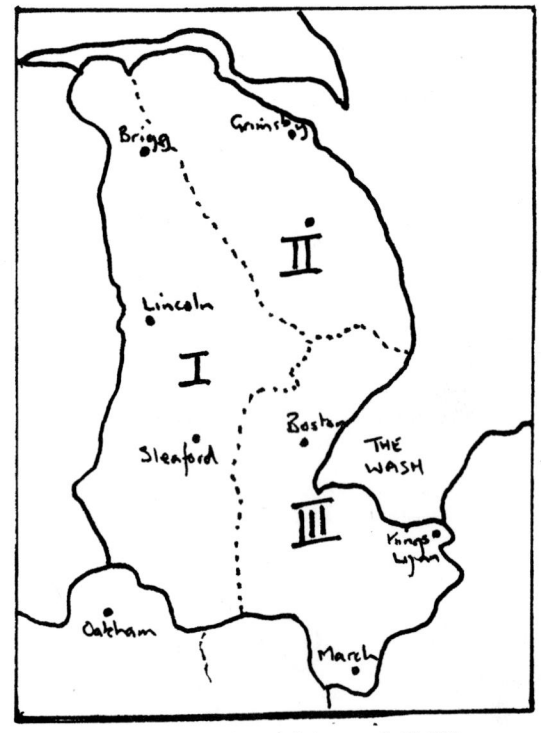

FIG.6. Location of Areas I II III

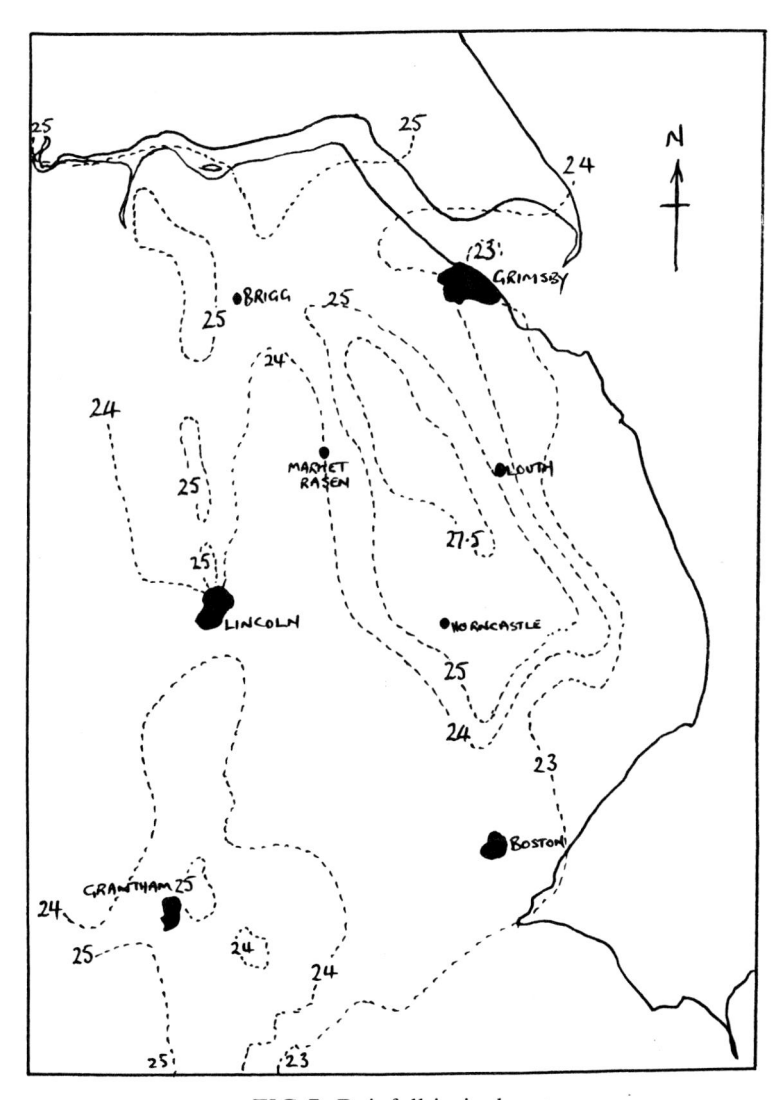

FIG.7. Rainfall in inches.
Taken from Standard Average Rainfall
Anglian Water Authority
Lincolnshire River Division
1916 – 1950

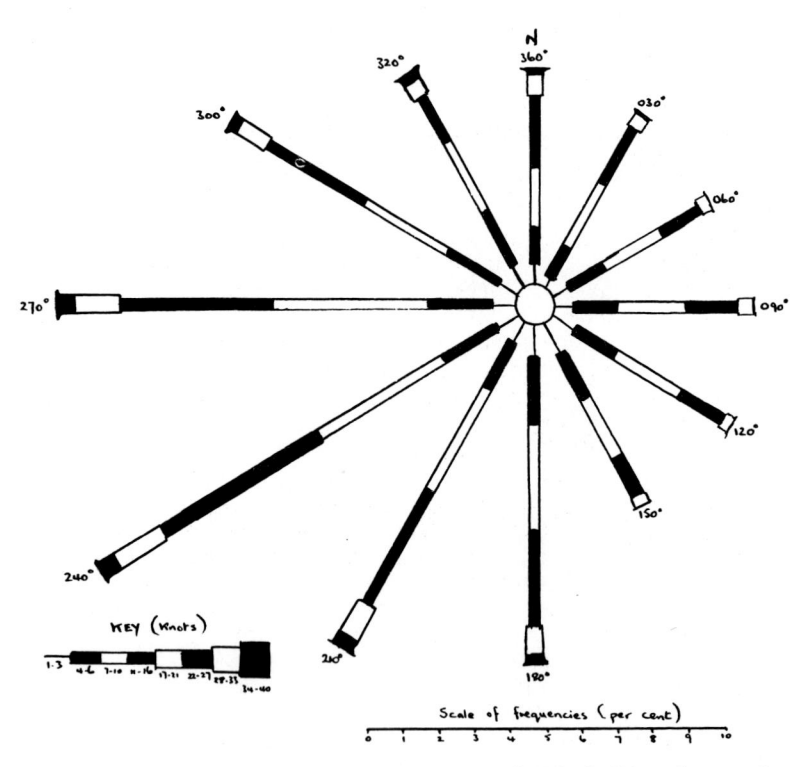

FIG.8. Annual Percentage Frequency of Wind Direction and Speed. 1962 – 1971 at Manby (53deg. 21' N, 00deg. 05' E).

WILDLIFE AND CONSERVATION IN LINCOLNSHIRE
by Mrs. A. Goodall.

It would be pointless to pretend that Lincolnshire's countryside has not been affected by the post-war agricultural revolution; the quest for ever-increasing yields from ever more arable acres. Anyone with eyes to see has a tale of coppices cleared, hedgerows gone, a favourite stream dredged and straightened or – quite illegally – verges ploughed.

There are areas of the county which are less affected, like the great estates which have largely retained their woodlands and shelter belts. In other areas the land use has been ironstone mining, as in the north-west, or gravel extraction, mainly in the river valleys, or commercial forestry. But we live in an agricultural county, and by and large we have the landscape our farmers give us.

Against this background the diversity of habitat and scenery which Lincolnshire has to offer comes to many people as a very real and pleasant surprise.

In this county we have ancient limewoods, gravelly streams, sandy heathlands, marshy meadows, limestone grasslands, reed-fringed lakes and glacial valleys. On the coast, away from the holiday spots, you may choose to visit shingle beaches, marram-clad sand-dunes, bird-rich mudflats or miles of open saltmarsh. Some of these areas are protected as nature reserves, national, local or private, or are covered by management agreements, (reserves owned, leased or managed by the Lincolnshire and South Humberside Trust for Nature Conservation are shown on the map, Fig. 3) but most of the land is privately owned. However, the owner of land is also its guardian and a most important function, both of the Trust and of the Nature Conservancy Council, is to offer advice and help to landowners on management to protect these vital habitats.

The N.C.C., an official body established by Parliament, is required by law to notify the local planning and water authorities of any area which is, in its opinion, "of special interest by reason of its flora, fauna, or geological or physiographical features". Areas notified in this way are known as **Sites of Special Scientific Interest** and the law is intended to ensure that they are not unwittingly destroyed. The planning authorities must consult the Conservancy before granting permission for any development affecting such a site, though unfortunately agricultural changes, such as ploughing up grassland, **do not** require planning permission. The sites remain private property and no other legal right, such as that of access, is conferred. In some cases this fact is itself a necessary protection for these areas; many habitats are fragile and even naturalists, especially in ever-increasing numbers, can damage them.

Fortunately there are many other areas, including some of the most lovely, where access is possible; some owners, including the Forestry Commission, allow reasonable access when sensible rules are observed; some areas are crossed by public rights of way; and some are Trust Nature Reserves, which members may visit by permit. Needless to say, right of access should **never** be assumed; permission must always be obtained if there is any doubt.

On any relief map of Lincolnshire one of the first regions to catch the eye is the broad flat plain of the fenlands. This is the Lincolnshire pictured by non-natives; reed-fringed dykes, far horizons, and grey geese against a wide, empty sky – the Lincolnshire of Peter Scott.

In reality the fenland is two regions. Circling the Wash, enmeshed by narrow roads and dotted with villages, are the silt fens. This is fertile land and most of it is under arable cultivation, but the fields are smaller, and often grow vegetables, fruit or bulbs. Around the villages especially there is still some grassland, and good hedges with hedgerow trees; ashes, oaks and willows. The same trees, with files of poplars, are also planted as shelter belts around isolated farms and the outskirts of villages, though larger areas of woodland are rare.

Outside the silt fens comes the wide circle of the peat fens, the fenland proper perhaps. This area was late coming into cultivation and even today it is sparsely populated. Hedges and trees are rare, the huge fields being divided mainly by drainage dykes, and the major land use is intensive arable. Late as it was in starting, once begun the drainage and conversion of the fens proceeded so fast and so completely that today we can catch only glimpses of its former state. Unfortunately much of the wildlife depended on a high water table, maintaining this against efficient drainage is an increasing problem. Compounding this, when wild populations become too small and scattered they may be unable to recolonise after such a natural setback as the long dry summers of 1975 and '76.

Among the most vulnerable groups are the marsh plants, which must once have been widespread and are now largely confined to reserves like Baston Fen. Once a 'wash' or flood storage area, this is today permanent pasture with borrow pits and counter drain. It also provides a home for another threatened group, the dragon and damselflies. These include the rare ruddy sympetrum, the bright blue southern aeshna, and occasionally most imposing of all the emperor dragonfly, with a body more than 3" long.

Less threatened are the wetland birds. In winter traditional areas are used by huge flocks of golden plover, and flooded fields are visited by wildfowl. Once these came in such numbers that wildfowling was big business and decoys sprang up throughout the fens. Increasing drainage and decreasing duck numbers put them out of use, but today the most famous of the East Fen decoys, at Friskney, is a nature reserve. Its open water still attracts ducks, but as one of the few pieces of woodland in the area it is at least as important to small woodland birds and animals.

Some insects are typical of the fens. Among the Orthoptera for instance, the lesser marsh grasshopper is widespread and the slender groundhopper is known from several sites. A rarer member of the same group, found only in the fens, is Roesel's bush-cricket. Its known sites are few, but where it occurs the colonies are strong and the males produce a continuous loud churr from dykeside grasses on sunny days.

North and west of the fenland there are areas of sand and gravel, deposited long ago on the margins of the great fen lake.

In the north, around Woodhall, it has produced poor farmland; dry sandy heath with marshy hollows. On reserves like those around Kirkby the dry areas have centaury, heath speedwell and shepherd's cress, with areas of ling and bracken used by nightjars. The wet pastures and boggy hollows grow bog pimpernel, devil's bit scabious and cottongrass. They also provide the county's only known site for the bog bush-cricket.

One of the most important land uses this century has been gravel and sand extraction. The new faces are quickly used by sand martins, and often also by kingfishers, which rear their young on bullheads from the Bain. Where reeds and rushes invade the flooded workings wainscot moths are common, their larvae providing food for reed and sedge warblers. From Domesday times this area has carried a lot of woodland, and still does, however, much of it is now conifer plantation.

On the western fen margin the gravels are not continuous; there is an extensive area in the south, more scattered patches to the north. Here most of the land is ploughed, though there is grassland around the villages and the Carr dyke is fringed by a number of tree belts and small woods. At Potterhanworth there is marshy woodland, with alders, water avens and purple loosestrife, and rides full of butterflies. The slightly higher woods at Nocton have speckled bush-crickets and nesting herons.

West of the gravels and overlying the dip slope of the limestone ridge the land is clay. The farming is mixed, with many small stream valleys and scattered fox-coverts. There are good hedges too, though "prairie-farming" is gaining ground in parts. The woods of Kesteven Forest lie on this dip slope clay, remnants again of extensive Domesday woodland. Again also many of them are now conifer plantations but even these have primroses and violets in the spring. All the woods hold fallow deer and red deer can also be seen, usually grazing the fields close to wood edges. The hardwoods of this area included much elm, now unfortunately suffering the ravages of Dutch Elm disease.

In the centre of the clay band, where glacial melt-water deposited sand on the limestone around Sleaford, it left poor land again. This was once used for rabbit warrening and later for sand extraction, and the reserve at Rauceby shows the effect of this mixed history.

Here plants of acid conditions flourish in the deeper sand pockets, with lime-loving species close by where the sand is shallower and their roots can reach the limestone.

West of the fen margin rises the limestone ridge, one of the county's two linear "uplands". Variously known as the Cliff, the Lincoln Edge and the heath, the ridge rises in south Kesteven to just under 500 feet. Here, at its widest part, it also has a capping of clay and much woodland. One of these woods, Tortoiseshell Wood in Witham parish, is now a nature reserve. Under its standard oaks the shrub layer includes coppiced ash and hazel, with field maple and the scarcer wild service. The woodland floor is carpeted in season with bluebells.

The limestone ridge itself is best known for its old and species-rich grasslands. Once much more widespread, remnants can still be seen on some stretches of the verge beside the Roman Ermine street and in small areas such as the reserves at Wilsford heath and Ancaster valley. These have some lovely plants and the same flora has developed also in many of the older limestone quarries scattered along the ridge. One such quarry also holds the county's only colony of the chalk carpet moth, and several harbour fallow deer and green woodpeckers.

Much of the ridge-top and dip slope has now been ploughed, and the large fields are bounded by dry-stone walls. On the scarp slope of the Edge there are marshy meadows where the springline runs.

The largest stretch of woodland on the ridge occurs at Broughton in the north where there were beechwoods, though they are being replaced with conifers.

Between the scarp slope of the cliff and the county boundary lie the western lowlands: the valley of the Trent and upper Witham and their tributaries. The soil is a patchwork of clays and gravels, producing similarly mixed farmland. Grass fields are common, well-hedged and often marshy. The Moor Closes reserve at Ancaster, lying on outwash gravels of the Ancaster gap, shows perhaps what the area once looked like.

East and south-east of Lincoln is a large area of old and current gravel workings which have left extensive reaches of open water. In winter the area holds many hundreds of waterfowl and some species also remain to breed. At the Burton Pits reserve these include great crested and little grebes, pochard, tufted duck, and kingfishers are also not uncommon. This area is a stronghold for the dragonflies; the brown aeshna is common and rarer species include the banded agrion and the emerald damselfly.

Further north the Trent valley provides a flyway for migrant birds on spring and autumn passage. The Trent carries many plant seeds down from the midlands and up from the sea, so unusual plants like the "policeman's helmets" often spring up on its banks. Seeds of saltmarsh plants are regularly carried far upstream, and every few years are joined by an unwelcome visitor like the thoroughly

LINCOLNSHIRE AS A HABITAT

poisonous hemlock water-dropwort.
West of the Trent in the northwest corner of the county lies the Isle of Axholme. The central ridge of clay, some 10 miles long, was literally an island in a fenny waste until the surrounding peat and silt lands were drained from the seventeenth century. The Isle and alluvial silt today are intensively arable, with small pockets of grassland and many deep wide drainage dykes; the peat is removed commercially for garden use.

As with the fens, the wildlife of the Isle depends on its watertable. Where there are low lying meadows, yellow flag, marsh marigolds and the rare fen sedge grow. The small remaining areas of peat, on Crowle Waste and the Haxey and Epworth turbaries, are protected as reserves and carry cottongrass, meadowrue, yellow and purple loosestrife and bog rosemary. On the peat wastes the black sympetrum dragonfly occurs, with moths such as the drinker and the last population of the large heath butterfly in the county.

Across the north of the county lie several deep tracts of cover sands. These occur in three main areas; the Trent valley north of Gainsborough, the scarp edge of the limestone cliff east of Scunthorpe, and the scarp edge of the wolds between Caistor and Market Rasen. In all three areas the poor sandy soil has proved unrewarding to farm, and was used as rough grazing or rabbit warrening; uses still reflected in many placenames. Today large areas lie under conifer plantations, but remnants of the original heathlands do exist.

The once extensive heath north of Gainsborough, for instance, is now largely covered by the Forestry Commission's Laughton Forest, but a small part has been preserved at Scotton Common. The dry areas here have ling and bell heather, with adders and common lizards and a high proportion of the county's breeding nightjars. In summer the roadside verges are alive with butterflies, and uncommon moths include the beautiful yellow underwing. The forest itself has much of importance, especially perhaps the county's last stronghold for the red squirrel.

Close to Scunthorpe the sand lying against the limestone ridge was used for warrening, and some of the old warrens still remain in private ownership. The mottled grasshopper is common here and moths found include the wood tiger and the latticed heath, while heather-covered areas still have the grayling butterfly. At the foot of the slope, pools in the peat overlying the sand are colonised by marsh cinquefoil and used by many dragonflies, including the black sympetrum. The scarce emerald damselfly was known and may still be here, since the commoner and very similar emerald damselfly abounds. Around Messingham the very pure silica sand is extracted for glass-making, leaving shallow lakes much used by water-birds and passage waders.

Around Market Rasen the sand is covered, as at Laughton, by the Willingham Forest, though part of one old warren, at Linwood, is now a nature reserve. Iceland moss – an alpine lichen which is very rare on lowland heath – and creeping willow can still be found, but the cottongrass and sundews may be casualties in a losing battle with the falling water table.

Between the cliff and the wolds in the north of the county lies the Ancholme valley. The river itself is canalised and serves now mainly to drain the valley farmland though it once carried barge traffic to Brandy Wharf and Caistor. Above Brigg the farming is mainly arable but to the north it is more mixed, with several old meadows in the carrlands.

South of Market Rasen the valley is occupied by the Barlings river, emptying south into the Witham. The Wragby clays, which are drained by the Barlings tributaries, were densely wooded in Domesday times and still carry much woodland today. Most of this comprises the Forestry Commission's Bardney Forest, but the treatment of the historic limewoods has been fairly sympathetic and large areas remain. In addition, conifers grow poorly here so mainly hardwoods have been planted. Native trees commonly occurring include wild service, dogwood, and guelder rose. Dark and oak bush-crickets are common and the white admiral butterfly is found.

Within one of the blocks of woodland lies the Little Scrubbs meadow reserve. This is maintained by regular mowing to preserve the meadow flora, which includes saw-wort, great burnet, dyer's greenweed and valerian, besides several orchids: green-winged, spotted and greater butterfly. The surrounding woods have nightingales, grasshopper warblers and woodcock.

East of the valley lies the broad upland of the wolds, which is not one habitat but several. The chalk of the wold tops once carried sheepwalks and is now under intensive arable, with very large fields and few internal hedges. Along the roadsides the typical hedgerow tree is ash. In autumn vast numbers of gulls sit on the burnt stubbles or follow the plough, and in winter flocks of golden plover feed on the fields.

This plateau is much dissected by stream valleys, carrying tributaries of the Bain, the Waring and the Steeping river to the south, the Hatcliffe beck to the North. These valleys are sometimes steepsided and often clay-bottomed and grassy, with alders and sallows, cuckooflower and ragged robin, kingfishers and snipe.

The wold edges are different again. To the west the scarp slope has much grassland, with small fields well-hedged, and shelterbelts around the villages. The dip slope onto the marsh carries several areas of parkland and the arable fields often have good hedges, with hedgerow oaks.

Towards the southern edge of the wolds are areas where sandstone outcrops. At Snipe Dales the stream has cut down through the sandstone into clay, so that the valley bottom is marshy, with ferns and marsh marigolds, while sand martins nest in the sandstone faces of the valley sides. The reserve also has badgers and deer, and

LINCOLNSHIRE AS A HABITAT 57

sparrowhawks hunt small birds along the edges of its plantations. From the eastern foot of the wolds, the marshes run down to the sea. The middle marsh, closest to the wolds, lies on clay. It's big regular fields carry mainly arable crops, but the dykes which bound them and the streams running off the slope have much wildlife interest. The dyke sides have colonies of lesser marsh grasshoppers and the holes of water voles, and several species of wainscot moths fly. Older woods on the clay, like the Hoplands wood reserve are species rich, with a ground flora including woodruff, herb paris, lady's mantle, common helleborine and butterfly orchids.

Beyond the middle marsh is the alluvial silt of the outmarsh, which once had smaller fields and much grassland where the Lincoln sheep were sent to fatten. Now the fields are arable like the middle marsh, with the pastures and shelter belts grouped around the villages and farms. The flora of the old grasslands is preserved at the Heath's Meadows reserve near Bratoft, where there are cowslips, devil's bit scabious, and green-winged orchids. A feature of this region is the 'blow-well' a group of ponds produced by springs bubbling at the surface. Some are now dry or overgrown, but the Tetney blow-wells have water violets and marsh marigolds, and the breeding birds include snipe and reed buntings. The willow carr provides food-plants for the larvae of many moths which fly there, while the closeness of the coast means that rarer migrants regularly arrive.

For many, perhaps most people, the Lincolnshire coast with its well-publicised holiday resorts probably does not automatically conjure up visions of solitude. Out of season though, or away from the holiday hotspots, the shore belongs to the fishermen and the birds.

To the north the county is bounded by the estuary of the River Humber, formed by the confluence of the Trent and Yorkshire Ouse. From Alkborough to Ferriby the river is edged by narrow grassy marshes outside the sea bank, inundated by the highest tides. The upper estuary also holds the Wildfowl Refuge, established in 1955, and in winter hundreds of pink-footed geese and many thousands of duck can be seen on the river and mudflats. Grazing ducks like the wigeon feed on Alkborough Flats, which also holds very large flocks of dunlin and ringed plover on spring and autumn passage.

From Barton to Goxhill the brick-making industry has left a string of old clay pits alongside the river. These provide a number of important wildlife habitats, from deep open water through to drying reed beds with invading scrub. On the bank grow teasels and sea wormwood, while the hawthorn scrub is often covered by the webs of broods of lackey moth caterpillars.

Between East Halton and Humberstone the coast is protected by a sea-wall but behind it the pits at Killingholme attract many birds, particularly migrant waders. Below Cleethorpes the shore becomes

58   THE BUTTERFLIES AND LARGER MOTHS

sandy, with prickly saltwort and sea rocket on the strand line and in sheltered areas saltmarshes form. The RSPB reserve at Tetney has the largest colony of little terns on the Lincolnshire coast – up to 200 pairs in some years, though between high tides and predation their breeding success is rarely high.

South of Donna Nook the coast is backed by sand-dunes, often colonised by sea buckthorn bushes. The fox moth flies here, with such butterflies as the Essex skipper and the green hairstreak. The natterjack toad was once common right down this coast, but the National Nature Reserve at Saltfleetby-Theddlethorpe is its last refuge.

Between Mablethorpe and Skegness the coast is often crowded, though in winter sanderlings chase the sea over the shingle and flocks of golden plover use the grass fields landward of the bank. Storms at sea often drive the rarer grebes, divers and sea ducks inshore, or over the bank on to the clay pits at Huttoft and Chapel. In autumn migrant birds follow the coast south until it suddenly turns west into the Wash, concentrating them on Gibraltar Point. For this reason the Bird Observatory was established here, and many thousands of birds are logged each year. The reserve is also important for its marsh and dune communities of plants and invertebrates.

For birds the Wash is the most important estuary on the east coast of Britain, and in terms of numbers using it each year it is second only to Morecambe Bay in the British Isles. Almost uncountable flocks of waders feed and roost here, and wildfowl to be seen in large numbers include pinkfooted and brent geese, shelduck and wigeon. Frampton marsh has a very large black-headed gull colony reputed to be the largest in the country, and in winter huge numbers of twite feed on the *Salicornia* and sea aster seeds here. Seals also use the Wash; common seals haul out on the mudbanks to calve, and grey seals can also be seen at some times of the year.

Perhaps few people would consider the Wash marshes as an area of outstanding natural beauty, but to those who love them they certainly are, and their importance to wildlife on a national, indeed international scale is being increasingly recognised. Here truly belongs the sense of distance and isolation to be found reflected in the fens – just over the bank.

THE ECOLOGY OF CERTAIN LINCOLNSHIRE BUTTERFLIES AND MOTHS.

The Authors.

Having an interest in the Lincolnshire lepidoptera need not mean that one's outside fieldwork is confined to the warmer months of the year when insect activity is at a peak. During colder winter days it is possible to go out "digging" for pupae. One can look for ova, or in the early spring when larvae emerge from hibernation and begin to feed up, they can be searched for at dusk by torch light. In one or another stage of their life-cycles, lepidoptera are everywhere throughout the year.

In his book "Leaves from a Moth-hunter's Notebook" (1980, p.169), P.B.M. Allan wrote "Clearly it must be in the nature of every lepidopteron to spread throughout the whole of an area which satisfies all its physiological requirements; for it is to the advantage biologically of every organism to occupy the whole area which is capable of being a habitat. In fact it will continue to spread. . . until the climatic factor becomes adverse, though the other ecological factors may still be congruous. For climate is ultimately the determining factor in the existence of an insect in any particular place."

Some of the Lincolnshire lepidoptera find the whole of the county area suitable to their physiological requirements, and they can be commonly seen in all areas. A read through some of the butterfly and moth records will illustrate how many are extensive in their range – perhaps because their larvae are not too specific about their foodplants. After all, there are grasses, nettles, dandelion and assortment of other "low plants" to be found almost everywhere. We are not concerned with these more "universal" species in this section of the chapter.

In his chapter entitled "Larva-hunting in Spring" (P.B.M. Allan, 1980, p.168), Mr. Allan wrote: "There are species which one finds only on chalk and sandy subsoils; there are some which inhabit dank woods on boulder clay. There are others which one meets with only on the sloping faces of cliffs, or can be collected only in fenlands by the man clad in waders. But whatever the quarry, provided one knows what one is about and one's plans are properly laid, larva-hunting by night in spring must come high on the list of the moth-hunter's preferences". The thinking naturalist – whether he deals with birds, bees or butterflies – not only wants to know "where" the object of his interest is sited. He will also ask "why" – as part of the process Allan calls knowing "what one is about" in the above quotation. To assist the reader in coming to this state of understanding concerning our County, we are pleased and grateful to have available the excellent articles in this book by our guest authors.

These authors have painted skilful cameos in words of the Lincolnshire geology, flora, weather, and range of habitats. What follows is the briefest of looks at how a number of our more restricted butterflies and moths fit into this landscape, and for ease of reference back to the section on flora, the same pattern of habitat types mentioned early in the chapter will be followed here.

A. Coastlands
(a) Saltmarshes

One of the most restricted moths in the County is the Scarce Pug which has its larvae feeding on sea wormwood. The flora section noted that this plant is not found "in quantity". In his Presidential Address to L.N.U. members (delivered March 18th 1978, and reported in L.N.U. Transactions, Vol. XIX, No. 3, Parts 1 and 2, 1978) Mr. J. H. Duddington commented that "This little moth is only found between the Humber and the Thames". (It has been recorded up to S. Yorkshire, found in a locality near Kilnsea). "This is the only region of Britain where the climate approaches the dry Continental type which possibly suits the moth."

Larvae of the Rosy Wave moth feed on sea lavender. This species used not to be uncommon around Gibraltar Point but after flooding in 1953 it became more scarce. Now, even though this is the extreme northern limit of its range in Britain, the population has recovered, and it is quite frequent again. (It would appear that the Rosy Wave is also an old inhabitant of the Fens. Dr. R. E. M. Pilcher, in L.N.U. Transactions Vol. VI. No. 1, 24, mentions it being a survivor at least of the warm period preceding the fourth and last glaciation of the last ice age).

Saltmarsh grasses support the Crescent Striped moth at places like North Somercotes, Holbeck and East Halton Skitter.

(b) Sand-dunes

Theddlethorpe, Gibraltar Point, Mablethorpe, Skegness, Holbeck, North Somercotes, Sandilands, Humberstone, Sutton-on-Sea, are all places frequently mentioned in the records for species depending on living on sand flora. The following examples (with larval foodplant bracketed) can be used in illustration: the Portland Dart (sea-fescue, marram etc.); the Shore Wainscot (marram); the Archer Dart (sea-sandwort, bedstraws); the White-line Dart (chickweed, bedstraws); the Coast Dart (sea-sandwort, sea spurge); the Marbled Coronet (sea and bladder campions etc.); the Sand Dart (sea rocket); the White Colon (sea bindweed); and the Narrow-bordered 6 spot Burnet (clover, birdsfoot trefoil).

The Portland Moth, named after the Duchess of Portland, has larvae which have adopted an unusual method of concealment (shared by a few other coastal species). The larva lives in a vertical tunnel in the sand, and the larval colour matches that of the sand over which it crawls to journey to its foodplant. The mouth of the tube in the sand collapses after the larva has entered, to make the hiding place almost invisible. Each evening the larva emerges to feed on its seaside plants. Incidentally, colonies of this moth do occur on sandy areas inland such as on the blown sand regions of South Humberside (the author has found it at Manton, and on Atkinson's Warren near Scunthorpe). Other "coastal" species are found, sometimes commonly on these sandy habitats, including the Archer Dart and Marbled Coronet mentioned above.

(c) **Shingle Beaches**

There is little of this type of habitat in Lincolnshire, but typical shingle species such as the Brown Tail moth, (sea buckthorn) and the Feathered Ranuncule (hound's tongue, sea plantain, and various low plants) have been recorded in the Gibraltar Point, Theddlethorpe and Saltfleetby areas.

B. **Woodlands**

Woodlands support the greatest number of our lepidoptera species. Some insects are specific to the dominant forest trees. Others frequent the shrub and hedgerow layers, and more find their niche in the very varied ground flora and grasses growing on the range of soil types within the County. Shrubby bushes like the Sallows growing along woodsides, woodland rides and road verges are extremely important to lepidoptera. A few years ago, an expanse of roadside sallows bordering a nature reserve near Scunthorpe were cut off at ground level as part of the verge "management" programme. A number of moth species once common in the lane had their populations decimated, and have not yet recovered. Similarly over the winter period 1980, a considerable length of blackthorn hedge besides a wood in the Bardney region was cut back most drastically. The twigs removed were carrying the overwintering eggs of the Brown Hairstreak butterfly. It remains to be seen whether this colony will be affected in this habitat.

(a) **Broadleaved Woodlands**

The White-letter Hairstreak is restricted to woods and lanes where elms (the larval foodplant) occur. The butterfly is attracted to feed at the blossoms of bramble and privet flowers. There has been concern for the future of this butterfly in the country over the last few years as a result of the ravages of the elm disease.

Another Hairstreak associated with forest trees is the Purple Hairstreak. The larvae feed on oak leaves, and the species is found in old oak woodlands like those near Skellingthorpe, Woodhall and Bardney.

The White Admiral is restricted to woods with a good growth of honeysuckle – the larval foodplant. This butterfly used only to be found in the New Forest, and in a few isolated colonies in the southern counties of England early this century. From about 1920 it started to extend its range, and it moved into a number of Midland and Eastern woodlands. Its population in Lincolnshire has expanded and contracted over the post-war period. It seemed to disappear entirely from Lincolnshire woodlands in the 1960s, but it has come to light again in the late 1970s in small numbers in restricted localities.

The Purple Emperor – now probably extinct in the county – is a typical woodland butterfly. This most attractive insect was to be found in a number of Lincolnshire woods in the nineteenth century (for example Burwell, Market Rasen, Laughton, Scotton etc.). It does require a specialised habitat, however, involving the sallow (the larval foodplant) and mature oak trees in oak woodlands.

The Speckled Wood (foodplant grasses) flits about in sunny patches and shadowy areas along woodsides or in woodland glades. It has been seen in woodlands at Newball, Bardney, Limber and Skellingthorpe.

Fritillary butterflies favour the woodland habitat, but sadly, in recent years, none have been found. In the not too distant past we had the Silver-washed Fritillary, the High Brown Fritillary, the Pearl-Bordered and Small Pearl-Bordered Fritillaries – all of which species have larvae feeding on violets. The High Brown Fritillary occurred at Woodhall and Scawby in the 1950s, the Pearl-bordered Fritillary was near Market Rasen and in a number of other South Lincolnshire woodlands in the 1960s; the Small Pearl-bordered Fritillary was at Scotton and Laughton in the 1950s... we hope that one day they may return to their former haunts.

Another butterfly which has disappeared is the Chequered Skipper. It was common in woods in Lincolnshire just a few years ago. Following a rapid decline it disappeared from all known localities in the 1960's. The authors, in company of Mr. R. E. M. Pilcher, made a special search for it in 1980, but to no avail. The species flew in flowery woodland rides (foodplant grasses) and a good number of these continue in existence, so it is most likely that climatic factors led to this decline. It may well have totally gone from England, but a Scottish race has been discovered in Western Inverness. Here, however, the butterfly lives in more open scrubland – similar to the habitat favoured on the continent.

The Holly Blue and Duke of Burgundy Fritillary have also altered in status over the last few decades. The former is now scarce, and the latter has had its few habitats so mauled that it is probably now extinct.

Considering the woodland moths, oak woods provide food and shelter for some hundreds of species. The Great Oak Beauty is found in only a few areas – it is now scarce. Another localised oak

LINCOLNSHIRE AS A HABITAT

feeder is the Frosted Green.
  Other trees are, of course, associated with other moth species. The beeches in the Limber area, for example, are the home of the Barred Hook-tip and the Clay Triple Lines.
  A further local and interesting moth is Blomer's Riplet (Rivulet) named after Captain Blomer in the early 1800s. The larval foodplant is wych elm, though the moth only seems to be in areas where these are found in company with beech trees. The right conditions occur in the Limber Woodlands, and this attractive little species has been found there.

(b) Coniferous Woodlands
  There are numerous acres of conifer plantation within the County, which, as reported in Mrs. Weston's article, "have little or no ground flora," except where there are rides or areas of scrub or mixed woodland.
  There are, however, a number of moth species which are specific to pine, spruce, fir and larch, etc. and recorded from such places as Laughton, Scotton, Linwood, Scawby, Market Rasen etc. Examples of these species are the Pine Beauty, the Reddish Pine Carpet, the Bordered White Beauty, the Grey Pine Carpet, and the Barred Red moths. The authors have watched the first of these species feeding in April on sallow catkins near Scotton on many occasions in the 1970's.

C. Grassland Communities
(1) Calcareous Grassland
  The most northerly colony of the Chalkhill Blue butterfly in the British Isles (larval foodplant vetch, birdsfoot trefoil) occurred in the south of the County on a wide road verge in a chalk area until very recently. Unfortunately, so far as is known, the colony has now failed.
  The local Chalk Carpet moth (foodplants trefoils, vetch, clover, black medick) is another typical insect of limestone-chalk hills. There is one small but thriving colony of this species in an area near Kirton Lindsey. The Pretty Chalk Carpet (foodplant traveller's joy) has been recorded near Boston, and the Dusky Sallow (larvae feed on chalkland grasses) has turned up at Ancaster, the Kirton Lindsey quarries and at Honington. This latter species, however, seems to be extending its range into other areas of Lincolnshire, and has been found recently in other types of habitat.
  The Marbled White butterfly, with larvae feeding on cock'sfoot, meadow and brome grasses has occurred on chalklands around Caistor, Ancaster, Colsterworth, Market Deeping and Stamford.

64 THE BUTTERFLIES AND LARGER MOTHS

(b) **Acid Heath**

Many interesting species are associated with plants thriving on acid soils, and Lincolnshire is rich in heathland pockets. The entomologist is enticed to areas like Laughton, Risby, Messingham, Manton, Holton-le-Moor, Wragby, Elsham, Epworth, Kirkby Moor, Linwood and Roughton.

The Grayling butterfly (larvae feed on various grasses like fescue-grass) can be found in local haunts, and the Green Hairstreak (furze, broom, ling, heath) is in a number of suitable locations. Sadly the Silver-studded Blue (heather, ling, furze etc) was lost near Laughton in the 1940s, after a thriving colony had been known there for many years.

The Wood Tiger (plantains) occurs on a number of moorland sites, and the Grey Rustic moth (bell heather, sallow) is at Epworth Turbary. A number of moth species have larvae specific to ling and heather. Examples of these are the Common Heath Beauty, the Grass-waved, the Heath Rustic, and the day flying Beautiful Yellow Underwing, which can be disturbed from the heather as one walks along.

(c) **Scrub**

Blackthorn, buckthorn, alder, birch, hawthorn, sallow, privet, bramble etc., are all important larval foodplants in shrubby scrub areas. The local Brown Hairstreak butterfly uses blackthorn as its foodplant, and the Brimstone butterfly, buckthorn. The Lappet moth is also a buckthorn, blackthorn, hawthorn feeder, and it has been recorded from Haxey, Tetney, Linwood and Sleaford etc. The Brown Scallop moth (buckthorn) is at Legsby, Ancaster and in the Pelham's Pillar wood area. The White waved Silver (birch, alder) occurs at Manton, Woodhall and Linwood, and the Privet Hawk at Holbeach, and other suitable localities in the south of the County.

D. **Peatlands**

(1) **Acid Bog**

Many of Lincolnshire's wet areas are now drained and under cultivation, but small areas of acid peatland survive in the north of the County in the region of Crowle Waste/Thorne Moors. One of our most valued butterflies lives in this type of habitat. This is the Large Heath butterfly – more really an alpine species, and in Lincolnshire found in one of its most southerly habitats, so far as the British Isles is concerned. The larvae feed on cotton grasses and sedges. These types of plants also feed the larvae of the Silver Hook moth, and the Haworth's Crescent moth. Both species occur in this region.

The Clouded Ermine and Emperor Moth are other typical acid heath dwellers.

(a) Fens

Dr. R. E. M. Pilcher has already been mentioned as a valued contributor to the records over a long period of years. In Vol. XV of the L.N.U. Transactions he published a paper on the larger moths to be found in the Lincolnshire Fens and Marshes, with notes on the species' ecological requirements and the factors influencing their distribution. He was, he wrote, prompted to write the work by the realisation that the earlier lepidopterists in the County had tended to devote their attention to the woods and the heathlands at the expense of the wetlands. Since this original paper was written, some twenty years ago, his continued recording has indicated that Lincolnshire is still rich in wetland species. His work has added many new species to the County list, and he has rediscovered other species which had been considered probably "lost" to us.

In Vol. XX, No. 1. 1980, of the Transactions of the Lincolnshire Naturalists' Union, he has published a second part to this paper, to re-assess the status of the fenland species. For a fuller coverage than is intended in this chapter, the reader can consult these articles.

The Wainscot family are typical fen inhabitants, with larvae feeding in and on the common reed, reedmace etc. The Obscure Wainscot, the Southern Wainscot, the Bulrush Wainscot, the Fen Wainscot have been recorded from fen districts like those at Saltfleetby, Kirkby Moors, Lincoln Fen and the South Thoresby wet areas.

A pretty geometrid moth, the Marsh Carpet (larvae feed on the flowers of yellow meadow rue) has existed locally in South Lincolnshire earlier this century, though it has not been seen in more recent years.

The Butterbur moth (larvae feed on roots of the butterbur plant) is another wet area resident. It is infrequently sighted in the North of the County, but more often recorded by R.E.M.P. in the South Thoresby region.

E. Gardens

Gardening is our most popular outdoor leisure activity. Garden centres and mail order organisations increasingly supply houseowners, and it thus becomes more and more important for gardeners to understand how important their activities are to wild life. It is said that perhaps there are a million acres of gardens in existence – a larger area than the 85,000 or so acres of nature reserves in the whole of England and Wales.

Parks and gardens increasingly have to be suitable refuges for many insects as other human activities (e.g. farming and urban development) clear these creatures from their natural haunts. Lepidoptera can be attracted to gardens by the planting of suitable shrubs and flowering plants. Particularly important are the nectar-bearing buddleias, valerian, nicotiana, sedums, sweet william and petunias.

Orange-tip butterflies breed in gardens on sweet rocket and honesty, and on the rock plant Arabis albida, as will the Green-veined White butterfly. A bed of bladder campion will attract species such as the Marbled Coronet, the Sandy Carpet, the Netted Pug, the Lychnis, the Campion etc. Larvae of these species can be found feeding in the seed capsules.

The Red Admiral, Peacock and Small Tortoiseshell butterflies, attracted to the shrubs and plants mentioned earlier, will lay eggs on small clumps of nettles if these are permitted to thrive in a "wild" corner of a garden.

Over the past fifty years or so, the Varied Coronet and the Golden Plusia moths have both colonised Britain from the continent, and have spread widely as they have adapted to the garden environment. The caterpillars of the Varied Coronet feed on sweet william, and those of the Golden Plusia feed on delphinium. This latter species can be a bit of a nuisance as the larvae consume the growing flower spurs – but this is the price an entomologist is happy to pay to promote a suitable garden habitat.

F. Migration of Species

Many butterfly and moth species are great wanderers, and a large number of species regularly visit us in Lincolnshire and South Humberside. The phenomenon must be determined by complex genetic as well as by environmental factors, however, since while certain species are notorious migrants, others seem never to stray at all.

A number of moths and butterflies constantly found in Britain are maintained either entirely or to a great extent by immigration. Were this to cease, they would therefore almost or completely disappear from these islands. As the ornithologist has to be concerned with the comings and goings of large numbers of bird species, the Lepidopterist similarly has to be aware of the movement of, at times, vast numbers of insects entering our air space.

The common Silver Y moth reaches us in the spring from southern Europe. It breeds here and populates Britain with myriads of the moth in late summer. It is one of our most "universal" species.

Three species of migrant Hawkmoths have been recorded in the County in recent years. The Convolvulus Hawk is observed nearly every year, after flying in from the Mediterranean area. In 1976, eleven moths were taken to the Scunthorpe Museum. Some were found in the Humberside villages and seemed to have been following the line of the Humber and Trent rivers.

If eleven were actually apprehended and handed over to the authorities, it is likely that many more got through customs undetected.

LINCOLNSHIRE AS A HABITAT

A sudden invasion of the Bedstraw Hawk moth occurred in the summer of 1973. This lovely moth reaches us from northern Europe. In the last 200 years the species has rarely been recorded in Britain in any number, but on this occasion several moths were taken in Lincolnshire, and the species larvae were in fair profusion on the coast. Dr. R. E. M. Pilcher described finding nearly 50, mostly feeding on small clumps of rosebay willow-herb. Others were found in Lincoln, and three were discovered devouring a fuchsia bush in Grimsby.

Caterpillars of our largest hawk moth, the Deaths Head Hawk are frequently found in Lincolnshire potato fields, and pupae have been turned out of the soil as the crops have been harvested in the late autumn. This species roams all over Europe and Africa as well as throughout Asia.

A rare visitor is the large and striking Camberwell Beauty butterfly which invaded Britain from Scandinavia in 1976. Approaching from this direction, the species is more commonly found, when it does migrate, along the eastern coast than along the southern shores of England. They were widespread during this above period, and three, for example, were taken in the Scunthorpe district.

The Red Admiral and Painted Lady butterflies are also migrant species – some years they occur in large numbers, and other years, when wind and weather conditions at the migration time are most adverse – they are totally absent. 1980 was a "Painted Lady" year, and the species was commonly seen in gardens and buddleia flowers in the north of the County area.

To the person interested in our lepidoptera, the phenomena of migration is a great blessing. There is always the thrill of meeting the "unexpected" species oneself, or of reading in the literature of the experiences of others. There is always the feeling that if "internal" species are not thriving too well, something of great "interest" could move into the area from "outside". There is always the hope that the next moth found, or the next larva turned up under the next leaf will be a county record, or better still, a species new to the British Isles. . .

CHAPTER 2
SYSTEMATIC CHECK LIST OF ALL RECORDED SPECIES

LEPIDOPTERA
Super-family PAPILIONOIDEA

PAPILIONIDAE
PAPILIONINAE
(Swallowtails)
1 *Papilio machaon* L.
  Common Swallowtail

PIERIDAE
PIERINAE
(Whites)
4 *Aporia crataegi* L.
  Black-veined White
5 *Pieris brassicae* L.
  Large Garden White
6 *Pieris rapae* L.
  Small Garden White
7 *Pieris napi* L.
  Green-veined White
8 *Pontia daplidice* L.
  Bath White
9 *Anthocharis cardamines* L.
  Orange-tip White
10 *Leptidea sinapis* L.
  Wood White
RHODOCERINAE
(Redhorns or Sulphurs)
11 *Colias hyale* L.
  Pale Clouded Yellow
13 *Colias croceus* Fourc.
  (edusa F.)
  Common Clouded Yellow
14 *Gonepteryx rhamni* L.
  Brimstone

SATYRIDAE
SATYRINAE
(Satyrs)
16 *Pararge megera* L.
  Wall Brown
17 *Pararge aegeria* L.
  Speckled Wood
18 *Eumenis semele* L.
  Grayling
20 *Erebia aethiops* Esp.
  (blandina F.)
  Northern Brown
22 *Maniola jurtina* L.
  (janira L.)
  Meadow Brown
23 *Maniola tithonus* L.
  Gatekeeper
24 *Coenonympha pamphilus* L.
  Small Heath
25 *Coenonympha tullia* Mull.
  (tiphon Rott.)
  Large Heath
26 *Aphantopus hyperantus* L.
  Common Ringlet
27 *Melanargia galathea* L.
  Marbled White

NYMPHALIDAE
APATURINAE
(Emperors)
28 *Apatura iris* L.
  Purple Emperor

THE BUTTERFLIES AND LARGER MOTHS

LIMENITINAE
(Sibyls or Wood-admirals)
29 *Limenitis camilla* L.
 (sibylla L.)
 White Admiral
NYMPHALINAE
(VANESSIDS or Angle-wings)
30 *Vanessa atalanta* L.
 Red Admiral
31 *Vanessa cardui* L.
 Painted Lady
33 *Nymphalis io* L.
 Peacock
34 *Nymphalis antiopa* L.
 Camberwell Beauty
36 *Nymphalis polychloros* L.
 Large Tortoiseshell
37 *Aglais urticae* L.
 Small Tortoiseshell
38 *Polygonia c-album* L.
 Comma
ARGYNNINAE
(Fritillaries)
39 *Argynnis paphia* L.
 Silver-washed Fritillary
40 *Argynnis cydippe* L.
 (adippe L.)
 High Brown Fritillary
42 *Argynnis aglaia* L.
 Dark Green Fritillary
44 *Clossiana euphrosyne* L.
 Large Pearl-bordered Fritillary
45 *Clossiana selene* Schiff.
 Small Pearl-bordered Fritillary
48 *Melitaea cinxia* L.
 Glanville Fritillary
49 *Euphydryas aurinia* Rott.
 (artemis Schiff.)
 Marsh Fritillary

NEMEOBIIDAE
NEMEOBIINAE
(Dukes)
50 *Haemearis lucina* L.
 Duke of Burgundy

LYCAENIDAE
THECLINAE
(Hairstreaks)
51 *Thecla betulae* L.
 Brown Hairstreak
52 *Thecla quercus* L.
 Purple Hairstreak
54 *Strymonidia w-album* Knoch
 White-letter Hairstreak
55 *Callophyrys rubi* L.
 Green Hairstreak
LYCAENINAE
(Coppers)
56 *Lycaena dispar* Haw.
 Large Copper
58 *Lycaena phlaeas* L.
 Small Copper
POLYOMMATINAE
(Blues)
61 *Plebejus argus* L.
 (aegon Schiff.)
 Silver-studded Blue
63 *Aricia agestis* Schiff.
 (astrarche Bergst.)
 Brown Argus Blue
64 *Polyommatus icarus* Rott.
 Common Blue
65 *Lysandra coridon* Poda
 Chalk-hill Blue
67 *Cyaniris semiargus* Rott.
 (acis Schiff.)
 Mazarine Blue
68 *Celastrina argiolus* L.
 Holly Blue
69 *Cupido minimus* Fuessl.
 Small Blue
HESPERIIDAE
PYRGINAE
(Black-and-white Skippers or Grey Skippers)
71 *Pyrgus malvae* L.
 Grizzled Skipper
72 *Erynnis tages* L.
 Dingy Skipper
HESPERIINAE
(Brown Skippers)

CHECK LIST OF RECORDED SPECIES

73 *Thymelicus sylvestris* Poda
 (thaumas Hufn.)
 Common Small Skipper
74 *Thymelicus lineola* Ochs.
 New Small Skipper
76 *Ochlodes venata* Br. &
 Grey *(sylvanus Esp.)*
 Large Skipper
78 *Carterocephalus palaemon* Pall.
 (paniscus F.)
 Chequered Skipper

Super-family SPHINGOIDEA

SPHINGIDAE
SMERINTHINAE
79 *Mimas tiliae* L.
 Lime Hawk
80 *Laothoe populi* L.
 Poplar Hawk
81 *Smerinthus ocellata* L.
 Eyed Hawk
SPHINGINAE
82 *Acherontia atropos* L.
 Death's-head Hawk
83 *Herse convolvuli* L.
 Convolvulus Hawk
86 *Sphinx ligustri* L.
 Privet Hawk
87 *Hyloicus pinastri* L.
 Pine Hawk
DEILEPHILINAE
91 *Celerio galii* Rott.
 Bedstraw Hawk
92 *Celerio livornica* Esp.
 (lineata F.)
 Striped Hawk
93 *Hippotion celerio* L.
 Silver-striped Hawk
94 *Daphnis nerii* L.
 Oleander Hawk
95 *Deilephila porcellus* L.
 Small Elephant Hawk
96 *Deilephila elpenor* L.
 Large Elephant Hawk

MACROGLOSSINAE
97 *Macroglossum stellatarum* L.
 Humming-bird Hawk
98 *Hemaris fuciformis* L.
 Broad-bordered Bee Hawk
99 *Hemaris tityus* L.
 (bombyliformis Esp.)
 Narrow-bordered Bee Hawk

Super-family BOMBYCOIDEA

NOTODONITIDAE
CERURINAE
101 *Harpyia bifida* Brahm
 (hermelina auct.)
 Poplar Kitten
102 *Harpyia furcula* Clerck
 Sallow Kitten
103 *Cerura vinula* L.
 Puss
NOTODONTINAE
104 *Stauropus fagi* L.
 Lobster Prominent
106 *Drymonia dodonaea* Schiff.
 (trimacula Esp.)
 Light Marbled Brown
107 *Chaonia ruficornis* Hufn.
 (chaonia Hubn)
 Lunar Marbled Brown
108 *Pheosia tremula* Clerck
 Greater Swallow Prominent
109 *Pheosia gnoma* F.
 (dictaeoides Esp.)
 (Lesser Swallow Prominent)
110 *Notodonta ziczac* L.
 Pebble Prominent
111 *Notodonta dromedarius* L.
 Iron Prominent
114 *Notodonta trepida* Esp.
 (anceps auct.)
 Great Prominent

72 THE BUTTERFLIES AND LARGER MOTHS

117 *Lophopteryx capucina L.*
    (camelina L.)
    Coxcomb Prominent
118 *Odontosia carmelita Esp.*
    Scarce Prominent
120 *Pterostoma palpina Clerck*
    Pale Prominent
121 *Phalera bucephala L.*
    Buff-tip
122 *Clostera curtula L.*
    Large Chocolate-tip
124 *Clostera pigra Hufn.*
    (reclusa F.)
    Small Chocolate-tip

THYATIRIDAE
    THYATIRINAE
125 *Habrosyne pyritoides Hufn. (derasa L.)*
    Buff Arches
126 *Thyatira batis L.*
    Peach Blossom
127 *Tethea ocularis L.*
    (octogesima Hubn)
    Figure of Eighty
128 *Tethea or Schiff*
    Poplar Lutestring
129 *Tethea duplaris L.*
    Least Satin Lutestring
131 *Asphalia diluta Schiff*
    Lesser Lutestring
132 *Achlya flavicornis L.*
    Yellow-horned Lutestring
133 *Polyploca ridens F.*
    Frosted Green Lutestring

LYMANTRIIDAE
    LYMANTRIINAE
134 *Orgyia recens Hubn.*
    (gonostigma auct.)
    Scarce Vapourer
135 *Orgyia antiqua L.*
    Common Vapourer
136 *Dasychira fascelina L.*
    Dark Tussock
137 *Dasychira pudibunda L.*
    Pale Tussock
138 *Euproctis chrysorrhoea L.*
    (phaeorrhoea Don.)
    Brown-tail
139 *Euproctis similis Fuessl*
    Gold-tail
140 *Laelia caenosa Hubn.*
    Reed Tussock
142 *Leucoma salicis L.*
    White Satin
144 *Lymantria monacha L.*
    Black-arched Tussock

LASIOCAMPIDAE
    LASIOCAMPINAE
145 *Malacosoma neustria L.*
    Common Lackey
146 *Malacosoma castrensis L.*
    Ground Lackey
147 *Trichiura crataegi L.*
    Pale Eggar
148 *Poecilocampa populi L.*
    December Eggar
149 *Eriogaster lanestris L.*
    Small Eggar
150 *Lasiocampa quercus L.*
    Oak Eggar
152 *Macrothylacia rubi L.*
    Fox
154 *Philudoria potatoria L.*
    Drinker

   GASTROPACHINAE
156 *Gastropacha quercifolia L.*
    Common Lappet

SATURNIIDAE
    SATURNIINAE
159 *Saturnia pavonia L.*
    (carpini Schiff)
    Empress

DREPANIDAE
    DREPANINAE
161 *Drepana binaria Hufn.*
    Oak Hook-tip

CHECK LIST OF RECORDED SPECIES 73

162 *Drepana cultraria* F.
Barred Hook-tip
163 *Drepana falcataria* L.
Pebble Hook-tip
164 *Drepana lacertinaria* L.
Scalloped Hook-tip

CILICINAE
165 *Cilix glaucata* Scop.
Chinese Character

NOLIDAE
NOLINAE
166 *Nola cucullatella* L.
Short-cloaked Black Arches
168 *Nola albula* Schiff.
(albulalis Hubn)
Kent Black Arches
169 *Celama confusalis* H.S.
Least Black Arches

ARCTIIDAE
LITHOSIINAE
171 *Atolmis rubricollis* L.
Red-necked Footman
172 *Nudaria mundana* L.
Muslin Footman
173 *Comacla senex* Hubn.
Round-winged Footman
174 *Miltochrista miniata* Forst.
Rosy Footman
176 *Cybosia mesomella* L.
Four-dotted Footman
177 *Lithosia quadra* L.
Large Footman
178 *Eilema deplana* Esp.
(depressa Esp.)
Buff Footman
179 *Eilema griseola* Hubn.
(stramineola Doubl.)
Dingy Footman
180 *Eilema lurideola* Zinck.
Common Footman
181 *Eilema complana* L.
Scarce Footman
185 *Eilema sororcula* Hufn.
Orange Footman

ARCTIINAE
189 *Utetheisa pulchella* L.
Crimson-speckled Flunkey
191 *Callimorpha jacobaeae* L.
Cinnabar
192 *Spilosoma lubricipeda* L.
(menthastri Esp.)
White Ermine
193 *Spilosoma urticae* Esp.
Water Ermine
194 *Spilosoma lutea* Hufn.
(lubricipeda auct.)
Buff Ermine
195 *Cycnia mendica* Clerck
Muslin Ermine
196 *Diacrisia sannio* L.
(russula L.)
Clouded Ermine
197 *Phragmatobia fuliginosa* L.
Ruby Tiger
199 *Parasemia plantaginis* L.
Wood Tiger
200 *Arctia caja* L.
Garden Tiger
201 *Arctia villica* L.
Cream-spot TIger

HYPSINAE
203 *Panaxia dominula* L.
Scarlet Tiger

Super-family PSYCHOIDEA

LIMACODIDAE
HETEROGENEINAE
208 *Apoda avellana* L.
(limacodes Hufn)
Festoon

ZYGAENIDAE
ZYGAENINAE
215 *Zygaena trifolii* Esp.
Broad-bordered Five-spot Burnet

74 THE BUTTERFLIES AND LARGER MOTHS

217 *Zygaena lonicerae* Scheven
Narrow-bordered Five-spot Burnet
218 *Zygaena fillipendulae L.*
Narrow-bordered Six-spot Burnet
221 *Procris statices L.*
Common Forester

SESIIDAE
SESIINAE
223 *Sesia apiformis* Clerck
Poplar Hornet Clearwing
224 *Sphecia bembeciformis* Hubn.
(crabroniformis Lew.)
Osier Hornet Clearwing

AEGERIINAE
229 *Aegeria tipuliformis* Clerck
Currant Clearwing
232 *Aegeria vespiformis L.*
(cynipiformis Esp.)
Yellow-legged Clearwing
233 *Aegeria myopaeformis* Borkh.
Small Red-belted Clearwing
234 *Aegeria culiciformis L.*
Large Red-belted Clearwing
235 *Aegeria formicaeformis* Esp.
Red-tipped Clearwing.

COSSIDAE
ZEUZERINAE
264 *Zeuzera pyrina L.*
Wood Leopard

COSSINAE
265 *Cossus cossus L.*
(ligniperda F.)
Goat

Super-family HEPIALOIDEA
HEPIALIDAE
HEPIALINAE
266 *Hepialus humuli L.*
Ghost Swift
267 *Hepialus sylvina L.*
Wood Swift
268 *Hepialus fusconebulosa* Deg. *(velleda Hubn)*
Map-winged Swift
269 *Hepialus lupulina L.*
Common Swift
270 *Hepialus hecta L.*
Golden Swift

Super-family NOCTUOIDEA
NOCTUIDAE
AGROTINAE
272 *Euxoa cursoria* Hufn.
Coast Dart
272 *Euxoa nigricans L.*
Garden Dart
274 *Euxoa tritici L.*
White-line Dart
276 *Euxoa obelisca* Schiff
Square-spot Dart
277 *Agrotis segetum* Schiff
Turnip Dart
278 *Agrotis vestigialis* Hufn
(valligera Schifff)
Archer Dart
280 *Agrotis clavis* Hufn
(corticea Schiff)
Heart and Club
281 *Agrotis denticulatus* Haw
(cinerea auct)
Light Feathered Rustic
282 *Agrotis puta* Hubn
(radius Haw)
Shuttle-shaped Dart
285 *Agrotis exclamationis L.*
Heart and Dart
286 *Agrotis ipsilon* Hufn
(suffusa Schiff)
Dark Dart

CHECK LIST OF RECORDED SPECIES 75

287 *Agrotis ripae* Hubn
   Sand Dart
289 *Lycophotia varia* Vill.
   (strigula Thunb)
   True Lovers' Knot
290 *Actebia praecox* L.
   Portland Dart
292 *Peridroma porphyrea Schiff (saucia Hubn)*
   Pearly Underwing
294 *Rhyacia simulans Hufn (pyrophila Schiff)*
   Dotted Rustic
295 *Spaelotis ravida Schiff (obscura Brahm)*
   Stout Dart
297 *Graphiphora augur* F.
   Double Dart
298 *Diarsia brunnea Schiff*
   Purple Clay
299 *Diarsia mendica* F.
   (festiva Schiff)
   Common Ingrailed Clay
301 *Diarsia dahlii* Hubn
   Barred Chestnut Clay
302 *Diarsia rubi* View
   (Bella Borkh)
   Small Square-spot
304 *Ochropleura plecta* L.
   Flame Shoulder
305 *Amathes agathina* Dup.
   Heath Rustic
309 *Amathes glareosa* Esp.
   Autumnal Rustic
310 *Amathes castanea* Esp.
   (neglecta Hubn)
   Grey Rustic
311 *Amathes baja Schiff*
   Dotted Clay
313 *Amathes c-nigrum* L.
   Setaceous Hebrew Character
315 *Amathes triangulum* Hufn.
   Double Square-spot
316 *Amathes stigmatica* Hübn.
   (rhomboidea Treits)
   Square-spotted Clay

317 *Amathes sexstrigata* Haw.
   (umbrosa Hubn)
   Six-striped Rustic
318 *Amathes xanthographa Schiff*
   Square-spot Rustic
319 *Axylia putris* L.
   Flame Rustic
320 *Anaplectoides prasina Schiff.*
   (herbida Hubn.)
   Green Arches
321 *Eurois occulta* L.
   Great Brocaded Rustic
322 *Gypsitea leucographa Schiff.*
   White-marked
323 *Cerastis rubricosa Schiff.*
   Red Chestnut Rustic
324 *Naenia typica* L.
   Gothic Type

NOCTUINAE

327 *Euschesis comes* Hubn.
   (orbona F.*)*
   Lesser Yellow-underwing
328 *Euschesis orbona* Hufn.
   (subsequa Hubn.*)*
   Lunar Yellow-underwing
329 *Euschesis janthina Schiff.*
   Lesser-bordered Yellow-underwing
330 *Euschesis interjecta* Hubn.
   Least Yellow-underwing
331 *Noctua pronuba* L.
   Common Yellow-underwing
332 *Lampra fimbriata Schreber (fimbria* L.*)*
   Broad-bordered Yellow-underwing

HELIOTHINAE

334 *Pyrrhia umbra* Hufn.
   (marginata F.*)*
   Bordered Orange

335 *Heliothis viriplaca* Hufn.
    (dipsacea L.)
    Marbled Clover
340 *Heliothis peltigera* Schiff.
    Dark Bordered Straw
341 *Heliothis armigera* Hübn.
    Scarce Bordered Straw

ANARTINAE

342 *Anarta myrtilli* L.
    Beautiful Yellow
    Underwing.

HADENINAE

345 *Mamestra brassicae* L.
    Cabbage Dot
346 *Melanchra persicariae* L.
    White Dot
347 *Polia hepatica* Clerck
    (tincta Brahm)
    Silvery Arches
348 *Polia nitens* Haw.
    (advena Schiff.)
    Pale Shining Arches
349 *Polia nebulosa* Hufn.
    Grey Arches
351 *Diataraxia oleracea* L.
    Bright-line Brown-eye
353 *Ceramica pisi* L.
    Broom Brocade
354 *Hada nana* Hufn.
    (dentina Esp.)
    Light Shears
355 *Scotogramma trifolii* Hufn.
    (chenopodii Schiff.)
    Small Nutmeg
357 *Hadena w-latinum* Hufn.
    (genistae Borkh.)
    Light Brocade
358 *Hadena suasa* Schiff.
    (dissimilis Knoch)
    Dog's-tooth
359 *Hadena thalassina* Hufn.
    Pale-shouldered Brocade
361 *Hadena bomycina* Hufn.
    (glauca Hubn)
    Glaucous Shears
363 *Hadena bicolorata* Hufn.
    (serena Schiff.)
    Broad-barred
    White Gothic
366 *Hadena conspersa* Schiff.
    (nana Rott.)
    Common Marbled
    Coronet
367 *Hadena compta* Schiff.
    Varied Coronet
368 *Hadena bicruris* Hufn.
    (capsincola Hubn.)
    Lychnis Coronet
370 *Hadena rivularis* F.
    (cucubali Schiff.)
    Campion Coronet
371 *Hadena lepida* Esp.
    (carpophaga Borkh.)
    Tawny Shears
373 *Anepia irregularis* Hufn.
    (echii Borkh.)
    Viper's-bugloss Gothic
374 *Heliophobus albicolon* Hubn.
    White Colon
375 *Heliophobus reticulata* Vill. *(saponariae* Borkh.)
    Bordered Gothic
376 *Tholera popularis* F.
    Feathered Gothic
377 *Tholera cespitis* Schiff.
    Hedge Gothic
378 *Cerapteryx graminis* L.
    Antler
380 *Xylomyges conspicillaris* L.
    Silver Cloud

ORTHOSIINAE

382 *Orthosia gothica* L.
    Common Hebrew
    Character
383 *Orthosia miniosa* Schiff.
    Blossom Underwing
384 *Orthosia cruda* Schiff.
    (pulverulenta Esp.)
    Small Quaker

CHECK LIST OF RECORDED SPECIES

385 *Orthosia stablis Schiff.*
   Common Quaker
386 *Orthosia populeti F.*
   Lead-coloured Drab
387 *Orthosia incerta Hufn.*
   (instablis Schiff.)
   Clouded Drab
388 *Orthosia munda Schiff.*
   Twin-spot Quaker
389 *Orthosia advena Schiff.*
   (opima Hubn)
   Northern Drab
390 *Orthosia gracilis Schiff.*
   Powdered Quaker
391 *Panolis flammea Schiff.*
   (piniperda Panz.)
   Pine Beau

   LEUCANIINAE
392 *Meliana flammea Curt.*
   Flame Wainscot
393 *Leucania pallens L.*
   Common Wainscot
395 *Leucania impura Hubn.*
   Smoky Wainscot
396 *Leucania straminea Treits.*
   Southern Wainscot
397 *Leucania pudorina Schiff.*
   (impudens Hubn.)
   Striped Wainscot
398 *Leucania obsoleta Hubn.*
   Obscure Wainscot
399 *Leucania litoralis Curt.*
   Shore Wainscot
400 *Leucania comma L.*
   Shoulder-striped Wainscot
404 *Leucania vitellina Hubn.*
   Delicate Wainscot
407 *Leucania lythargyria Esp.*
   Clay Wainscot
408 *Leucania conigera Schiff.*
   Brown-line Wainscot

   NONAGRIINAE
411 *Rhizedra lutosa Hubn.*
   (crassicornis Haw.)
   Large Wainscot

413 *Arenostola pygmina Haw.*
   (fulva Hubn.)
   Small Wainscot
414 *Arenostola extrema Hubn.*
   (concolor Guen.)
   Concolorous Wainscot
415 *Arenostola fluxa Hubn.*
   (hellmanni Ev.)
   Mere Wainscot
417 *Arenostola elymi Treits.*
   Lyme-grass Wainscot
419 *Arenostola phragmitidis Hubn.*
   Fen Wainscot
421 *Nonagria algae Esp.*
   (cannae Ochs.)
   Reed Wainscot
422 *Nonagria sparganii Esp.*
   Webb's Wainscot
423 *Nonagria typhae Thunb.*
   (arundinis F.)
   Bulrush Wainscot
424 *Nonagria geminipuncta Haw.*
   Twin-spot Wainscot
425 *Nonagria dissoluta Treits.*
   Brown-veined Wainscot
427 *Coenobia rufa Haw.*
   (despecta Treits.)
   Rufous Wainscot
428 *Chilodes maritima Tausch.*
   (ulvae Hubn.)
   Silky Wainscot

   CARADRININAE
429 *Meristis trigrammica Hufn. (trilinea Schiff.)*
   Treble-line
430 *Caradrina morpheus Hufn.*
   Mottled Rustic
431 *Caradrina alsines Brahm*
   Uncertain
432 *Caradrina blanda Schiff.*
   (taraxaci Hubn.)
   Smooth Rustic

435 *Caradrina clavipalpis*
Scop. *(quadripunctata F.)*
Pale Mottled Willow
436 *Laphygma exigua Hubn.*
Small Mottled Willow
438 *Dypterygia
scabriuscula L.
(pinastri L.)*
Bird's-wing
441 *Apamea lithoxylaea
Schiff.*
Common Light Arches
442 *Apamea sublustris Esp.*
Reddish Light Arches
444 *Apamea monoglypha
Hufn. (polyodon L.)*
Dark Arches
446 *Apamea epomidion Haw.
(hepatica L.)*
Large Clouded Brindle
447 *Apamea crenata Hufn.
(rurea F.)*
Cloud-bordered Brindle
448 *Apamea sordens Hufn.
(basilinea Schiff.)*
Rustic Shoulder-knot
449 *Apamea unanimis Hubn.*
Small Clouded Brindle
450 *Apamea pabulatricula
Brahm (connexa Borkh.)*
Union Rustic
451 *Apamea oblonga Haw.
(abjecta Hubn.)*
Crescent Striped
452 *Apamea infesta Ochs.
(sordida Borkh.)*
Large Nutmeg
453 *Apamea furva Schiff.*
Confused Brindle
454 *Apamea remissa Hubn.
(obscura Haw.)*
Dusky Brocade
455 *Apamea scolopacina Esp.*
Slender Brindle
456 *Apamea secalis L.
(didyma Esp.)*
Common Rustic

457 *Apamea ophiogramma
Esp.*
Double-lobed
458 *Apamea ypsillon Schiff.
(fissipuncta Haw.)*
Dismal Brindle
461 *Eremobia ochroleuca
Schiff.*
Dusky Sallow Rustic
462 *Procus strigilis Clerck*
Marbled Minor
463 *Procus latruncula Schiff.*
Tawny Minor
464 *Procus versicolor Borkh.*
Rufous Minor
465 *Procus Fasciuncula Haw.*
Middle-barred Minor
466 *Procus literosa Haw.*
Rosy Minor
467 *Procus furuncula Schiff.
(bicoloria Vill.)*
Cloaked Minor
468 *Phothedes captiuncula
Treits. (expolita Staint.)*
Least Minor
469 *Luperina testacea Schiff.*
Flounced Rustic
472 *Euplexia lucipara L.*
Small Angle-shades
473 *Phlogophora
meticulosa L.*
Large Angle-shades
476 *Thalpophila matura Hufn
(cytherea F.)*
Straw Underwing

AMPHIPYRINAE

478 *Petilampa minima Haw.
(arcuosa Haw.)*
Small Dotted Buff
480 *Hydrillula palustris Hubn.*
Marsh Buff
481 *Celaena haworthii Curt.*
Haworth's Crescent
482 *Celaena leucostigma
Hubn. (fibrosa Hubn.)*
Brown Crescent

CHECK LIST OF RECORDED SPECIES 79

484 *Hydraecia oculea L.*
 (nicitans Borkh.)
 Common Ear
485 *Hydraecia paludis* Tutt
 Saltern Ear
486 *Hydraecia lucens* Freyer
 Large Ear
488 *Gortyna micacea Esp.*
 Rosy Ear
489 *Gortyna petasitis Doubl.*
 Butterbur Ear
490 *Gortyna flavago Schiff.*
 (ochracea Hubn.)
 Orange Ear
493 *Cosmia pyralina Schiff.*
 Lunar-spotted Pinion
494 *Cosmia affinis L.*
 Lesser-spotted Pinion
495 *Cosmia diffinis L.*
 White-spotted Pinion
496 *Cosmia trapezina L.*
 Dun-bar
497 *Enargia paleacea Esp.*
 (fulvago Hubn.)
 Angle-striped
499 *Zenobia retusa L.*
 Double Kidney
500 *Zenobia subtusa Schiff.*
 Olive Kidney
501 *Panemeria tenebrata Scop.*
 (arbuti F.)
 Small Yellow Underwing
502 *Amphipyra pyramidea L.*
 Copper Underwing
502a *Amphipyra berbera* Rungs
 Drab Copper Underwing
503 *Amphipyra tragopoginis*
 Clerck
 Mouse
504 *Rusina tenebrosa Hubn.*
 (umbratica auct.)
 Brown Feathered
505 *Mormo maura L.*
 Old-lady

APATELINAE
506 *Cryphia perla Schiff.*
 Marbled Beau

511 *Moma alpium* Osbeck
 (orion Esp.)
 Scarce Merveille-du-jour
512 *Apatele leporina L.*
 Miller
513 *Apatele aceris L.*
 Sycamore Dagger
514 *Apatele megacephala Schiff.*
 Poplar Dagger
515 *Apatele alni L.*
 Alder Dagger
517 *Apatele tridens Schiff.*
 Dark Dagger
518 *Apatele psi L.*
 Grey Dagger
520 *Apatele menyanthidis* View
 Light Knot-grass Dagger
523 *Apatele rumicis L.*
 Dusky Knot-grass Dagger
524 *Craniophora ligustri Schiff.*
 Crown
525 *Simyra venosa Borkh.*
 (albovenosa auct.)
 Powdered Dagger.

CUCULLIINAE
527 *Cucullia umbratica L.*
 Common Shark
528 *Cucullia asteris Schiff.*
 Starwort Shark
529 *Cucullia chamomillae Schiff.*
 Chamomile Shark
531 *Cucullia absinthii L.*
 Pale Wormwood Shark
533 *Cucullia verbasci L.*
 Mullein Shark

XYLENINAE
537 *Lithomoia solidaginis Hubn.*
 Bilberry Brind

80 THE BUTTERFLIES AND LARGER MOTHS

538 *Lithophane semibrunnea Haw.*
   Tawny Pinion
543 *Lithophane ornitopus Hufn. (rhizolitha F.)*
   Grey Shoulder-knot
544 *Xylena exsoleta L.*
   Cloudy Sword-grass
545 *Xylena vetusta Hubn.*
   Red Sword-grass
546 *Xylocampa areola Esp. (lithorhiza Borkh.)*
   Grey Early

DASYPOLIINAE
550 *Brachionycha sphinx Hufn. (cassinia Schiff.)*
   Common Sprawler
552 *Bombycia viminalis F.*
   Minor Shoulder-knot
553 *Aporophyla lutulenta Schiff.*
   Deep Brown Rustic
555 *Aporophyla lunula Stroem (nigra Haw.)*
   Black Rustic
557 *Allophyes oxyacanthae L.*
   Green-brindled Crescent
559 *Griposia aprilina L.*
   Common Merveille-du-jour
561 *Eumichtis satura Schiff. (porphyrea Esp.)*
   Beautiful Arches
562 *Eumichtis adusta Esp.*
   Dark Brocade
563 *Eumichtis lichenea Hubn.*
   Feathered Ranuncule
564 *Parastichtis suspecta Hubn.*
   Suspected
565 *Dryobotodes aremita F. (protea Schiff.)*
   Brindled Green Mottle
567 *Dasypolia templi Thunb.*
   Brindled Ochre

568 *Antitype flavicincta Schiff.*
   Large Ranuncule
569 *Antitype chi L.*
   Grey Chi
571 *Eupsilia transversa Hufn. (satellitia L.)*
   Satellite
572 *Jodia croceago Schiff.*
   Orange Upperwing
574 *Omphaloscelis lunosa Haw.*
   Lunar Underwing
575 *Agrochola lota Clerck*
   Red-Line Quaker
576 *Agrochola macilenta Hubn.*
   Yellow-line Quaker
577 *Agrochola circellaris Hufn. (ferruginea Esp.)*
   Brick
578 *Agrochola lychnidis Schiff. (pistacina F.)*
   Beaded Chestnut
579 *Anchoscelis helvola L. (rufina L.)*
   Flounced Chestnut
580 *Anchoscelis litura L.*
   Brown-spot Chestnut
581 *Atethmia xerampelina Esp.*
   Centre-barred Sallow
582 *Tiliacea citrago L.*
   Orange Sallow
583 *Tiliacea aurago Schiff.*
   Barred Sallow
584 *Citria lutea Stroem (flavago F.)*
   Pink-barred Sallow
585 *Cirrhia icteritia Hufn. (fulvago L.)*
   Common Sallow
586 *Cirrhia gilvago Schiff.*
   Dusky-lemon Sallow
590 *Conistra vaccinii L.*
   Common Chestnut
591 *Conistra ligula Esp. (spadicea Staint.)*
   Dark Chestnut

CHECK LIST OF RECORDED SPECIES 81

HYLOPHILIDAE

WESTERMANNIINAE
592 *Bena prasinana* L.
 (fagana F.)
 Green Silver-lines
593 *Pseudoips bicolorana*
 Fuesal. (quercana Schiff.)
 Scarce Silver-lines
594 *Earias clorana* L.
 Cream-bordered Green

NYCTEOLINAE
595 *Nycteola revayana* Scop.
 (Undulana Hubn.)
 Large Marbled Tort

PLUSIIDAE

EUSTROTIINAE
603 *Lithacodia fasciciana* L.
 White-spot Marbled
606 *Eustrotia uncula* Clerck
 (uncana L.)
 Silver Hook

CATCOCALINAE
608 *Catocala fraxini* L.
 Clifden Nonpareil
610 *Catocala nupta* L.
 Red Underwing
613a *Clytie illunaris* Hubn.
 Trent Double Stripe
615 *Euclidimera mi* Clerck
 Mother Shipton
616 *Ectypa glyphica* L.
 Burnet Companion
617 *Colocasia coryli* L.
 Nut-tree Tuffet
619 *Episema caeruleocephala* L.
 Figure of Eight
621 *Polychrisia moneta* F.
 Silver Eight
623 *Plusia chrysitis* L.
 Common Burnished Brass
626 *Plusia Bractea Schiff*
 Gold Spangle

627 *Plusia festucae* L.
 Gold Spot
627a *Plusia gracilis Lampke*
 Lampke's Gold Spot
630 *Plusia jota* L.
 Plain Golden Y.
631 *Plusia pulchrina* Haw.
 Beautiful Golden Y
632 *Plusia ni Hubn.*
 (brassicae Ril.)
 Silver V(ni moth)
635 *Plusia gamma* L
 Common Silver Y
636 *Plusia interrogationis* L.
 Scarce Silver Y
638 *Unca triplasia* L.
 Dark Spectacle
639 *Unca tripartita Hufn.*
 (urticae Hubn.)
 Light Spectacle

OPHIDERINAE
641 *Acontia luctuosa Schiff.*
 Four-spot
644 *Lygephila pastinum* Triets.
 Plain Black Neck
648 *Rivula sericealis* Scop.
 Straw point
649 *Phytometra viridaria* Clerck *(aenea Hubn.)*
 Small Purple Bars

GONOPTERINAE
651 *Scoliopteryx libatrix* L.
 Herald

HYPENINAE
652 *Bomolocha fontis Thunb.*
 (crassalis Treits.)
 Beautiful Snout
653 *Hypena proboscidalis* L.
 Common Snout
656 *Hypena rostralis* L.
 Buttoned Snout
658 *Schrankia costaestrigalis* Steph.
 Pinion-streaked Snout

661 *Zanclognatha tarsipennalis* Treits.
Brown Fanfoot
662 *Zanclognatha nemoralis* F.
(grisealis Schiff.)
Small Fanfoot
663 *Zanclognatha cribrumalis* Hubn. *(cribralis* Hubn.)
Dotted Fanfoot
665 *Herminia barbalis* Clerck
Common Fanfoot
666 *Laspeyria flexula* Schiff.
Beautiful Hook-wing

Super-family GEOMETROIDEA

GEOMETRIDAE
ARCHIEARINAE
667 *Archiearis parthenias* L.
Common Orange-underwing
668 *Archiearis notha* Hubn.
Light Orange-underwing

OENOCHROMINAE
669 *Alsophila aescularia* Schiff.
March Usher

GEOMETRINAE
671 *Pseudoterpna pruinata* Hufn. *(cytisaria* Schiff.)
Greater Grass Emerald
672 *Geometra papilionaria* L.
Large Emerald
673 *Comibaena pustulata* Hufn.
(bajularia Schiff.)
Blotched Emerald
674 *Hemithea aestivaria* Hubn. *(strigata* Mull)
Common Emerald
675 *Chlorissa viridata* L.
Small Grass Emerald
679 *Hemistola immaculata* Thunb.
(chrysoprasaria Esp.)
Lesser Emerald

680 *Jodis lactearia* L.
Little Emerald

STERRHINAE
681 *Calothysanis amata* L.
Large Blood-vein
682 *Cosymbia albipunctata* Hufn.
(pendularia auct.)
Birch Mocha
684 *Cosymbia annulata* Schulze
(omicronaria Schiff.)
Maple Mocha
687 *Cosymbia punctaria* L.
Maiden's Blush
688 *Cosymbia linearia* Hubn.
(trilinearia Borkh.)
Clay Triple-lines
689 *Scopula ternata* Schrank
(fumata Steph.)
Smoky Wave
692 *Scopula promutata* Guen.
(marginepunctata auct.)
Mullein Wave
694 *Scopula imitaria* Hubn.
Small Blood-veined Wave
695 *Scopula emutaria* Hubn.
Rosy Wave
698 *Scopula immutata* L.
Lesser Cream Wave
699 *Scopula lactata* Haw.
(remutaria Hubn.)
Greater Cream Wave
702 *Sterrha interjectaria* Boisd.
(fuscovenosa auct.)
Dwarf Cream Wave
704 *Sterrha dilutaria* Hubn.
(holosericata Dup.)
Silky Wave
707 *Sterrha dimidiata* Hufn.
(scutulata Schiff.)
Single-dotted Wave
710 *Sterrha seriata* Schrank
(virgularia Hubn.)
Small Dusty Wave
711 *Sterrha subsericeata* Haw
Satin Wave

CHECK LIST OF RECORDED SPECIES 83

712 *Sterrha sylvestraria Hubn.*
 (straminata Treits)
 Ringed Wave
716 *Sterrha straminata Borkh.*
 (inornata Haw.)
 Plain Wave
717 *Sterrha aversata L.*
 Riband Wave
719 *Sterrha biselata Hufn.*
 (bisetata Rott.)
 Small Fan-footted Wave
720 *Sterrha emarginata L.*
 Small Scallop Wave

LARENTIINAE

723 *Xanthorhoe quadrifasiata Clerck*
 (quadrifasicaria L.)
 Large Twin-spot Carpet
725 *Xanthorhoe ferrugata Clerck (unidentaria Haw.)*
 Dark Twin-spot Carpet
726 *Xanthorhoe spadicearia Schiff.*
 (ferrugata Staud, non Clerck.)
 Red Twin-spot Carpet
728 *Xanthorhoe designata Hufn. (propugnata Schiff.)*
 Flame Carpet
729 *Xanthorhoe montanata Schiff.*
 Silver-ground Carpet
730 *Xanthorhoe fluctuata L.*
 Garden Carpet
731 *Nycterosea obstipata F.*
 (fluviata Hubn.)
 Narrow-barred Carpet
733 *Colostygia pectinataria Knoch (viridaria F.)*
 Spring Green Carpet
734 *Colostygia salicata Hubn.*
 Striped Twin-spot Carpet
735 *Colostygia multistrigaria Haw.*
 Grey Mottled Carpet

736 *Colostygia didymata L.*
 Small Twin-spot Carpet
737 *Pareulype berberata Schiff*
 Barberry Carpet
738 *Earophila badiata Schiff.*
 Shoulder-striped Carpet
739 *Anticlea derivata Schiff.*
 (nigrofasciaria auct.)
 Streamer Carpet
740 *Mesoleuca albicillata L.*
 Beautiful Carpet
743 *Perizoma sagittata F.*
 Marsh Carpet
746 *Perizoma affinitata Steph.*
 Large Rivulet
747 *Perizoma alchemillata L.*
 (rivulata Schiff.)
 Small Rivulet
748 *Perizoma flavofaciata Thunb.*
 (decolorata Hubn.)
 Sandy Carpet
749 *Perizoma albulata Schiff.*
 Grass Rivulet
750 *Perizoma bifaciata Haw.*
 (unifasciata Haw.)
 Barred Rivulet
751 *Perizoma minorata Treits.*
 Heath Rivulet
758 *Euphyia bilineata L.*
 Yellow Shell
759 *Melanthia procellata Schiff.*
 Pretty Chalk Carpet
760 *Mesotype virgata Hufn.*
 (lineolata Schiff.)
 Oblique-striped
761 *Lyncometra ocellata L.*
 Purple Bar Carpet
762 *Lampropteryx suffumata Schiff.*
 Water Carpet
764 *Electrophaes corylata Thunb.*
 Broken-barred Carpet
765 *Ecliptopera silaceata Schiff.*
 Small Phoenix

THE BUTTERFLIES AND LARGER MOTHS

767 *Lygris prunata L.*
Large Phoenix
768 *Lygris testata L.*
Common Chevron
769 *Lygris populata L.*
Northern Spinach
770 *Lygris mellinata F. (associata Borkh.)*
Currant Spinach
771 *Lygris pyraliata Schiff. (dotata L.)*
Barred Straw Chevron
772 *Cidaria fulvata Forst.*
Barred Yellow
773 *Plemyria rubiginata Schiff. (bicolorata Hufn.)*
Blue-bordered Carpet
774 *Chloroclysta siterata Hufn. (psittacata Hubn.)*
Red-green Carpet
775 *Chloroclysta miata L.*
Autumn Green Carpet
776 *Dysstroma truncata Hufn. (russata Borkh.)*
Common Marbled Carpet
778 *Dysstroma citrata L. (immanata Haw.)*
Dark Marbled Carpet
779 *Thera obeliscata Hubn*
Grey Pine Carpet
781 *Thera cognata Thunb. (simulata Hubn.)*
Chestnut-coloured Carpet
782 *Thera firmata Hubn.*
Reddish Pine Carpet
783 *Thera juniperata L.*
Juniper Carpet
784 *Hydriomena furcata Thunb. (sordidata F.)*
July Highflyer
785 *Hydriomena caerulata F. (impluviata Schiff.)*
May Highflyer
786 *Hydriomena ruberata Freyer*
Ruddy Highflyer
787 *Philereme vetulata Schiff.*
Brown Scallop
788 *Philereme transversata Hufn. (rhamnata Schiff.)*
Dark Scallop
789 *Triphosa dubitata L.*
Common Tissue
790 *Rheumaptera cervinalis Scop. (certata Hubn.)*
Scarce Tissue
791 *Theumaptera undulata L.*
Shell Scallop
792 *Theumaptera hastata L.*
Large Argent-and-sable
794 *Epirrhoe rivata Hubn.*
Wood Carpet
795 *Epirrhoe alternata Mull (sociata Borkh.)*
Common Bedstraw Carpet
796 *Epirrhoe tristata L.*
Small Argent-and-sable
797 *Epirrhoe galiata Schiff.*
Galium Carpet
800 *Chesias legatella Schiff. (sparitata Fuessl.)*
Streaked Carpet
802 *Odezia atrata L. (chaerophyllata L.)*
Chimney-sweeper
803 *Anaitis plagiata L.*
Slender Treble-bar
805 *Carsia sororiata Hubn. (paludata Thunb.)*
Manchester Treble-bar
807 *Horisme vitalbata Schiff.*
Umber Waved Carpet
809 *Horisme tersata Schiff.*
Fern Carpet
810 *Lobophora halterata Hufn. (hexapterata Schiff.)*
Large Seraphim
811 *Mysticoptera sexalata Retz. (sexalisata Hubn.)*
Small Seraphim
812 *Acasis viretata Hubn.*
Brindle-barred Yellow

CHECK LIST OF RECORDED SPECIES

813 *Trichopteryx polycommata Schiff.*
Barred Tooth-striped
814 *Trichopteryx carpinata Borkh. (lobulata Hubn.)*
Early Tooth-striped
815 *Orthonama lignata Hubn. (vittata Borkh.)*
Oblique Carpet
816 *Ortholitha mucronata Scop. (umbrifera Prout)*
Common Lead-belle
818 *Ortholitha chenopodiata L. (limitata Scop.)*
Shaded Broad-bar
821 *Ortholitha bipunctaria Schiff.*
Local Chalk Carpet
822 *Larentia clavaria Haw. (cervinata Schiff.)*
Mallow Carpet
823 *Pelurga comitata L.*
Dark Spinach
824 *Oporinia autumnata Borkh.*
Large Autumnal Carpet
825 *Oporinia filigrammaria H.—S.*
Small Autumnal Carpet
826 *Oporinia dilutata Schiff. (nebulata Thunb. non Scop.)*
November Carpet
827 *Oporinia christyi Prout*
Christy's Carpet
828 *Operophtera brumata L.*
Common Winter
829 *Operophtera fagata Scharf. (boreata Hubn.)*
Northern Winter
830 *Asthena albulata Hufn. (candidata Schiff.)*
White Waved Carpet
831 *Minoa murinata Scop. (euphorbiata Schiff.)*
Drab Carpet
(Drab Looper)

832 *Hydrelia flammeolaria Hufn. (luteata Schiff.)*
Yellow Waved Carpet
833 *Hydrelia testaceata Don. (sylvata Schiff.)*
Sylvan Waved Carpet
834 *Euchoeca nebulata Scop. (obliterata Hufn.)*
Dingy Shell
836 *Discoloxia blomeri Curt*
Blomer's Ripplet
837 *Antiocollix sparsata Treits*
Dentated Pug
838 *Eupithecia pini Retz. (togata Hubn.)*
Cloaked Pug
839 *Eupithecia subumbrata Schiff. (scabiosata Borkh.)*
Shaded Pug
840 *Eupithecia subnotata Hubn.*
Plain Pug
842 *Eupithecia distinctaria H.S.*
Thyme Pug
843 *Eupithecia tenuiata Hubn.*
Slender Pug
844 *Eupithecia inturbata Hubn. (subciliata Doubl.)*
Maple Pug
845 *Eupithecia haworthiata Doubl. (Isogrammaria H.S.)*
Haworth's Pug
846 *Eupithecia plumbeolata Haw.*
Lead-coloured Pug
847 *Eupithecia linariata Schiff.*
Toadflax Pug
848 *Eupithecia pulchellata Steph.*
Foxglove Pug
849 *Eupithecia irriguata Hubn.*
Marbled Pug
850 *Eupithecia exiguata Hubn.*
Mottled Pug

851 *Eupithecia insigniata* Hubn. (*consignata* Borkh.)
   Pinion-spotted Pug
852 *Eupithecia valerianata* Hubn.
   Valerian Pug
853 *Eupithecia pygmaeata* Hubn. (*palustraria* Doubl.)
   Marsh Pug
854 *Eupithecia venosata* F.
   Netted Pug
855 *Eupithecia centaureata* Schiff. (*oblongata* Thunb.)
   Lime-speck Pug
856 *Eupithecia trisignaria* H.—S.
   Triple-spotted Pug
858 *Eupithecia satyrata* Hubn.
   Satyr Pug
859 *Eupithecia tripunctaria* H.S. (*albipunctata* Haw. non *Hufn.*)
   White-spotted Pug
860 *Eupithecia absinthiata* Clerck (*minutata* Schiff.)
   Wormwood Pug
861 *Eupithecia goossensiata* Mab. (*minutata* Doubl. non Schiff.)
   Ling Pug
862 *Eupithecia expallidata* Doubl.
   Bleached Pug
863 *Eupithecia assimilata* Doubl.
   Currant Pug
864 *Eupithecia vulgata* Haw.
   Common Pug
865 *Eupithecia denotata* Hubn. (*campanulata* H.S.)
   Bell-flower Pug

866 *Eupithecia castigata* Hubn.
   Grey Pug
867 *Eupithecia icterata* Vill.
   Tawny Speckled Pug
868 *Eupithecia succenturiata* L.
   Bordered Pug
869 *Eupithecia indigata* Hubn.
   Ochreous Pug
870 *Eupithecia pimpinellata* Hubn. (*denotata* Guen.)
   Pimpinel Pug
871 *Eupithecia extensaria* Freyer
   Scarce Pug
872 *Eupithecia nanata* Hubn.
   Narrow-winged Pug
873 *Eupithecia innotata* Hufn.
   Angle-barred Pug
874 *Eupithecia fraxinata* Crewe
   Ash Pug
876 *Eupithecia virgaureata* Doubl.
   Goldenrod Pug
877 *Eupithecia abbreviata* Steph.
   Brindled Pug
878 *Eupithecia dodoneata* Guen.
   Oak-tree Pug
880 *Eupithecia sobrinata* Hubn. (*pusillata* auct.)
   Juniper Pug
882 *Eupithecia lariciata* Freyer
   Larch Pug
883 *Eupithecia tantillaria* Boisd. (*pusillata* Schiff.)
   Dwarf Pug
884 *Chloroclystis coronata* Hubn.
   V Pug
855 *Chloroclystis debiliata* Hubn.
   Bilberry Pug

CHECK LIST OF RECORDED SPECIES 87

886 *Chloroclystis rectangulata L.*
　　Green Pug
887 *Gymnoscelis pumilata Hubn.*
　　Double-striped Pug
887a *Chloroclystis chloerata (mabille 1870)*
888 *Abraxas sylvata Scop. (ulmata F.)*
　　Clouded Magpie
889 *Abraxas grossulariata L.*
　　Common Magpie
891 *Lomaspilis marginata L.*
　　Clouded Border
892 *Ligdia adustata Schiff.*
　　Scorched Silver
894 *Bapta bimaculata F. (taminata Schiff.)*
　　White-pinion Spotted
895 *Bapta temerata Schiff. (punctata Hubn.*
　　Clouded Silver
896 *Deilinia pusaria L.*
　　White Waved Silver
897 *Deilinia exanthemata Scop.*
　　Common Waved Silver
898 *Ellopia fasciaria L. (prosapiaria L.)*
　　Barred Red
899 *Campaea margaritata L. (margaritaria L.)*
　　Barred Light-green
900 *Angerona prunaria L.*
　　Orange Thorn
901 *Semiothisa notata L.*
　　Blunt Peacock Angle
902 *Semiothisa alternaria Hubn. (alternata Schiff.)*
　　Sharp Peacock Angle
903 *Semiothisa liturata Clerck*
　　Tawny-barred Angle
904 *Theria rupicapraria Schiff.*
　　Early Umber
905 *Erannis leucophaearia Schiff.*
　　Spring Umber
906 *Erannis aurantiaria Hubn.*
　　Scarce Umber
907 *Erannis marginaria F. (progemmaria Hubn.)*
　　Dotted Border
908 *Erannis defoliaria Clerck*
　　Mottled Umber

ENNOMINAE

909 *Anagoga pulveraria L.*
　　Barred Umber Thorn
910 *Ennomos autumnaria Wernb. (alniaria Schiff. non L.)*
　　Large Thorn
911 *Ennomos quercinaria Hufn. (angularia Hubn.)*
　　August Thorn
912 *Deuteronomos alniaria L. (Tiliaria Borkh.)*
　　Canary-shouldered Thorn
913 *Deuteronomos fuscantaria Steph.*
　　Dusky Thorn
914 *Deuteronomos erosaria Schiff.*
　　September Thorn
915 *Selenia bilunaria Esp. (ilunaria Hubn.)*
　　Early Thorn
916 *Selenia lunaria Schiff.*
　　Lunar Thorn
917 *Selenia tetralunaria Hufn. (illustraria Hubn.)*
　　Purple Thorn
918 *Aperia syringaria L.*
　　Lilac Thorn
919 *Gonodontis bidentata Clerck*
　　Scalloped Hazel Thorn
920 *Colotois pennaria L.*
　　Feathered Thorn
921 *Crocallis elinguaria L.*
　　Scalloped Oak Thorn
922 *Plagodis dolabraria L.*
　　Scorched-wing

THE BUTTERFLIES AND LARGER MOTHS

923 *Opisthographtis luteolata L. (crataegata L.)*
 Sulphur Thorn
924 *Epione repandaria Hufn. (apiciaria Schiff.)*
 Common Bordered-beauty
927 *Pseudopanthera macularia L.*
 Speckled Yellow

OURAPTERYGINAE
928 *Ourapteryx sambucaria L.*
 Swallow-tailed Elder

BISTONINAE
929 *Phigalia pedaria F. (pilosaria Schiff.)*
 Pale Brindled-beauty
930 *Apocheima hispidaria Schiff.*
 Small Brindled-beauty
933 *Lycia hirtaria Clerck*
 London Brindled-beauty
934 *Biston strataria Hufn. (prodromaria Schiff.)*
 Oak Brindled-beauty
935 *Biston betularia L.*
 Pepper-and-salt
 (Peppered moth)

BOARMIINAE
936 *Menophra abruptaria Thunb.*
 Waved Umber Beauty
938 *Cleora rhomboidaria Schiff. (gemmaria Brahm)*
 Willow Beauty
939 *Cleorodes lichenaria Hufn.*
 Brussels Lace

940 *Deileptenia ribeata Clerck (abietaria Schiff.)*
 Satin Beauty
941 *Alcis repandata L.*
 Mottled Beauty
944 *Boarmia roboraria Schiff.*
 Great Oak Beauty
945 *Pseudoboarmia punctinalis Scop. (consortaria F.)*
 Pale Oak Beauty
946 *Ectropis biundularia Borkh. (bistortata auct.)*
 Early Engrailed
947 *Ectropis crepuscularia Schiff. (biundularia Esp.)*
 Small Engrailed
950 *Aethalura punctulata Schiff. (punctularia Hubn.)*
 Grey Birch Beauty
952 *Pachycnemia hippocastanaria Hubn.*
 Horse-chestnut Longwing
958 *Ematurga atomaria L.*
 Common Heath Beauty
959 *Bupalus piniaria L.*
 Bordered White Beauty
961 *Itame wauaria L.*
 V. Looper
962 *Itame brunneata Thunb. (fulvaria Vill.)*
 Rannoch Looper
963 *Lithina chlorosata Scop. (petraria Hubn.)*
 Brown Silver-lined
964 *Chiasmia clathrata L.*
 Heath Lattice
965 *Dyscia fagaria Thunb. (belgaria Hubn.)*
 Grey Scalloped Bar
968 *Aspitates ochrearia Rossi (Icitraria Hubn.)*
 Yellow Belle
969 *Perconia strigillaria Hubn.*
 Grass-waved

CHAPTER 3.

RECORDS OF LINCOLNSHIRE SPECIES

The data covering all the species recorded within the county constitutes the most important part of the book. The listing follows the order of the late I. R. P. Heslop's revised, indexed "Checklist of the British Lepidoptera" (1964), with all available supplements (7 supplements up to 18th May 1970) published up to the date of writing being taken into account. This "checklist" was selected as being the most suitable to our needs on the strength of the way in which the species are numbered. (For example, Papilionidae, numbered 1 to 3; Sphingidae, 79 to 99; Noctuidae, 271 to 591; Geometridae, 667 – 969, etc.). Kloet and Hinck's "Check List of British Insects" (1972) was considered, but rejected on the grounds that its number system was not compatible with easy indexing. This can be seen in "A Recorder's Log Book of British Butterflies and Moths", J. D. Bradley and D. S. Fletcher, (1979) (Papilionidae, numbered 1536 to 1540; Sphingidae, 1971 to 1993; Noctuidae, 2080 to 2495; Geometridae, 1661 to 1970 etc.).

The Psychidae (Psychinae and Talaeporiinae – some 24 species) and Syntomidae (Syntominae, Thyretinae and Antichlorinae – 4 species) have been omitted, as they have traditionally been included in the Microlepidoptera, and are still widely treated as such, so that reliable records are not available.

Each species, then, is numbered according to Heslop's Systematic list, and is introduced by its Latin name and its vernacular name. Generally, Heslop's revised 'English' name has been used, but when it has been considered that this differs so much from the previous name as to cause confusion, then the previous English name has been additionally printed.

For each species the intention has been to provide information to cover the following aspects.

Heslop's Number   **Latin Name**   **English Name**
1. **The Typical Habitat** (Seashore/Coastal, Fen/Marshland, Heath/Moorland, Chalk Downs, Woodland etc.)
2. **Statement as to Status** (Rare, Migrant, Occasional Visitor, Local, Common, Widespread etc.)

3. **Date on wing**
4. **Any Important General Comment** (Number of Generations, Population Trends, Variation, Melanism etc.)
5. **Specific Recorded Sites** (dated with identification of the **Contributor**)
6. **Larval Food Plant**

    Generally speaking, the butterflies, and the larger moths (e.g. Hawk moths) and their larvae, are more frequently noticed by naturalists without any special entomological interest. This can lead to a lepidoptera recording secretary receiving more records for these species, and one can be tempted in a work of this kind to give the "popular" species an undue emphasis. An effort has been made in the following section to give a fair coverage to each recorded species. Where a species is widespread or abundant, only a fraction of the recorded sites may have been quoted as typical localities. Where a species is of special interest, as a rarity, for example, a fuller coverage of the sightings will have been given. In a few instances, the naming of exact locations has been deliberately avoided, to avert the possibility of undue interference with a breeding population, as there is always the chance that this could lead to its extinction. As a rule, it has been considered that the general locality without being too specific, is adequate information for most species in this book. Where no abbreviations have been used and names have been printed in the records in full, this practice has been retained. Abbreviations used for names or contributors are those which have been devised by various L.N.U. lepidoptera recording secretaries, and are quoted as found in the records. A full list of L.N.U. and other contributors can be found in Chapter 4.

    For each species recorded, a statement has been made as to the larval food plant. It was felt that this would be useful because the entomologist can use this item as a handy point of reference, and the non-entomological naturalist can find an answer to the question he or she invariably asks when finding an adult insect, or a wandering larva out in the field, – i.e. "What does it eat?"

    To date, records are available for 57 species of butterfly (out of a total in Heslop of 78) and for 599 moths (out of a possible 866). This gives a total of 656 species – over two-thirds of the total number of Macrolepidoptera on the British list. It is an impressive total, and must to some extent be accounted for by the diverse nature of the habitats available within the county, as described in the previous chapter. It must however, be remembered that this number includes a significant number of immigrant species that occur from time to time on the East coast and along the river valleys, as well as a few species now extinct, or those which have declined to the point where no recent records are available. In "A Guide to the Butterflies and Larger Moths of Essex", Firmin et al.

(1975) the authors comment "the decline – or eclipse – (of a species can be) attributable to the destruction or major disturbance of the environment by man, which unfortunately in Essex has been particularly marked of late years. On the other hand quite a few species have actually increased since the turn of the Century due mainly to the spread of their food plant, notably as a result of the creation of man-made habitats, e.g. reservoirs, and several others have colonised the county". Any reader familiar with the development in agriculture, forestry, industry etc. in Lincolnshire this decade, would realise that this statement is equally applicable to Lincolnshire.

The vast majority of records can be safely vouched for by the authors. It goes without saying, however, that it is impossible to verify some of the very early ones. Where they have a special interest value, early records have been used, but in most cases recent records have been selected as being more valid. Where doubts have developed as to recent records of rare or very local species, considerable effort has been made to satisfy the authors that no mistaken identity has occurred.

In spite of this effort, however, it is accepted that there may well be an occasional (it is hoped – very rare) mistake, or omission, for which the authors alone are responsible. There are considerable areas within the county which are not worked regularly, and a number of entomologists must occasionally work within the county without passing on their records to a source accessible to the L.N.U. lepidoptera secretary. Any information in the future as to omissions or errors will be gratefully received, and amendments will be filed for future publication in the appropriate LNU Transactions or reprints of this work.

THE BUTTERFLIES AND LARGER MOTHS
Species Recorded in Lincolnshire.

Super-family PAPILIONOIDEA
PAPILIONIDAE
PAPILIONINAE (Swallowtails)

1. *Papilio machaon* L. **Common Swallowtail**
This butterfly is one of Britain's more local insects, restricted to the fens and marshes of Cambridgeshire and Norfolk, where its chief food-plant occurs. From time to time it has been recorded in most Southern and Midland Counties, and Lincolnshire is no exception. One specimen was recorded at Bourne Fen in 1872 (S.S.). In 1890 it was noted at Woodhall by Rev. Peacock – as reported in "The Natural History of Woodhall". On 10th August 1958 one female was sighted at Cowbit, in the Welland Washes near Spalding. (E.J.R.). This was assumed to be a windblown vagrant from Wicken, as the wind was blowing strongly from this direction at the time.
L.F.P. Milk Parsley, Fennel.

PIERIDAE
PIERINAE (Whites)

4. *Aporia crataegi* L. **Black-veined White**
This species is now extinct in this country. According to the Lincs. Butts., in the late 1800s, "several specimens were taken by Mr. Baines, who states that it used to be fairly common in a locality near Gainsborough". The locality was later identified in the records as White's Wood Lane.
L.F.P. Hawthorn

5. *Pieris brassicae* L. **Large Garden White**
Found commonly across the County. In some years numbers increase considerably due to migration. Abundant immigrants were noted crossing the Humber in 1870 and 1876 by Mr. J. Cordeaux. In August 1936, between Mablethorpe and Chapel St. Leonards, odd specimens were recorded flying towards the land from a boat about 2½ miles from the shore. (W. R. Withers, L.N.U. trans. Vol.9, p.119). August 1947 saw a "large invasion" of Large Whites at Mablethorpe and "Much damage resulted". (F.T.B.). During the latter year, this species was recorded as being "drowned" in small numbers along the Lincolnshire coast, (F.L.K.). There are mainly two generations per year – the first in late May/June, the second in August.
L.F.P. Brassicae and various other Cruciferae.

6. *Pieris rapae* L.   **Small Garden White**
Common throughout the County, in two broods from March/April to August/September. Migrations were seen crossing the Humber in 1870 and 1876 (J.C.). "There was a big migration from the continent and extensive defoliation of the crops occurred in 1947", (A. Roebuck, L.N.U. Trans. Vol. 12, p.66).
This migratory habit was also recorded in "some notes on the Macro-lepidoptera of Gibraltar Point", (R. E. M. Pilcher, L.N.U. Trans. Vol.XIII, No.4 July 1955). It was noted "Although it is not a local phenomenon, it is essentially a coastal one, and mention must therefore be made of the occasional mass immigration of Pieridae, for it is likely to recur and to present to the fortunate observer an astonishing spectacle, as when in the late hours of a June afternoon in 1944, a snow storm of white butterflies crossed the sea, and landed with its right flank on Gibraltar Point. For a short time every blade of grass appeared to hold its resting small white, (Pieris rapae)".
                                    L.F.P. Various Cruciferae.

7. *Pieris napi* L.   **Green-veined White**
Found widely in damp areas and along hedgesides – and in recent years has adapted to life in household gardens. In good years three generations occur, the first in April/May, and the second and third across August/September. Examples of records are: Market Deeping, May 1961 (J.H.D.); Normanby Park, Scunthorpe, "plentiful, well marked, dark yellow", 9th May 1964, (G.A.T.J.); Lincoln, 21st September 1969, (R.G.).
                                    L.F.P. Hedge or Garlic Mustard,
                                        Horse-radish etc.,

8. *Pontia daplidice* L.   **Bath White**
Only an occasional migrant from the continent. One was taken on 17th July 1894 by Mr. G. Skelton in his garden at Bargate, Grimsby, (Lincs. Butts.). A second was recorded 25th August 1913 at Kirton-in-Lindsey by the Rev. E. A. Woodruffe Peacock. This male was seen in his father's garden at Wickenstree House (L.N.U. Trans. Vol.11, p.97).
                                    L.F.P. Garden and wild
                                        Mignonette and many
                                            Cruciferae.

9. *Anthocharis cardamines* L.   **Orange-tip White**
Found along woodsides and road verges throughout the county in May/June. Numbers have fluctuated considerably over the years. In the early 1960s numbers declined, but an excellent recovery has happened in the 1970s, with breeding occurring in gardens (on

Honesty and Sweet Rocket) in the North of the County. An example from the records: May 1971, Old Leake and Wyberton "Definitely spreading range and building up numbers." (A.E.S.)
L.F.P. Cuckoo-flower, Hedge Mustard and other Cruciferae.

10. *Leptidea sinapis L.*   **Wood White**
This is a very "localised butterfly, found mainly in the south and just into the midlands", and it frequents "the edges of woodland and along rides." (Goodden, 1978). It flies mainly in May and June, but it is known that in certain years, in certain localities, there can be a second brood in August. The only mention of it in L.N.U. records is for a reputed sighting of it many years ago in Bourne Woods by Mr. S. Smith (Lincs. Butts.).
L.F.P. Leguminosae-trefoils and vetches.

RHODOCERINAE — (Redhorns or Sulphurs)
11. *Colias hyale L.*   **Pale Clouded Yellow**
An uncommon migrant to the County. The first record (1858) was for Louth when 4 were taken (V. Crow). One specimen was recorded on the railway (2.6.1892) by W. Houlden of Aby. In 1933, (6 Sept), a male was captured by H. J. J. Winter, (A.E.M.). At Spanby, (Aug 1941), two specimens were captured by Capt. W. A. Cragg, (F.L.K.).
L.F.P. Lucerne, Clovers.

13. *Colias croceus Fourc.*   **Common Clouded Yellow**
    (edusa F.)
A migrant to the County, but quite frequently recorded. In the 1920s for instance, there were 11 L.N.U. records, and in the 1940s some 16. The first County record was at Louth, 1877, when the butterfly was "very common" (R.W.G., and H.W.K.).
Twenty-one were taken at Market Rasen, 5.9.82, (W.D.C., F.A.L. and W.L.). Examples of localities in the 1940s were Skellingthorpe, Limber, Fiskerton and Sleaford. One record quotes F.L. Kirk as saying "They appeared in large numbers all over the County up to the end of October". A more recent record was 20th August 1966 at the Haven, Boston, (N.G.).
L.F.P. Clovers, Lucerne etc.

14. *Gonepteryx rhamni L.*   **Brimstone**
More common in woodland rides and along lanes in the South of the County, where the food plant occurs. In the 1970s however, the range has been extended in the North, and the butterfly has even been found in gardens. After hibernation as an imago, the

RECORDS OF LINCOLNSHIRE SPECIES 95

males and females are recorded from March/April. They were reported by F.L.K. (1948) as "becoming active in the latter half of March". On 28th August 1972, the species was common at Ancaster. (A.E.S.). Other good localities in L.N.U. records are at Woodhall, Skellingthorpe, Linwood and Holywell. Early in the 1970s the species extended its range and was frequently seen in the South Humberside region.

L.F.P. Buckthorn.

SATYRIDAE
SATYRINAE — (Satyrs)

16. *Parage megera* L. **Wall Brown**
Common and widespread, generally with 2 generations (late May and August/September). In the 1970s it appears to be less plentiful in the North of the County. Examples from the records: 25th May 1965 – Newball, (P.C.H.); and 6th September 1969, Gibraltar Point, (G.P.).

L.F.P. Various Grasses.

17. *Pararge aegeria* L. **Speckled Wood**
A local insect, found mainly in the larger woodlands to the South and Centre of the County (Newball, Bardney, Limber, Twyford Forest, Aswarby, Skellingthorpe etc.). Two, and occasionally three generations occur between May and August. Examples: Skellingthorpe, 23rd April 1957 – Rev. G. Houlden; "In Lincs. appear to be extending their range" – (F.L.K. 1948); Twigmoor – "2 specimens, August 1948," and Bourne Wood, 9th September 1978, (Both J.H.D.). Some lepidopterists consider that the species has declined in numbers in the late 1970s.

L.F.P. Various Grasses.

18. *Eumenis semele* L. **Grayling**
Now only found on restricted localities in the County where sandy heathlands survive (e.g. Risby and Messingham in the North). There are early L.N.U. records from Ashby (Brigg), and Scotton Common (28.7.05 by Dr. C.) In August 1936, it was recorded at Scotton by J. Mason. It was common at Manton in 1950 (L.N.U. Meeting), but it has now disappeared from this locality. After the species disappeared at Scotton, an attempt was made to re-introduce it from Messingham where the habitat was in danger (early 1970s). This was during a period of flooding at Scotton and the experiment was not successful. Recent records show it at Risby (Aug. 1975 – L.N.U. Meeting), and Messingham (1973 – R.E.M.P.). In August 1978, a few were recorded at Risby by D. Burton, so the species is still maintaining itself successfully in this area.

L.F.P. Grasses

20. *Erebia aethiops* Esp.   **Northern Brown**
    (blandina F.)   **(Scotch Argus)**
A few specimens of this butterfly were taken at Hubbard's Hills near Louth in 1856 by Mr. T. W. Wallis (R.W.G. and H.W.K.). According to South (1956), this butterfly only occurs in the "North of England and in Scotland", being found in the Highlands, and in England in "Durham, Westmorland, Cumberland, Lancashire and Yorkshire". Further Lincolnshire specimens were taken near Louth in 1859 or 1860 (F.C.) according to the Lincs. Butts.
                                                L.F.P. Purple Moor-grass
                                                Tufted Hair-grass etc.

22. *Maniola jurtina* L.   **Meadow Brown**
    (janira L.)
Widespread in all grassy localities from the end of June to August. Examples: 13th July 1969 – Haxey, (R.J.); 8th August 1971 – Jericho Wood, (R.W.B.). Other typical localities are New Park Wood – Bardney, Bennington Marsh, Gibraltar Point, and Stamford.
                                                L.F.P. Meadow-grasses etc.

23. *Maniola tithonus* L.   **Gatekeeper**
A butterfly of the hedgerows and woodsides, and along grass verges. Basically it can be found throughout the county, but it may be only locally common. There is a single generation in July and August. In the 1970s numbers seemed to be on the increase. Records include: 31st July 1970 – Butterwick, (A.E.S.); and July – August 1967 – Ancaster, Boston, Horsington, (J.B.).
                                                          L.F.P. Grasses

24. *Coenonympha pamphilus* L.   **Small Heath**
Abundant on grassy heathlands, sandhills, warrens, rough grassland etc. throughout the County. The first brood occurs May/June, with the second through to September. Localities include: Gibraltar Point, (27th Sept. 1976, R.G.); Stapleford Moor (23rd July 1964 – L.N.U. – G.A.T.J.); and Holywell, (13th June, 1957 – F.L.K.).
                                                          L.F.P. Grasses

RECORDS OF LINCOLNSHIRE SPECIES 97

25. *Coenonympha tullia* Mull. **Large Heath**
Limited to wet, peaty heathland, mosses, where the food plant occurs. The earliest L.N.U. record was for Alford where 2 were taken in 1888 (E.W. – Lincs. Butts.). In June 1953 three males were noted at Manton Common (approaching form philoxenus – J.H.D.). On 25th June 1961, freshly emerged males and females were found at Epworth Turbary, (J.H.D.) and on 11th July, on a peaty area some six miles South of Scunthorpe females were found (J.H.D.). It has also been recorded from Crowle Waste within the Lincs. boundary by R.E.M.P. (June 1970), and by R.J. (June 1971), when it was flying in numbers. The colony still exists (1979) in this latter area, as two specimens were seen on 29th June 1980 (R.J. and J.H.D.). It is missing from all other previously known haunts. Subsequent to a number of these previous records the areas at Manton and Messingham have been turned over to agriculture, and it is therefore extinct in these localities.
L.F.P. Various Sedges, Purple Moor-grass etc

26. *Aphantopus hyperantus* L. **Common Ringlet**
A locally common species of the hedgerow and woodside. Typical localities: 12th July 1967 – Ancaster, (G.H.); 21st July 1968 – Scotgrove Wood, "very numerous", (A.D.T.); 17th July 1971 – Bourne Wood, "Quite common", (A.E.S.). This species is noted for the variation in size of eye-spots on the underwing. Var.*lanceolata* (enlarged spots), and Var.*caeca* (very diminished spots), have been found at Black Walk Nook, Manton (by J.H.D. and Geo. Hyde), and *caeca* has been found on the Brumby Common Museum Nature Reserve at Scunthorpe (R.J.).
L.F.P. Various grasses

27. *Melanargia galathea* L. **Marbled White**
L.N.U. records infer that this was once a common butterfly in Lincolnshire (in the late 1800s) on limestone grasslands. Early records have it at Caistor and Newball Wood between 1873 and 1881, (J.C.). Canon Fowler recorded it on 15th July 1888 at Wickenby and Linwood Common, (Lincs.Butts.). More recently it was found at Ancaster (1919 L.N.U. Meeting), and Holywell near Stamford (1938 F.L.K.). On July 19th, 1959, it was recorded at Scotgrove Wood as "the most interesting thing I have seen this year – the first time I have ever seen it in Lincolnshire", by the Rev. P. C. Hawker. It was also recorded in Lincs. in 1960 in the Stamford area by R.E.M.P., where it was claimed to have a "slender hold". Its last sighting was at Glebe Quarry, Colsterworth, when a single specimen was found at an L.N.U. field meeting, 2nd August 1980.
L.F.P. Various meadow Grasses

NYMPHALIDAE
APATURINAE — (Emperors)

    28. *Apatura iris* L.   **Purple Emperor**
This butterfly is naturally a "high flying" species, and it generally inhabits the larger Oak woodlands. It is found in clearings where the food plant, the Broad Leaved Sallow occurs, though it is a long time since it has been seen flying at all in Lincolnshire, as a result of the disappearance of our Oak woodland. It was first recorded near Louth in 1858 (F.C.) where it was claimed to be "rather rare". It turned up in Burwell Wood, (R.W.G. and H.W.K.), in 1882, and was recorded near Market Rasen, (W.L.), on 18th August 1893. Other areas mentioned in the records include Newball, (G.H.), 7th July 1896; Willingham, (F.A.L.), 1877-1879; Welton Wood, "one male and 3 others seen", (E.W. and J.E.M.), 25.7.1890. Mr. Baines mentioned its occurrence in the Gainsborough district, and Canon Fowler recorded Hartsholme and Doddington in the late 1800s.
                                                      L.F.P. Sallows (particularly Broad-leaved Sallow), and Willow

LIMENITINAE — (Wood Admirals)

    29. *Limenitits camilla* L.   **White Admiral**
       (sibylla L.)
Frequents woodland rides where honeysuckle hangs. Early this century Baines recorded this species as occurring in two localities near Gainsborough. It was also recorded at Langton, near Horncastle by J. Conway Walter. Early in the 1940s the species began to extend its range, and in 1941 it was found at Skellingthorpe Old Wood by C. Taylor. There were then a considerable number of records for many of Lincolnshire's central and southern woodlands in the 1940s and 1950s – including Lynwood, Newball, Fiskerton, Woodhall, Bleasby, Bardney, Legbourne, etc. The population went into a decline in the mid 1960s, with the species disappearing from most haunts. Colonies have, however, been seen in the Bardney, woodlands in the 1970s by R.E.M.P., A.E.S., J.H.D. and others.
                                                          L.F.P. Honeysuckle

NYMPHALINAE — (Vanessids)

    30. *Vanessa atalanta* L.   **Red Admiral**
A migrant species which is observed most years – sometimes in large numbers, other years seemingly absent. Spring migrants breed in this county successfully, leading to imagines being seen in the Autumn. Odd specimens have been recorded very early in the year, so occasionally the adult insect is able to overwinter here,

though usually it is thought that the Autumn adults perish in our coldest months. It has been recorded almost everywhere in the county. As examples, it was noted in Boston, (R.E.M.P.), in 1960, where it "swarmed in summer and autumn", and in Scunthorpe, (P. Riley), in 1973, when in August and September it was "abundant – 40 in garden on fallen apples".

L.F.P. Stinging Nettle.

31. *Vanessa cardui* L. **Painted Lady**
Again a migrant, though much more irregular than the previous species. Some years it is very abundant – and it can be seen anywhere, any time between May and September. The spring visitors breed in this county. As typical records: "1969 – Lincoln, Bardney Ponds, Snakesholme Wood – very common" (A.D. Townsend); Boston, 11th August 1974, (A.E.S.); Scunthorpe, August 1977, (J.H.D.); Humberstone, 11th November 1978, (J. Owen). There was a large migration across the county in 1980, when specimens were known to cross the coast in July, and spread to all areas in the following weeks.

L.F.P. Thistle

33. *Nymphalis io* L. **Peacock**
Widely distributed across the whole county. This is an indigenous species, but some years it appears to be more common than others.

Imagines can be seen quite early in the year after hibernation, and the butterfly can then be seen on the wing at any time between March and September/October. It can often be found in gardens, visiting nectar bearing flowers. As an example from the records, E. Mason, 14th September 1977 recorded "150 observed" at Lynwood.

L.F.P. Stinging Nettle

34. *Nymphalis antiopa* L. **Camberwell Beauty**
A very rare Autumn migrant, arriving on the East Coast of the County from Scandinavia. It was first recorded in September 1858 near Louth (C. Clayton). 1898 was apparently a good year, as one was seen at Croxby (8th April) by Rev. W. Cooper, and it was recorded at Gainsborough (F.M.B.) as "not uncommon some years – 7 examples having been seen in Tillbridge Lane, and one in my garden". After an absence of some 78 years, there were then 10 L.N.U. records in the county for 1976. These covered Gibraltar Point, Sutton-on-Sea, Cherry Willingham, Lincoln, South Gulham, Brumby Common Nature Reserve, Scunthorpe, Garthorpe and Ousefleet (all in August, September and October).

One wintered in Scunthorpe in confinement (c/o Mrs. Rees), and it fed in the early spring of 1977, before it died in March.

L.F.P. Willows

36. *Nymphalis polychloros L.* **Large Tortoiseshell**
A butterfly of more lightly wooded areas – very rare with no recent records in the county. There are records for the late 1800s e.g. 1877/78, (August, F.A.L.), Market Rasen; and 1897, "3 seen near Newball Wood" (W.L.). More up to date, E. J. Redshaw reported 2 specimens, (20th April 1958), near Cowbit.
L.F.P. Elm, Willow

37. *Aglais urticae L.* **Small Tortoiseshell**
Common everywhere. It can be seen flying in early spring sunshine after hibernation, but it is most common from July and August into the late Autumn. Occasional fine varieties have been taken in the county – e.g. 1945 in a Scunthorpe garden by Geo. Hyde.
L.F.P. Stinging Nettle

38. *Polygonia c-album L.* **Comma**
A woodland butterfly which hibernates as imago. It can be seen in the spring visiting Sallow bloom, and its next generation emerges in August/September. A number of Central Lincs. haunts are mentioned in the records for the late 1800s and early 1900s. After a short absence, it next turned up at Grantham in August 1934, (R. Benn and A. E. Musgrave). Sightings then increased every year through to the 1950s, as the range extended to cover virtually the whole county up to the Humber. The species then went into decline, and became generally scarce. Since then, small numbers have been recorded each year to the South e.g. South Thoresby – 22nd September 1971, (R.E.M.P.); Billingborough – "taken in confinement and returned to wood, April 2nd, after accepting honey solution", 10th Feb. 1976, Miss E. M. Goom.
    A striking albinistic variety was taken at Raventhorpe near Scunthorpe on 11.4.48, (J.H.D., Entomologist Vol.81, p.149).
L.F.P. Stinging Nettle

ARGYNNINAE — (fritillaries)

39. *Argynnis paphia L.* **Silver-washed Fritillary**
A woodland butterfly, on the wing in July and August, and often seen visiting Bramble blossom, Valerian or Thistle heads. It was last noted in the records in 1960, when it was seen to be "strong over a wide area around Boston" (R.E.M.P.). Earlier this century, it was classed as a locally common butterfly around the Market Rasen area particularly, but the population seemed to decline around 1950. Examples of earlier records are: 1925 and 1935, Market Rasen "plentiful", (T. H. Court); 1945 Linwood, "Common", (G.H.); 1947, near Alford, "plentiful", (F.L.K.).
L.F.P. Dog Violet

RECORDS OF LINCOLNSHIRE SPECIES

40. *Argynnis cydippe* L. **High Brown Fritillary**
This woodland species, usually spotted in July and August, has always been scarce within the county. The earliest record for it was for Greenfield (near Alford), for 1890 (E.W.). It was next recorded three times in the 1940s. Firstly in 1940, records (by A.E.S.) covered Swinnwood, Motherwood, Tothill Wood, Muckton Wood and Well Vale. Secondly, it was seen on 4th August 1940, in Grasby Bottom (Else), and thirdly it was taken at Scawby (1948) by J.H.D. There are also three records for the 1950s, the last being for Martin, Horncastle, where two were on the wing, (4th July 1959), and seen by G.A.T.J. It was recently reported from Temple Wood, (25th July 1981) by Mr. G. Posnett.
L.F.P. Dog Violet.

42. *Argynnis aglaia* L. **Dark Green Fritillary**
Frequents flowery meadows and rough heathland in July and August, where various species of Violet occur. It has been recorded over a very wide area of the county at such localities as Mareham Lane (5th July 1941, F.L.K.), when the species provided "a wonderful sight, many females just emerging", and at Market Rasen (T. Court 1941), when it "swarmed in the district in woods and fields". Other areas mentioned in the records include Sleaford, Grantham, Skellingthorpe and Louth, where the species was common. To the North it was common at Scotton and Manton in the mid 1950s. It was last recorded at Kirkby Moor in 1970, (M.T.), and in August 1975, by R.E.M.P.
L.F.P. Dog violet, Marsh Violet, Heartsease.

44. *Clossiana euphrosyne* L. **Large Pearl-bordered Fritillary**
This butterfly was recorded as a locally common insect in May and June, in a number of areas in the late 1800s. It was still claimed to be "common in its usual haunts" (F.L.K.) up to 1939. These included Hartsholme, Newball Wood, Doddington and Stainton Wood. In 1942 it was recorded at Grantham, (F.L.K.), and in 1943 at Hurn Wood, (Chambers). In 1945 it turned up at Gainsborough, (9th May, T.H.C.), and it was common around Woodhall in the middle 1950s. After being seen at Linwood, (J.H.D.), in 1958, it declined rapidly, and there are no records after the early 1960s.
L.F.P. Many common Violet species

THE BUTTERFLIES AND LARGER MOTHS

45. *Clossiana selene Schiff.* **Small Pearl-bordered Fritillary**
A local butterfly which was found in the county on damp heathlands, and along woodland margins. There is a first generation on the wing in June/July, and a second in the Autumn. This can be illustrated from the records – the butterfly was seen on 8/7/1909 at Peacock's Hole, Morton Carrs, near Gainsborough by G.W.M., and A.E.M. recorded a second brood on the wing in August of 1923.
More recently it was at Bleesby Wood, (G.M.), in 1946, and it was last recorded at Scotton Common in 1951 and 1955 by M.P.G.

L.F.P. Violets

48. *Melitaea cinxia L.* **Glanville Fritillary**
Lincs. Butts. stated that this insect "Formerly occurred in Lincolnshire, and county records from 1702 downwards are given in C. W. Dale's "History of our British Butterflies".

L.F.P. Ribwort Plantain

49. *Euphydryas aurinia Rott.* **Marsh Fritillary**
(artemis Schiff.)
Recorded in the county in damp, marshy meadow habitats in late May and June. Its two main areas were in the Market Rasen and Skellingthorpe districts. In the former it was recorded between 1925 and 1943, (T.H.C.). In the latter, there was a small colony between 1943 and 1947, and from 1948 it seemed to disappear completely, after a very rapid decline. More recently, however, it has been claimed that several have been seen (mid 1970s) near Cranwell.

L.F.P. Devil's-bit Scabious, Honeysuckle

NEMEOBIIDAE

NEMEOBIINAE — (Dukes)

50. *Hamearis lucina L.* **Duke of Burgundy**
Seen on the wing from the end of May, into June, in woodland clearings. It was claimed in Lincs. Butts. that it "has been found in the 1800s in extremely restricted localities in Lincolnshire", (T. H. Allis).
In 1960, R.E.M.P. reported that it was "on its last legs in the county as its habitat has been dreadfully mauled".
This "last station" was in fact near Stamford, and this area was destroyed by 1961.

L.F.P. Cowslip and Primrose

LYCAENIDAE

THECLINAE — (Hairstreaks)

51. *Thecla betulae* L. **Brown Hairstreak**
Found along woodland borders and rides in July and August, and even later in some years. It was only infrequently recorded in the county in the early part of this century, but in the 1940s and 1950s it turned up in Newball Wood, (T.C.T.), and Southrey Wood, Woodhall Spa (F.L.K.). It was seen at Stainfield Wood on 15th September 1962 (E. E. Steel), and since 1970 it has been found each year in the Bardney Woodlands by R.E.M.P. Ovae and larvae were discovered in this latter region in 1979 and 1980, by A.E.S., J.H.D. and R.J., and it was seen in Temple Wood on 25th July 1981 by Mr. G. Posnett.
L.F.P. Blackthorn

52. *Thecla quercus* L. **Purple Hairstreak**
This butterfly is fairly widely distributed across the county, but its colonies are quite local, and restricted to a number of Oak woodlands. In 1916 it was said to be common in woodland at Skellingthorpe (larvae found on 3.6.1916 by Rev. F. L. Blathwayt, and imagines later in the season). On 22nd July 1937, it was seen flying at a L.N.U. meeting at Well Vale, near Alford. In the 1960s and 1970s it has been recorded from Woodhall, Kirkby Moor, Bardney, Laughton Common, and at Holme Plantation (near Messingham). The latter specimen was a female, seen Sept. 1977 (by J.H.D.).
L.F.P. Oak

54. *Strymonidia w-album Knock* **White-letter Hairstreak**
Mainly found (July and August) along the margins of woodlands where Elms or Wych Elms occur. Some years it has been recorded in numbers. Other years it is seemingly missing, but it can re-occur in its previous haunts. Records include: 1929, Belton, "noticed on several occasions by Rev. H. E. Stancliffe, visiting flowers in his garden". It was recorded in Rauceby High Wood by F.L.K. in 1940. More recent records are for Hendale Wood (1950 – P.F.) and Scawby (1953 – J.H.D.). In the 1970s, it was found on 8th August 1971 in Jericho Wood, (R. W. Bacon), and larvae were found, (by A.E.S.), at Castle Bytham, (June 1973).
L.F.P. Elm and Wych Elm

55. *Callophrys rubi* L. **Green Hairstreak**
Found among rough scrub land in May/June, it was frequent in many different localities up to the 1960s. It then disappeared from N. Lincs, and many of its previous haunts, and it is now mainly

found along the belt of scrub dune lands along the East Coast (e.g. Theddlethorpe and Gibraltar Point).

Examples of records are Mablethorpe, 1952-60, "fairly common until 1960," (T. R. New); and Gibraltar Point, 12th June 1971, "locally common", (A.E.S.).

L.F.P. Furze and Broom

LYCAENINAE — (Coppers)

56. *Lycaena dispar* Haw. **Large Copper**

In most of the standard works on lepidoptera, (e.g. "Butterflies", E. B. Ford, 1967, and "The Butterflies of the British Isles", Richard South, 1956) reference is made to the fact that the Large Copper was first known to Lewin (1795) with specimens being taken in Huntingdonshire. Ford wrote (p.166) "for a time it was found commonly in some parts of fenland, but by 1840 it had become rare, and the last specimens seem to have been caught in 1847 or 1848". South wrote (p.150) "Haworth (1803) mentions its occurrence in the fens of Cambridgeshire". In these two localities South described how "it appears to occur in great profusion, as several hundred specimens have been captured within these last ten years by the London collectors who have visited Whittlesea and Yaxley Meres, during the month of July, for the sole purpose of obtaining specimens of this insect". On the same page (now p.151) it is recorded how "the last capture, consisting of five specimens, appears to have been made in Holme Fen, by Mr. Stretton either in 1847 or 1848".

The English race of the Large Copper (sub-species *dispar*) is therefore extinct, though continental sub-species (*rutilus* and *batavus*) have been successfully introduced to a number of regions (the latter form, from Holland, for example, to Wicken Fen in Cambridgeshire).

According to L.N.U. records, "at a L.N.U. Field Meeting at Gainsborough, (8.7.1909) Mr. Baines exhibited a number of *Lycaena dispar* taken by a Mr. Forrington in the mid 1800s at Morton Carrs (in the Trent Valley) near Gainsborough".

This was thought to be the only record of the species in Lincolnshire. The authors were delighted, and greatly surprised, however, when Mr. John Redshaw of Pinchbeck, Spalding, Lincs. wrote to Mr. Duddington on 15th February 1982, giving details of a further record which he had come across by chance. Mr. Redshaw had been reading in the early minute books of the Spalding Gentlemen's Society, which was founded in 1710, and which is still active – Mr. Redshaw being a member.

In the Fifth Minute Book, under the entries for 28th September, 1749, appears the following report:

"The Orange Argus of Elloe, an Elegant Butterfly depicted and discovered by the Secr'y. The Secretary Shew'd the Company a

Butterfly which he took on the Dozen's Bank of which he sent a limming to Mr. Da Costa to whom and to the Aurelian Society 'tis entirely unknown. Therefore the President Desired it might be Scetched into the Book and at his Instance the Secr. was so obligeing to make these very exact pictures thereof above."

This "limming" is a coloured painting showing the upper and under surfaces of a large orange butterfly, to which is appended the scientific name "Argus Aurantius Elloensis". Although the wings are shown set unnaturally, with the front edge at right angles to the thorax, the upper and under surfaces are painted in detail, and excellent colour, leaving no doubt that the specimen was a male Large Copper. This is a most striking discovery, since this note and clear painting pre-dates other records of the butterlfy by some 46 years, and it extends its known range.

Mr. Redshaw, in L.N.U. Transactions XX,3 explains that Mr. Da Costa was Emanuel Mendes Da Costa, a notable naturalist of his time, Foreign Secretary to the Royal Society, and a Member of the Spalding Gentlemen's Society.

He also writes "Dozen's Bank, where the specimen was taken, presumably in the preceding July or early August when this species is on the wing, is situated in the parish of Pinchbeck on the western edge of the 10km grid square TF/22, and in 1749 formed the eastern boundary bank of the old Counter Drain Washes. These washes were drained about 1775 when the old Counter Drain was realigned to join the Vernatt's Drain at Pode Hole".

The article continues "Although these washes would have been flooded in winter it is likely that similar conditions persisted in the Cambridgeshire fens prior to drainage".

Mr. Redshaw's article gives further detail of the pattern of drainage in this Pinchbeck region and he speculates that "as the known 19th century colonies of this species appeared to be confined to comparatively small areas, it seems possible that the Pinchbeck specimen was taken within the range of a local colony".

The authors are most grateful to Mr. Redshaw for permission to quote his article and findings at length. A discovery of this nature is obviously most exciting to a butterfly enthusiast.

L.F.P. Great Water-dock

58. *Lycaena phlaeas L.* **Small Copper**

Flies commonly across the whole county. In good years three generations occur between April and late October. A number of interesting varieties have been taken including var.*obsoleta* (with the copper bands on the hindwing missing) – 1952. (J.H.D.), at Ashbyville near Scunthorpe. A Most unusual cream variety was captured (by J.H.D., Ent. Vol. 89., p304), from the same colony in 1956. This particular haunt has now disappeared under the Anchor Steelworks. L.F.P. Sorrel and Dock

THE BUTTERFLIES AND LARGER MOTHS

POLYOMMATINAE — (Blues)

    61. *Plebejus argus* L.   **Silver-studded Blue**
      (aegon Schiff.)
This is a heathland butterfly, and its flight period is in May and June. Lincs. Butts. reported that the species occured Near Laughton Common, Gainsborough, (F.B.M.), in the Owston Ferry district (A.R.) and at S. Hartsholme and Doddington, (J.F.M.)", in the late 1800s and early 1900s. It vanished from most of these haunts, but a strong colony (seen by Geo. Hyde in the early 1940s, and T. C. Taylor (1944) among other entomologists) maintained itself at Laughton up to 1945 – before it finally disappeared.
                                          L.F.P. Gorse and Birds-foot Trefoil

    63. *Aricia agestis Schiff.*   **Brown Argus Blue**
      (astrarche Bergst.)
Another heathland butterfly – found on rough, open ground in two generations (May/June and again in August).
    Lincs. Butts. claimed it was "found rarely – in the Louth district" and was "included in the collection of butterflies found near Louth, 1863, in the Mechanics Institute, but is not now found, (R.W.G. and H.W.K.)". From 1926, however, it was found at Grantham in "fair numbers" (A. E. Musgrave) "where the form occurring has the orange spots in the margin very prominent. This form is more usually associated with districts further South than our own". It continued to thrive, being recorded at Ancaster, Stamford, Woodhall Gibraltar Point, Market Rasen and Scunthorpe, and in the period of the late 1940s and mid 1950s it was at its best. It was common at Mablethorpe for a few years from 1952 onwards, but then it went into rapid decline everywhere, and there are no L.N.U. records for the species after 1960, (T. R. New), when it was last seen in its Mablethorpe locality.
                                                L.F.P. Rock-rose, Stork's-bill

    64. *Polyommatus icarus Rott.*   **Common Blue**
Found quite commonly across the County on open, rough ground on the banks of railway cuttings and even along road verges, in two generations (June and August/September).
    Records seemed to decrease slightly in the late 1960s, but by the mid 1970s a recovery had been made. Recent records include: Gibraltar Point, (G. Posnett), Sept. 6th 1969; Bourne Wood, "a fairly strong colony", (A.E.S.), June 20th 1970; and Burton Pits, Lincoln, (E. Mason), July and August 1975.
                                                      L.F.P. Trefoils

65. *Lysandra coridon* Poda **Chalk-hill Blue**
A butterfly of limestone grassland areas, on the wing in July/August. Lincs. Butts. had the species "common in Lincolnshire on chalk, (T. H. Allis), in the 1890s". Early records locate the butterfly at Ancaster, (Aug. 1919, H. C. Bee). There are some twenty L.N.U. records for Ancaster up to the 1950s when the colony was particularly flourishing. It continued to thrive into the '60s, (A typical record being "Common, males and females – R. W. Bacon – 7th August 1965"). After 1970, the colony began to tail off and on 10th August, 1973, Sir W. P. Parker, Bart. reported only "4 males – a deplorable year". It was last seen in 1974, (M.T.), and since the latter date there have been no further sightings from this colony – the colony the furthest North in the British Isles – of the Chalk-hill Blue.
L.F.P. Horseshoe Vetch

67. *Cyaniris semiargus* Rott. **Mazarine Blue**
  (acis Schiff.)
In his "Butterfly Book", (1955), Edmund Sanders wrote "once an established resident in England, this butterfly has been extinct here since 1876". According to Lincs. Butts. it was found "near Epworth" as late as 1864. Additionally, it is stated "Mr. S. Hudson continued to find it, though not commonly, at Epworth in the North of Lincolnshire". He wrote in the Zoologist for that year. "I find it in meadows, but they are large in extent; and the insect appearing just before the grass is ready for the mower, prevents a proper search from being made for it." In a letter to E.A.W.P. written by the late Mr. S. Hudson in 1903, it was stated that it was not yet extinct. (Nat. 1904, p224).
L.F.P. Kidney Vetch

68. *Celastrina argiolus* L. **Holly Blue**
This butterfly occurs in shrubberies and gardens where the foodplants of the two generations are found. The spring generation lays eggs on Holly buds (in May) and the autumn generation lays on Ivy. The species has a fairly wide distribution across the south of the county, and is found as far North as Lincoln. The range was extended across the county in the early 1940s, and specimens reached the Humber (Barton – C. G. Else, 1951). After this, the species seemed to withdraw to its former territory.
In 1960 it was abundant at Boston, (R.E.M.P.). Numbers have decreased in some localities, but small numbers or odd ones have been spotted in some old haunts (e.g. Bourne Woods). Examples of previous records are: Lincoln, 20th August, 1972, "small numbers", (E.M.); and 4th September 1971, "one only", Holywell, (A.E.S.).
L.F.P. Holly & Ivy

69. *Cupido minimus* Fuessl. **Small Blue**
Generally a species found on grassy slopes on chalk – on the wing in June. Lincs. Butts. had the status as "rare", the only record being "Near Glentham, 1877-1879, (F.A.L.)".

L.F.P. Kidney Vetch

HESPERIIDAE

PYRGINAE — (Black and White Skippers or Grey Skippers)

71. *Pyrgus malvae* L. **Grizzled Skipper**
Frequents meadows, commons or woodland clearings in May and June, and sometimes into July. Exceptionally there can be a second generation (e.g. Ferriby Cliff, 14th September, 1943, G. Else).

According to Lincs. Butts., it was scarce around Market Rasen, Gainsborough and Skellingthorpe in 1901. It became more plentiful in the 1920s, (e.g. Little Ponton, "plentiful" 1925, (L.M.)). It was more scarce in the 1930s but again became "abundant, 1943, Market Rasen", (T. H. Court). There was a rapid increase in the 1950s, when records included: Manton Warren, 12th May 1956, (L.N.U. meeting), and specimens at Woodhall, Knowles Wood, Aswarby Thorns, etc. In the later 1960s numbers reduced to "odd ones" again, though in June 1971 it was fairly common at Sedgebrook, (R.W.B.), and in the Bardney area in 1971 it was "numerous (A.D.T.).

Up to 1979, it has continued to be found near Bardney, and small numbers have been recorded in woodlands in the South West of the county in 1980.

L.F.P. Bramble & Wild Strawberry

72. *Erynnis tages* L. **Dingy Skipper**
Found in open spaces, on rough ground and hillsides in May/June, with exceptionally, a second generation in the Autumn. In 1900 it was classed as "locally common in some districts" in Lincs. Butts.

Records up to the 1960s include Market Rasen, Newball Wood, Ancaster, Woodhall, and Knowles Wood.

In the 1960s most records were for single specimens – e.g. Temple Wood, 1967, (A.E.S.); and Wilsford Heath, near Kelby (old quarry), (G.N.H.).

From 1970 onwards one colony has maintained itself near Haxey, (last L.N.U. record, June 4th 1978 (R.J.)) and a colony still exists at Bardney.

L.F.P. Bird's-foot Trefoil.

RECORDS OF LINCOLNSHIRE SPECIES 109

HESPERIINAE — (Brown Skippers)

73. *Thymelicus sylvestris Poda* **Common Small Skipper**
(thaumas Hufn.)
Widespread across the county – could be found anywhere in July and August on rough ground, along waysides and in woodland clearings.
Records include: Normanton Thorns, 17th July 1971, (R.W.B.); Kirkby Moor, "Fairly plentiful", 20th August 1972, (E. Mason); Scunthorpe area, (found at Messingham Sand quarries and on Brumby Common Nature Reserve etc.) July, August 1978, (R.J.).
L.F.P. Soft Grasses.

74. *Thymelicus lineola Ochs.* **New Small Skipper**
(Essex Skipper)
Found on rough ground in July, this species is very similar to the Common Small Skipper, and it is difficult to separate the two species without close examination. It was mainly found in the S.E. of the country, but it has now extended its range into Lincolnshire. L.N.U. records show it as "spreading fast in the Boston area in Lincs. in 1960", (R.E.M.P.).
In 1971 it was "not uncommon within 5 miles of Boston", (23rd July – A.E.S.), and in 1974/75 it was "abundant at Saltfleetby and Theddlethorpe", (R.E.M.P.). It was recorded in 1979 in Snakesholme Wood, Chamber's Wood (Little Scrubbs), and in the Burton Pits, (M.T.).
L.F.P. Longer coarser Grasses–
e.g. Couch Grass

76. *Ochlodes venata Br. & Gray* **Large Skipper**
(sylvanus Esp.)
Inhabits woodland margins and rides, and rough ground in June/July, and found everywhere.
Typical sites include New Park Wood, "common" 6th July 1968, (A.E.S.); Normanton Thorns, 17th July 1971, (R.W.B.); Great Scrubbs Wood, Bardney, 14th July 1979, (J.H.D. and R.J.). etc..
L.F.P. False-Brome Grass

78. *Carterocephalus palaemon Pall.* **Chequered Skipper**
(paniscus F.)
This species **was** found in woodland rides and clearings in late May/June. Lincs. Butts. gave its status as "very local" – e.g. near West Rasen, 27.5.1856, (Rev. W. W. Cooper), and Skellingthorpe, 3.6.1900, (J.E.M.). In 1893, in Newball Wood, it was

abundant from May 4th to May 25th, when "hundreds could be seen" (Rev. F. L. Blathwayt). After 1900 its main localities were Market Rasen (1925 – "not too plentiful" T. H. Court), and Skellingthorpe (1953 – "I saw between 30 and 40 – it was the commonest butterfly" – Canon G. Houlden). The Market Rasen colony disappeared after 1925. That at Skellingthorpe thrived up to 1960, and then declined until the last specimen was recorded in 1967. (G.H.). A survey by a number of lepidopterists in 1980 failed to make any sighting of this species in any area of the County.

L.F.P. Grasses – e.g. False-Brome Grass

Super-family SPHINGOIDEA

SPHINGIDAE

SMERINTHINAE

79. *Mimas tiliae L.* **Lime Hawk**
This species mostly occurs where Lime and Elm trees line roadsides and drives. In the main it is an urban species in this county, and its range is widespread though colonies tend to be localised. During the last ten years there has been an extension in range to the North of the county. It is on the wing in late May, through June, and is very variable – 'pink' banded varieties have been taken. Typical localities: Grimsby, (M.V. light, G.A.T.J.), 22nd May 1956: Sleaford, (N.H.), 23rd June 1960; Boston, (Several, A.E.S.), 30th June 1972; Goxhill, (S. Van Den Bos), June 1976; Scunthorpe, (4 larvae on elm, P. Walter), September 1976.

L.F.P. Lime, Elm

80. *Laothoe populi L.* **Poplar Hawk**
The poplar hawk is as common and as widespread throughout the county as the occurrence of its food plant. There are often two generations, with a staggered emergence of the adults between May and September. The moth shows considerable colour variation from dark winged, through grey, to occasional pink winged specimens. In Lincoln in 1943 it was recorded that "larvae were defoliating young poplars", by A. Roebuck (L.N.U. Trans. Vol.2, p.39). A female was noted ovipositing on 24th May 1952 at Alford, (F.L.K.). Other records include: Saltfleetby, (G.A.T.J.), 18th July 1958, and Grimsby, (M.V. light, G.A.T.J.), 31st May 1958. In Mablethorpe it was reported as "common" every year between 1952-65 (T. R. New). etc.,

L.F.P. Poplars, Aspen, and occasionally Sallow or Willow

RECORDS OF LINCOLNSHIRE SPECIES 111

81. *Smerinthus ocellata* L. **Eyed Hawk**
This species favours fairly damp places where Sallows and Willows abound. It is very widespread, though is not too common in some areas of large woodland. It is frequently recorded in gardens where the larvae feed on Apple or Weeping Willow foliage. The imago is on the wing in June in the main, with larvae being found up to September. There is little variation in this species. Records include: Boston, (B.H.) 1951; Skegness, (B.W.), June 1965; Scunthorpe area, "larvae very numerous," (J.H.D.), 1966; widespread across Lincolnshire, 1977, (J.H.D.); etc.,
L.F.P. Sallow, Willows, Apple etc.,

SPHINGINAE

82. *Acherontia atropos* L. **Death's head Hawk**
This migrant species has been recorded from most parts of the county this century, and some years the larvae have been quite common in agricultural areas growing Potatoes. A few adult insects have been recorded as early as May, and from this month larvae, pupae and adult insects have been recorded through to October during the Potato harvest. Some of the later adult insects may be migrant specimens, though occasionally, it is thought that a number of the early larvae may breed through to pupate and emerge successfully in the Autumn. As examples of recent records, "in 1950 large numbers were found over a wide area, including Snitterby, Waddingham, Louth, Alford, Lincoln, Saxilby, Doddington, Washingborough, Nocton, and in the Fens surrounding Boston. Larvae were found from early August and pupae in October. Altogether over 100 seen", (H.M.S.). At Barton-on-Humber 3 larvae were found on 19th August 1952 (M.P.G.), and at Lincoln "Three adults were found within the city during the week ending 21.9.56" (F.T.B.). It has been seen almost annually in the 1960s and 1970s, the most up to date record being for a specimen brought to the Scunthorpe Museum, 18th July 1979. (E. Moody).
L.F.P. Potato, and occasionally Snowberry

83. *Herse convolvuli* L. **Convolvulus Hawk**
A migrant species – some years recorded in large numbers, other years absent. Most sightings occur in September, though August and October specimens are occasionally noted. Usually it is the adult moth which is recorded – only rarely are larvae discovered. Localities include: Barton-on-Humber, (E.P.H.B.), 1915; Gibraltar Point, "One in Heligoland bird trap", (L.N.T.), 5th Sept. 1950; Boston, "usually pretty plentiful around here", (R.E.M.P.), 1959; Scunthorpe, "11 taken to museum found in

Humber villages between August and October", (J.H.D.); Lincoln, (G. Posnett), September 1976.

L.F.P. Convolulus.

86. *Sphinx ligustri* L. **Privet Hawk**
Most frequently this moth is recorded in the South of the County, where it occurs each year, but never seems to be very common. It is an indigenous British moth, the population occasionally being augmented by migrants. Its main locality seems to be around the Boston area where it was reported to be "common in garden" in 1961 and 1962, (R.E.M.P.). Other localities include Holbeach, "one moth end of June", (B.H.); Lincoln, (Mr. R. Wood Powell), 15th Sept. 1945; Market Deeping, "one larva 1971", (J.H.D.); and larvae were found near Boston (by M.P.G.) in the mid 1970s.

L.F.P. Privet and more rarely Lilac

87. *Hyloicus pinastri* L. **Pine Hawk**
This moth is fairly common in parts of Norfolk (e.g. around Thetford) and is extending its range from East to West across the country. Lincolnshire is probably on the very edge of its range, and only three records are available – Boston 30th June 1955, S. Thoresby, 20th July 1976, (Both R.E.M.P.), and Boston 1979, (A.E.S.).

L.F.P. Pine

DEILEPHILINAE

91. *Celerio galii* Rott. **Bedstraw Hawk**
A rare migrant. The bedstraw hawk was first recorded on 2nd August 1888 "near Alford at a clump of Sweet William" by R. Garfit. Only 4 recordings were made between this date and 1972 when Mr. A. D. Townsend found "several larvae on Rosebay Willow Herb at Lincoln". Ten further records occur in the 1970s, covering Grimsby, Gibraltar Point, Scunthorpe, Scotton and Alkborough, and suggest that some pupae may have survived the mild winters. The most recent record is for a larvae being found at Ashbyville (Scunthorpe) in 1979 (reported to J.H.D.).

L.F.P. Bedstraw, Willow Herb Fuchsia

92. *Celerio livornica* Esp. **Striped Hawk**
 (lineata F.*)*
A very rare migrant. It was recorded at Lincoln in August 1906, (Arnold), when "two examples were discovered in a wash house". A fresh specimen was sighted in a garden at St. Catherine's,

RECORDS OF LINCOLNSHIRE SPECIES 113

Lincoln, at Phlox blooms, on the evening (7.15 p.m.) of August 21st 1907, (J.F.M.). Only two other records exist, the latest being for Gainsborough, (captured by Mr. E. J. Scott, and reported by Mr. J. W. White), 20th April 1945.
L.F.P. Vine, Fuchsia, Dock, Antirrhinum etc

93. *Hippotion celerio* L. **Silverstriped Hawk**
Very rare. Two records exist for the 1800s, and a single record for 1926 – near Gainsborough in 1859, (in C. W. Dale's "History of British Hawkmoths") (E. Tearle); Burton Road, Lincoln, (J.F.M.), 9th October 1883; and near Louth, "taken at light in a cottage" on 24th September 1926, (A. E. Musgrave).
L.F.P. Vine, Yellow Bedstraw, Virginia Creeper, Fuchsia

94. *Daphnis nerii* L. **Oleander Hawk**
A very rare migrant with only 3 L.N.U. records. "A specimen was taken at South Somercotes on 10th October 1903, and is now in the possession of the Louth Naturalists' Society" (Lep. Lincs. 1.). Another occurred at Holbeach (7th Sept. 1926, Mr. Curtis). On 22nd October 1949, one was "picked up on a dredger in Grimsby Docks". It was reported in the Daily Mail, and investigated by G.A.T.J.
L.F.P. Oleander

95. *Deilephila porcellus* L. **Small Elephant Hawk**
Locally, this moth is quite common, and its range extends across the county from the coast, inland to wooded areas. The species prefers a drier type of habitat to the elephant hawk, and can be found on the wing from the end of May, through June to July. It is often seen being attracted to nectar bearing flowers such as Honeysuckle and Valerian at dusk. Localities include Market Rasen, (T. H. Court), 14th August 1950; Mablethorpe, "common every year 1952-65", (T. R. New); Holbeck, "very comon", (G.A.T.J.), June 1976.
L.F.P. Bedstraws, Willow-herb

96. *Deilephila elpenor* L. **Large Elephant Hawk**
Generally common and widespread throughout the county – seeming to increase in numbers as its food plant has extended its territory. The imago can be found over a long period with fresh specimens emerging from June into the Autumn. Typical records are: Holton-Le-Moor, (Miss J. Gibbons) 1950; Brocklesby, "Plentiful at light" (G.A.T.J.), 25th June 1957; Lincoln, (R. Goy),

30th June 1969, and Boston, (A.E.S.), "quite common 25th-30th June 1972" etc.

L.F.P. Willow-herb

MACROGLOSSINAE

97. *Macroglossum stellatarum L.*   **Humming-bird Hawk**
A common migrant, sometimes abundant in numbers, this species can be seen visiting nectar bearing flowers like Phlox, Rhododendron and Tobacco plant at any date between July and October. 1947 was a particularly good year – "Lincolnshire – very common", (F.L.K.). 1959 also saw large numbers recorded at Cleethorpes, Lincoln, Sleaford, Swaby, Louth and Scunthorpe, where the moth was "common in parks and gardens on flowers during October – the last one seen on Oct. 30th", (J.H.D.). There are numerous later records for such places as: Willoughby, Goxhill, Scawby, S. Thoresby, Coningsby, Boston etc.,

L.F.P. Bedstraws

98. *Hemaris fuciformis L.*   **Broad-bordered Bee Hawk**
This species mainly flies in woodland areas where its food plant occurs, and it can be locally common. It is on the wing from the end of May into June, and is fond of visiting blossom, such as Rhododendron flowers. It was recorded at Holywell, (L.N.U.), 16th June 1951; Woodhall Spa, "Common", (F.L.K.), 1953; Willingham Forest, (L.N.U.), 26th May 1956; "larvae and ova were very common at Twigmoor," (R.J.), 13th July 1967; and at Boston, (A.E.S.), 1975.

L.F.P. Honeysuckle

99. *Hemaris tityus L.*   **Narrow-bordered Bee Hawk**
   (bombyliformis Esp.)
This is mainly a woodland species, keeping to damp places where its food plant occurs, and its emergence is in May and June. It is a very local species, and has not been frequently recorded in the County. Early records include Market Rasen, (F.A.L.), 1877-1879, and Skellingthorpe, (J.F.M.), 3rd June 1901. More recently it was found at Market Rasen, "Feeding on Aubretia in late May, and another on 31st July", (T. H. Court) in 1943. It was last recorded at Kirkby Moor, (R.E.M.P.), in 1970 and 1971.

L.F.P. Devils'-bit Scabious.

RECORDS OF LINCOLNSHIRE SPECIES 115

Super-family BOMBYCOIDEA
NOTODONTIDAE
CERURINAE

101. *Harpyia bifida* Brahm   **Poplar Kitten**
     (hermelina auct.)
A distinctly local moth within the county – mostly recorded in July where the food plant occurs.
It was recorded as "fairly common" at Haverholme Priory in June 1902, (J.D.C.), but there are only a handful of L.N.U. records for the species this century. These include Market Rasen, 1st July 1950, (T.M.C.); Grimsby, "at M.V. lamp" 1951, (G.A.T.J.); and Osgerby Moor, "larvae on Aspen", July 1951 (L.N.U. Meeting). It was last recorded from Holme Plantation, 6th July 1979, (A.J.).
L.F.P. Aspen and Poplars

102. *Harpyia furcula* Clerck   **Sallow Kitten**
This species is found through June and into early July where Sallows occur in hedgerows or on heathland. L.N.U. records give its status as "frequent", and it is widespread over the whole of Lincolnshire. Early records placed it at Holbeach, Hammeringham, Alford, etc., and later records include Grimsby 6th July 1956, and Holbeck 5th July 1975, (both G.A.T.J.).
L.F.P. Sallow or Willow

103. *Cerura vinula* L.   **Puss**
Commonly found across the county from the end of May, into June, where Poplars occur. Typical records are: Sleaford, May 1959, (G.N.H.); Boston, 28th May 1965, (J.B.); and Twigmoor, (Ermine Street), 20th May 1975, (R.J.).
L.F.P. Poplars, Sallows

NOTODONTINAE

104. *Stauropus fagi* L.   **Lobster Prominent**
The moth is on the wing during the period May to July (depending on the season) – where Beech occurs. There is a single record for the county, for N. Somercotes, 1917, for "larva on Birch in garden," (Rev. S. Proudfoot).
L.F.P. Beech and occasionally Birch, Oak or Hazel

106. *Drymonia dodonaea* Schiff.   **Light Marbled Brown**
    (trimacula Esp.)
Restricted to Oak woodlands, and found only locally in the month

of June. Localities include Linwood, 16th June 1968, "plentiful at light", (M.P.G., G.A.T.J. and J.H.D.); Moor Farm (Trust Reserve) where it was recorded in 1973, (R.E.M.P.), and in 1974, (M.C.T.). It was at Linwood again more recently on 4th July 1981, (J.H.D.).

L.F.P. Oak

107. *Chaonia ruficornis* Hufn.   **Lunar Marbled Brown**
(chaonia Hubn.)
Another very local moth, on the wing in Oak woods in May. It was found to be "abundant" at Boston in May 1961, (R.E.M.P. and G.A.T.J.), and was recorded at Holme Plantation (Messingham), on 25th May 1973, (J.H.D.).

L.F.P. Oak

108. *Pheosia tremula* Clerck   **Greater Swallow Prominent**
This species can be found wherever the foodplant occurs, in two generations some years (May and August). In the 1800s it was more plentiful than the Lesser Swallow Prominent, but now the opposite seems to be true as increased numbers of the latter have been recorded since 1945.

Records cover the whole of the county, and include Grimsby, Woodhall, Ruskington, Saltfleetby, Humberston and Skellingthorpe etc.,

L.F.P. Poplars, and Sallows etc

109. *Phoesia gnoma* F   **Lesser Swallow Prominent**
(dictaeoides Esp.)
Mostly found on heathland habitats, with a growth of Birch scrub, in May/June and July/August. According to the early records, it appeared to be scarce up to 1945, when numbers seemed to increase, but its range has always been fairly extensive. A typical record is that for Woodhall, 9th August 1956, (F.L.K.). Other heathland localities in the records are Linwood, Laughton, Twigmoor, Manton, Burton Pits and Kirkby Moor.

L.F.P. Birch

110. *Notodonta ziczac* L.   **Pebble Prominent**
Found commonly across the county, particularly where damp areas occur, with Sallows, Poplars and Willows. There are two broods a year – May/June and August. Early records include Haverholme Priory, "very common", 24th August 1900, (J.D.C.); and Lincoln Ballast Pits, 1923, (A.E.M.). More recent localities mentioned are Grimsby, Saltfleetby, Mablethorpe and Willoughby near Alford.

L.F.P. Sallows, Willows etc.,

111. *Notodonta dromedarius* L. **Iron Prominent**
A heathland moth where scrub Birch occurs – frequently found on sandy heaths in the North West of the county in May, and in August/September.
Lep. Lincs. 1 stated that this species was "apparently scarce up to 1900", but it was recorded from Gainsborough, Panton, Wrawby Moor and Haverholme Priory. It appears to have become more frequently seen since the early 1940s. Recent records for sandy heathland areas include Wrawby Moor, Laughton, Skellingthorpe, Manton, Holme, Scotton etc.,
L.F.P. Birch and Alder

114. *Notodonta trepida* Esp. **Great Prominent**
(anceps auct.)
This species inhabits Oak woodland or parkland, and flies early in the year (May).
Lep. Lincs. 1 noted its status as "rare", with "a few most years at Market Rasen", (W.L.). Another early record is for "Skellingthorpe – a pair in cop," (D. H. Pearson). In 1916, the Rev. F. L. Blathwayt found it again at Skellingthorpe. It "came to light" at Linwood, 8th May 1959, (G.A.T.J.), and again at Boston in 1961, when there were "12 on sheet at once", (R.E.M.P.).
The last record was for Bardney, New Park Wood, 14th May 1966, (A.D.T.).
L.F.P. Oak

117. *Lophopteryx capucina* L. **Coxcomb Prominent**
(came.ina L.*)*
This is the commonest of our prominents – widespread across the county in two generations (May/June and July/August), with the larvae feeding on a large variety of trees and bushes.
Typical records are for Limber, 17th June 1964, (G.A.T.J.); Gibraltar Point bird observatory, 14th June 1966, (Barry Wilkinson, Warden); Mablethorpe "fairly common", 1952 to 1965, (T.R.N.).
L.F.P. Oak, Beech, Hawthorn, etc

118. *Odontosia carmelita* Esp. **Scarce Prominent**
Found early in the year (April/May) in areas where Birch occurs. It has only once been recorded in the county at Boston, 10th April 1961, (R.E.M.P.).
L.F.P. Birch

120. *Pterostoma palpina* Clerck **Pale Prominent**
Can be seen all over the county where Poplars and Sallows occur, in May/June, and again in July/August. Early records mention

Cowbit, (M. Hufton), and the Holbeach district, where it was "common", (L.M.C.).
In the 1960s and '70s, records included Elsham Park, bred 29th April to 21st May, 1967, (G.N.H.); Boston 30th May 1970 (A.E.S.); and Scunthorpe, "plentiful 1972", (J.H.D.).

L.F.P. Poplars, Sallows

121. *Phalera bucephala* L. **Buff-tip**
Abundant across the county, as the larvae are polyphagous on forest trees. The moths are rarely seen by day, as they are difficult to detect at rest, but the larvae are gregarious until quite large, and are often found in large colonies. The imago is on the wing in June in a single generation.

Places mentioned in the records include Crowle, Revesby, Ruskington, Sleaford, Legsby and Gibraltar Point, etc.,

L.F.P. Many species of trees

122. *Clostera curtula* L. **Large Chocolate-tip**
Lincolnshire represents the most northerly point of the range of this species, and its distribution covers the area as far North as Bardney Forest in the county. It is a woodland species, mainly where Aspen occurs, and South (1961) gives two broods a year in April/May and July/August.

There are only a few L.N.U. records. Lep. Lincs. I put the moth at "N. Wickenby – J. A. Hardy (Naturalist's Chronicle 1896 p.l.)" and at Market Rasen (W.L.), and Linwood, "one larva and one pupa on Aspen, 5th August 1907," (J.P. and G.W.M.). The only later records are for Woodhall, and Bardney Forest in "two woods", 1969, (both R.E.M.P.).

L.F.P. Poplar and Aspen

124. *Clostera pigra* Hufn. **Small Chocolate-tip**
A species found in marshy heathland where Creeping Willow and Aspen occur. Generally there are two broods (May and August).

Lep. Lincs. claimed it was "very local, but common where it occurs", and recorded it at Haverholme Priory, "fairly common", (J.D.C.); and "Near Scotton Common – bred frequently, (F.M.B.)". It was also taken in the larval stage by L.N.U. members at a field meeting there on 28.7.1902.

Recent records for wet, sandy heathland areas in what is now the South Humberside region, include the following localities in the 1950s and 1960s: Twigmoor, Holme Plantation, Scotton, Manton, (all J.H.D.).

L.F.P. Creeping Willow, Aspen

THYATIRIDAE
THYATIRINAE

125. *Habrosyne pyritoides* Hufn. **Buff Arches**
Well distributed in all wooded areas across the county, it mainly flies in June and July.
Recorded from a large number of areas, typical haunts include Hubbard's Hills, Bourne Woods, Hartsholme, Barton, Grimsby, Revesby, Roughton Moor and Boston.
L.F.P. Bramble (Often on low growing strands under trees)

126. *Thyatira batis* L. **Peach Blossom**
Frequent in the county in all wooded localities in June, and will visit sugar patches, as well as a M.V. light. The numbers found each year seemingly fluctuate, and the species suddenly became particularly plentiful in the north of the county in the late 1960s. Wooded areas mentioned in L.N.U. records include Langworth, Mother Wood, Bourne Wood, Brocklesby, Broughton and Burton, etc.,
L.F.P. Bramble

127. *Tethea ocularis* L. **Figure of Eighty**
(octogesima Hubn.)
This species is on the wing in June/July where Poplar and Aspen occur. Lep. Lincs. 1 had it as "rare" and located it "near Theddlethorpe" in 1904 (A. E. Gibbs), and at Usselby, 1887, (F.A.L.).
Since 1956, it has been recorded more frequently. It turned up at Ruskington 10th June 1956, (G.N.H.); Grimsby, 11th July 1962, (G.A.T.J.); Boston, "abundant" 1964, (R.E.M.P.); and Goxhill, 23rd June 1973, (C. Potts).
L.F.P. Aspen, Poplar

128. *Tethea or Schiff.* **Poplar Lutestring**
The moth occurs where poplars grow, and is on the wing in the spring. Lep. Lincs. 1 stated that it was "locally common" in the late 1800s, but there are only four L.N.U. records for this century.
The Lep. Lincs. localities were North Langworth, 22nd May 1893, (J. W. Carr); Market Rasen "a few each year – larvae sometimes very common"; and Motherwood, "common at sugar", (both J.W.M.).
The records for this century are for the Scunthorpe area, 1917, "larvae taken by Rev. Proudfoot"; Swanpool and Boultham district, 1924, "at light", (A.E.M.); and Skellingthorpe, 1945, (H. M. Small).
L.F.P. Poplar

129. *Tethea duplaris* L. **Least Satin Lutestring**
(Lesser Satin Carpet, or Common Lutestring)
Found on wet heathland where Birch trees grow, in July and August. Lep. Lincs. 1 classed it as "local". L.N.U. records are mainly for the period from 1950 to the present date, and they include Laughton, Roughton Moor, Epworth Turbary and Woodhall (17th July 1964 – G.A.T.J.).
Specimens taken in Lincolnshire are the darker Northern form of this moth.

L.F.P. Birch

131. *Asphalia diluta* Schiff. **Lesser Lutestring**
(Oak Lutestring)
This species usually emerges in September, and is restricted to Oak woodlands. L.N.U. records claim that the moth was "scarce" in the early 1900s, and even more recently, it has only been found "locally". (Often at sugar in the late Autumn). In Lep. Lincs. 1 it is recorded that Barrett, speaking of it in connection with other counties, said "locally it seems to be much more frequent in Lincs."

It was in the Market Rasen district from 1876-80, (F.A.L.). More up to date, it has "visited sugar at Laughton" in September 1956, (J.H.D.), been recorded at Twigmoor, 27th August 1959, "at M.V. light", and has also been found at Scotgrove Wood in 1971, (both A.D.T.).

L.F.P. Oak

132. *Achlya flavicornis* L. **Yellow-horned Lutestring**
Widespread in the county with larvae being Birch feeders. Lep. Lincs. 1 gave it as "common". The imago emerges from the end of March into April, and can be found resting on Birch twigs.

Records include Roughton Moor, 11th April 1964; Bardney, 13th March 1965, "plentiful", (both G.N.H.); and Boston, March and April 1965 (J.B.). Ova were numerous at Scotton Common in the Spring of 1980 (J.H.D.).

L.F.P. Birch

133. *Polyploca ridens* F. **Frosted Green Lutestring**
Flies in late March, through April and into May, in oak woodland. Lep. Lincs. 1 had its status as "rare", and named S. Hartsholme, 13th March 1900, (J.F.M.), as one locality.

There are only a few other records, including Woodhall Spa, 1st May 1965, "one only to light, temperature 48deg. F., fair wind and new moon", (G.A.T.J.); and New Park Wood, Bardney, 1977, "larvae taken in June", (J.H.D.).

L.F.P. Oak

RECORDS OF LINCOLNSHIRE SPECIES

LYMANTRIIDAE
LYMANTRIINAE

134. *Orgyia recens* Hubn.   **Scarce Vapourer**
(gonostigma auct.)
A very local moth in the British Isles. Only a handful of habitats are known – the majority seeming to be around Lincolnshire – and at least 6 or 7 localities recorded in the Scunthorpe district. South (1961) states that these moths emerge in two generations, the first being on the wing in June, with resulting larvae then feeding up quickly to produce a second generation in August. L.N.U. records suggest that the main flight period in Lincolnshire is in July. It would seem that there is generally one main generation, with occasionally a partial emergence in the Autumn.
    Records include: Scotton Common, 17th July 1961, when it was noted that "males were flying", (no doubt in search for females which are apterous-wingless); Holme Plantation (Messingham) July 1969 and 1972, and Messingham Sand Quarries, June 1976 (larvae) – (all J.H.D.); Crowle Waste "larvae in June 1978" (R. Key). Also in the 1970s, a number of larvae have been found in gardens on the Berkeley Estate, Scunthorpe, feeding on rose and other cultivated shrubs, (R.J.), in the spring.
                                    L.F.P. Sallow, Oak, Bramble etc.,

135. *Orgyia antiqua* L.   **Common Vapourer**
Found in a large variety of habitats – chiefly where woodland occurs, but also in gardens and along hedgesides etc., The moths are out from July through to September, with the larvae emerging over a prolonged period during the following year – throughout the Spring and Summer months.
    It occurs everywhere in the county, and is common. Its habitats include: Saltfleetby, 1950, (L.N.U.): Sleaford, 7th August 1961, (G.N.H.); Austacre Wood, 17th August 1969, (A.D.T.), etc.,
                      L.F.P. Most native trees and shrubs

136. *Dasychira fascelina* L.   **Dark Tussock**
A species found in the wetter areas of Scotland and the more Northerly conties of England in June or July.
    There are no county records for the 1900s, but Lep. Lincs. 1 noted the following records: Near Gainsborough district (F.M.B.); Mablethorpe, "larvae taken feeding on Sea Buckthorn" (R. Garfit and E.W.); Skegness – "larvae exhibited at a meeting of the Leeds Naturalists' Club and Scientific Association on 24th August 1880 by John Grassam", (Nat. Vol. 6, p.45); Skegness, "bred from one larvae about 1890", (J.C.L.-C.).
                          L.F.P. Hawthorn and Sallows

122   THE BUTTERFLIES AND LARGER MOTHS

  137. *Dasychira pudibunda* L. **Pale Tussock**
Mainly a woodland, or bushy heathland species, common and frequently found across the county in the months of May and June. The moth is often seen at rest on branches, and comes in numbers to a M.V. light.
  Typical records include: Holton-le-Moor, 1951, (Miss J. Gibbons); Newball Wood, June 1963, (G.N.H.); Broughton High Wood, May 1978, (J.H.D.).
L.F.P. Most native trees

  138. *Euproctis chrysorrhoea* L. **Brown-tail**
    (phaeorrhoea Don.)
A hedgerow species, but very scarce in the county. According to South (1961), it is "essentially a coast species in Britain, and confined to Essex, Kent, Sussex, and Hampshire". Lep. Lincs. 1 had it – near Grimsby (E.H.F.); Newball Wood, 25th June 1905, (J.F.M.); West Ashby, 1902, (F.S.A.); Market Rasen district, 1876-80, (F.A.L.).
  The only subsequent records are for Mablethorpe, 1963, "one specimen", (T. R. New); and Gibraltar Point, 1.8.78, (R.E.M.P.).
L.F.P. Hawthorn, Rose, Sea Buckthorn

  139. *Euproctis similis* Fuessl. **Gold-tail**
(Yellow-tail)
The larvae of this species feed on a large variety of food plants. The moth can be found commonly in June or July right across the county.
  Typical records are: Waltham, 2nd July 1956, (Miss B. Hopkins); Linwood, 25th June 1959, (Rear Admiral A. D. Torlesse); Barrow Haven, 20th September 1969, "larvae", (R. Goy); etc..
L.F.P. Hawthorn, Sallows, etc.,

  140. *Laelia caenosa* Hubn. **Reed Tussock**
This moth was restricted to a fenland habitat, and was claimed by South (1961) to be extinct "somewhere about 1880" when "lamps would not draw a single specimen". Lep. Lincs. 1 had the following entry "Dr. F. Arnold Lees records having taken one specimen of this rarity near Market Rasen between Linwood and Lissington, on the 1st August 1878".
L.F.P. Fen Sedge, Common Reed.

142. *Leucoma salicis* L. **White Satin**
A very local moth, but the species can be common where Poplars occur. The imago emerges in July.
Localities in the records are: Grimsby, 24th July 1956, (G.A.T.J.); Scunthorpe, 24th July 1967, "common around street lights," (J.H.D.); Boston, 7th August 1972, (A.E.S.).
L.F.P. Poplars, Aspen, Willows

144. *Lymantria monacha* L. **Black-arched Tussock (Black Arches)**
A woodland moth which flies in August. Lep. Lincs. said it was "frequent but local" and recorded it at Burton – "at rest on tree boles" (F.M.B.); Market Rasen – "a few each year," (W.L.); Tumby, 1880, (L.N.U.); Tothill, 1880, (C. D. Ash); and South of Lincoln, 1881, (Canon Fowler).
The first record for this century was for Benniworth Haven, 13th July 1947, "one seen and captured resting on Oak tree", (C. G. Else). It was next seen at Skellingthorpe in 1949, (H. M. Small), and last recorded at Mablethorpe – "one specimen", (T.R.N.).
L.F.P. The foliage of various trees and shrubs.

LASIOCAMPIDAE

LASIOCAMPINAE

145. *Malacosoma neustria* L. **Common Lackey**
Common in the county, it occurs every year in July or August, particularly being found along the coast among the sand dunes. Occasionally the species can be a pest in urban housing areas, where it can defoliate hedges and fruit trees.
Records include Saltfleetby, 1950, (L.N.U.); Chapel St. Leonards, 20th August, 1962, (J.H.D.); and Boston, "larvae on apple", 14th June 1969, (A.E.S.), etc.,
L.F.P. Foliage of deciduous trees and shrubs

146. *Malacosoma castrensis* L. **Ground Lackey**
This very variable species inhabits salt marshes and estuaries, and it is seen in July and August. There is no mention of it in L.N.U. records. However, according to Mr. R. E. M. Pilcher, it was known to exist in the Gibraltar Point and Wainfleet regions in the 1850s.
L.F.P. Many plants growing in "Saltern" habitats –
e.g. Sea Wormwood, Sea Plantain

147. *Trichiura crataegi* L. **Pale Eggar**
Occurs in wooded districts in September – and is found in small numbers most years in a number of localities across the county. There are a number of records for Saltfleetby – e.g. 13th September 1963, (A.D.T.). Other habitats are Holbeck, 27th August 1975, (G.A.T.J.); South Kelsey, 29th August 1976, (S.Van den Bos); and Normanby-le-Wold, 20th August, 1978, (D. Brant).

L.F.P. Hawthorn, Birch etc

148. *Poecilocampa populi* L. **December Eggar**
Mainly this is a woodland species, but it can turn up in parks, gardens, and along hedgerows. The moth nearly always emerges in November, and it is observed almost every year in the Autumn. Localities mentioned in the records include: Holywell, "larvae", 3.6.1915, (G.W.M.); Bottesford, 21st November 1955, (M.P.G.); Holbeck, 18th October 1975, (G.A.T.J.).

L.F.P. Oak, Birch etc.

149. *Eriogaste lanestris* L. **Small Eggar**
The adult of this species emerges early in the year, and lays eggs on hawthorn or sloe in the hedgerows.
According to Lep. Lincs. it was found before 1900 at Legsby, Newball and Panton – "larvae common", (G.H.R.). There are only two recordings this century – for Thurlby, 5th June 1922, "larvae on sloe" and for Lincoln Ballast Pits, 1923, (both A.E.M.).

L.F.P. Hawthorn, Birch etc.,

150. *Lasiocampa quercus* L. **Oak Eggar**
Mainly this species occurs along the coast, but it can also be found inland on areas of wooded heath, in May/June, or July/August, depending on the form. The Northern form *(callunae)* is the earlier of the two, the Southern *(quercus)* the later.
Coastal records include Saltfleetby, July, 1959 "males flying along the sand dunes in numbers" (Canon G. Houlden), and Mablethorpe, "common on sand dunes", 1962-65, (T. R. New). Other localities include Laughton, 17th August 1961, "came to light", (A.D.T.); and East Halton Skitter, 16th July 1976, (S. Van den Bos).

L.F.P. Bramble, Heather, Hawthorn etc.,

152. *Macrothylacia rubi* L. **Fox**
Common some years on heathland in late May or June – and with larvae seen to be particularly plentiful in the autumn on the sand dunes at Saltfleetby and Gibraltar Point, and on the sandy

RECORDS OF LINCOLNSHIRE SPECIES 125

heathland in South Humberside. Typical records: Linwood and Scotton Common, 26th May 1960, (J.H.D.); Gibraltar Point, June 1965, (B. Wilkinson); Kirkby Moor, August 1967, "larvae common", (J.B.).
L.F.P. Heather, Birch, low plants etc.

154. *Philudoria potatoria* L. **Drinker**
Common in damp areas across the county in July and August. There is considerable colour variation in the wing tones of the males, from a deep buff to a purple colouration. Records include such localities as: Woodhall, Linwood, Scunthorpe, Mablethorpe, Gibraltar Point and Crowle Waste, etc..
L.F.P. Coarse grasses

GASTROPACHINAE

156. *Gastropacha quercifolia* L. **Common Lappet**
Found along wood margins and hedgerows in July and August, the species is quite widespread, but it appears to be more plentiful in the South of the county. In the 1970s the indications are that it is extending its range towards South Humberside, as recent records include a number of sites such as Haxey, Tetney and Linwood. Lep. Lincs. 1 had it as rare, but records in the 1950s and 1960s include: Near Sleaford, July 1957, "males and females", (G.N.H.); Near Osbournby, 15th August 1962, (W. M. Peat); Moor Farm, 1973, (R.E.M.P.).
L.F.P. Hawthorn, Apple etc.

SATURNIIDAE
SATURNIINAE

159. *Saturnia pavonia* L. **Empress**
(carpini Schiff.) **(Emperor)**
A moth of sandy heathland areas – on the wing in May. Lep. Lincs. stated that the species was "not common", and the same may be said of its status today. Larvae were found at Scotton Common on Sallows, 15th August 1951, (M.P.G.); the moth was seen at Roughton Moor, 12th May 1951, and Kirkby Moor, 5th May 1956, (Both F.L.K.); and Messingham Moor, 10th May 1962, (J.H.D.), etc. A more recent record is for a larva found in the Bardney Forest area, 1978, (A.E.S.).
L.F.P. Heather, Bramble, Sallow

DREPANIDAE
DREPANINAE

161. *Drepana binaria* Hufn. **Oak Hook-tip**
Flies in Oak woodland in May and again in August and

September. According to Lep. Lincs. "Its most northerly known locality is in Lincs.". Before 1945 its distribution was mainly across the South of the County. From this date it has extended its range Northwards into the South Humberside region (Twigmoor – J.H.D.). It has reached the River Trent having been found at Alkborough in 1977 (M.P.G.).
Other records include: Laughton, 7th September 1962, (A.D.T.), Holywell, 28th May 1956, (F.L.K.), and Twigmoor, 22nd May 1976, (R.J. and J.H.D.).
L.F.P. Oak

162. *Drepana cultraria* F. **Barred Hook-tip**
Found in areas where Beech trees occur, and according to South (1961), "preferably on chalky soil". It was at Limber, 6th June 1902, (F.S.A. and J.P.) and was said to be "rare" in Lep. Lincs.
It was found near Caistor, 1939, (C. G. Else); Limber, October 1971, (larvae – J.H.D.); and Broughton, May 1978, (M.P.G.).
L.F.P. Beech

163. *Drepana Falcataria* L. **Pebble Hook-tip**
Before 1900 this species was "rare, scarce in Lincs." Now, it is not common, but it occurs on most heaths were Birch is abundant, flying in May/June and August. Areas in the records include: Skellingthorpe (Rev. Blathwayt and Rev. Proudfoot), Messingham, Linwood, Manton and Epworth Turbary, etc.
L.F.P. Birch

CILICINAE

165. *Cilix glaucata* Scop. **Chinese Character**
Found commonly in the county on Hawthorn in hedgerows, it flies in May/June, and again from late July into August and September.
Localities in the records: near Holdingham, 30th July 1961, (G.N.H.); Grimsby, 14th June 1962, (G.A.T.J.); Boston, July and August 1967, (J.B.); Sleaford 12th May 1964, (G.N.H.), etc.
L.F.P. Hawthorn, Blackthorn

NOLIDAE
NOLINAE

166. *Nola cucullatella* L. **Short-cloaked Black Arches**
(Short-cloaked Moth)
On the wing in July mainly, in wooded localities, this species appears to be widespread in range, but local where it is found. The earliest record for it was for Wyberton, (J.C.), (Lep. Lincs. 1). After a gap of quite a few years, it turned up more recently at the Limber Woodlands (Pelham's Pillar), 15th July 1960, and was

recorded at Grimsby each year between 1952 and 1958 (both G.A.T.J.). It was also near Tattershall, June 1962, in the form of "larvae", (G.N.H.).

L.F.P. Blackthorn, Plum, Apple

168. *Nola albula Schiff.* **Kent Black Arches**
(albulais Hubn)
According to South (1961), this species was first observed in England in 1859, when four specimens were taken in N.E. Kent. Since that date it has appeared spasmodically, and been rarely recorded in habitats in the South of England in coastal counties. Lep. Lincs. 1 referred to "one example on the wing in the Grimsby district, 19.8.1906", (E.H.F.). The only other record this century is for Gibraltar Point, (where the food plant is prolific), 3rd July 1976, (R.E.M.P.).

L.F.P. Dewberry

169. *Celama confusalis H.—S.* **Least Black Arches**
A very local moth in woodlands in May and June – there have been five L.N.U. records this century. It was at Lynwode Warren, 5th June, 1911, (L.N.U. meeting); Pelham's Pillar Wood, 15th May 1954, and Bradley Woods (Grimsby), 6th June 1955, (both G.A.T.J.); Linwood Warren in May 1958, (J.H.D.), being seen in this latter locality into the mid 1960s; and at Gibraltar Point, 1975, (M.T.).

L.F.P. Oak, Beech, Blackthorn

ARCTIIDAE
LITHOSIINAE

171. *Atolmis rubricollis L.* **Red-necked Footman**
A member of an unusual group of moths, by virtue of the nature of the foodplant eaten by the larvae – this being for this particular species "Green Algae on Fir, Beech, Oak and Palings", (P.B.M. Allan, "Larval Foodplants", 1949).

The moth is out from the end of May to July. Lep. Lincs. 1 classed it as rare, but gave a number of localities for it – Legsby, Newball Wood, Ailby Wood, Mother Wood, Greenfield Wood, Market Rasen, Skellingthorpe, etc.

There is a record for it this century for Division 2 (Scunthorpe District), "larvae taken by Rev. S. Proudfoot", (No date). It was also seen in Welton Wood, 10th June 1950, (R.E.M.P.).

L.F.P. Algae as above

172. *Nudaria mundana L.* **Muslin Footman**
Another lichen and algae feeder – this time on material growing on walls, roofs and rocks. The moth flies in June. It used to occur on

the Risby Warren "flying about stone walls bordering the Warren at dusk", 5th July 1906, (G.W.M.); it was also recorded at Haverholme Priory (J.D.C.), and at Gainsborough "some years" (F.M.B.); The only recent records are for Normanby Park in July, in the 1970s, (J.H.D. and R.J.).

L.F.P. Algae and small Lichens on rocks

173. *Comacla senex Hubn.* **Round-winged Footman**
Flies in marshy places in July/August, with the larvae feeding on algae growing or. dead reeds, and on mosses growing on the ground.
It was classed as "locally common" in Lep. Lincs. 1 near Skegness, 16th July 1879, (G. T. Porritt) and at Theddlethorpe, (A. E. Gibbs – in "Entomologist" Vol. XXXVIII p.81). It was next recorded at Saltfleetby, 18th July 1958, "common at M.V. lamp" (G.A.T.J.); Gibraltar Point, 1974, (M.T.); Holbeck, 1975, (G.A.T.J.); and Epworth Turbary, August 1978, (J.H.D.).

L.F.P. Algae and Mosses

174. *Miltochrista miniata Forst.* **Rosy Footman**
A species of Oak woodlands – out in July. Lep. Lincs. 1 considered it "rare" and gave Linwood, Scotton Common and Horsington near Horncastle as localities. Rev. F. L. Blathwayt and Rev. S. Proudfoot found it at Skellingthorpe in 1915. More recently there are a number of records for both the Linwood and Woodhall areas. E.G. Linwood, 1972, (A.D.T.); Woodhall, 1964, "Fairly plentiful at light", (A.D.T., G.A.T.J.,).

L.F.P. Algae growing on rocks, and dead Oak leaves

176. *Cybosia mesomella L.* **Four-dotted Footman**
Locally found on Heather, and where heathland and woodland intermingle, in late June and July.
Lep. Lincs. only gave the Market Rasen area as "common" in 1857. There are subsequently a large number of records for Linwood from the 1950s to the 1970s (e.g. 25th June 1959 – A.D.T.). The only other localities in the records are Kirkby Moor (Woodhall), 20th June 1965, (G.N.H.), and Crowle Waste, 1972, (A.D.T.), and 1978, (R.K.).

L.F.P. Algae growing on Heather

177. *Lithosia quadra L.* **Large Footman**
(Four-spotted Footman)
Mainly a woodland species, out in July and August. There are no recent records. Lep. Lincs 1 had it as "rare", and recorded it at N.

Willingham, "one female", 7th July 1878, (F.A.L.); S. of Grantham, (Miss F. Woodward); and Hartsholme, "scarce", (W. D. Carr).

L.F.P. Goat's-horn Lichen, Dog Lichen.

178. *Eilema deplana* Esp. **Buff Footman**
(depressa Esp.)
A woodland species, flying in July, and very rare in the county. Lep. Lincs 1 had a single record: Near Market Rasen, one in July, 1894, (W.L.). The only records for this century are for Hangham Wood, 1973, (R.E.M.P.), and for Little Scrubs Wood, 1975, (M.T.).

L.F.P. Green algae growing on Oak and Birch

179. *Eilema griseola* Hubn. **Dingy Footman**
(stramineola Doubl.)
Found in damp areas in July/August, and "rare" in the county. Early records were for Scotton Common, Haverholme Priory (ab. *flava* Haw.-pale form), Willingham Park and South of Wyberton.
There are only a few later records: Laughton, 23rd August 1967; Scotton Common, 17th July 1976; and Walesby, August 1978, (All J.H.D. – and all localities including the pale form).

L.F.P. Lichen growing on Sallows

180. *Eilema lurideola* Zinck. **Common Footman**
Common throughout Lincolnshire, and on the wing from the end of June into July. Localities mentioned in the records include: Boultham District, Alford, Manton, Saltfleetby, Swallow Vale, Holdingham, Ancaster, Sleaford, Woodhall, Theddlethorpe, etc.
The last L.N.U. record is for East Halton Skitter, 16th July 1976, when "over 100 came to M.V. light", (S. Van den Bos).

L.F.P. Lichen and algae growing on Oak, and on many other trees and walls.

181. *Eilema complana* L. **Scarce Footman**
A rare species in Lincs. – mainly found in coastal regions. It was first recorded near Theddlethorpe, 1904 (A. E. Gibbs – in 'Entomologist' Vol.XXXVIII, p.81), and at Bracebridge, 6th July 1901, (J.E.M.). The only later records are for Saltfleetby, 14th August 1965, and Gibraltar Point, 1971, (both A.D.T.).

L.F.P. Various Algae and Bramble.

185. *Eilema sororcula Hufn.*   **Orange Footman**
This species occurs chiefly in Oak woods in May and June, in counties to the south of Lincolnshire.
There is an early record for "South of Holbeach district, where it is scarce" (L.M.C. in Lep. Lincs. 1.).
More recently it has been seen at Wyberton, 17th June 1953, Boston 1965, and Moor Farm, 1973, (All R.E.M.P.).
L.F.P. Algae and Lichens on Oak

186. *Peliosa muscerda Hufn.*   **Dotted Footman**
A very local species found chiefly in July and August in the fens of Norfolk (see South 1972 p.52) and in restricted localities in the New Forest and marshes of Sandwich, Kent. South writes that the larvae have "escaped detection" in their "fenny home" but in captivity have "thrived on a mixed diet of algae growing on alder and sallow, mosses, and withered leaves of bramble, sallow, and also on lettuce."
A fresh male was found by Dr. R. E. M. Pilcher at Saltfleetby on 17 July 1982.
L.F.P. see above

ARCTIINAE

189. *Utetheisa pulchella L.*   **Crimson-speckled Flunkey**
A migrant to the county, but when the migrants arrive at a suitable time, eggs can be deposited which can emerge to develop into imagines later in the season. The only records were in Lep. Lincs. 1 "Near Lincoln, 3 examples, one in a garden on the Burton Road, and one near the Barracks (C. P. Arnold); Linwood, 27.9.1877, (F.A.L.); and Boston, 1880, (Annie Dows).
L.F.P. Forget-me-not, Borage.

191. *Callimorpha jacobaeae L.*   **Cinnabar**
Abundant in the county in May and June, on sandy areas where ragwort flourishes. It has been recorded at Gibraltar Point "Thousands of larvae in 1950", Sleaford, Atkinson's Warren (Scunthorpe), Risby Warren, Ashbyville, Tattershall, Mablethorpe etc.,
L.F.P. Ragwort

192. *Spilosoma lubricipeda L.*   **White Ermine**
       (menthastri Esp.)
Widespread – found everywhere in the county in all types of habitat, including gardens, – June.
Found typically at Tetney Haven, Barton on Humber, Grimsby, Syston Park, Boston, Limber etc. etc.,
L.F.P. Larvae almost 'omnivorous' – eat most low growing plants

RECORDS OF LINCOLNSHIRE SPECIES 131

193. *Spilosoma urticae Esp.* **Water Ermine**
A moth of marshes and fenland, which has only occurred in the South of the county. Flies in June.
The only records are from Lep. Lincs. 1; "Near East Ferry Common, one example about 1892, (A.R.); South of Holbeach, "scarce"; Haverholme Priory, (J.D.C.)".
L.F.P. Yellow Loose strife, Watermint

194. *Spilosoma lutea Hufn.* **Buff Ermine**
(lubricipeda auct.)
This is a common species in the county, found in numbers in all types of habitat. The ground colour of the moth is variable – from a pale buff colour to a rich yellow ochre. The species has particularly interested lepidopterists as a result of this colour variance, and because of the varying intensity of the dark spotting found on the wings. In extreme cases, black marking can cover almost all the wing areas. A number of the varieties of this species have been named – and many have been found, bred, and selectively in-bred in Lincolnshire. The var. *radiata*, var. *fasciata* and var. *eboraci* are all dark forms leading up to var. *zatima* – the particularly black form.
Var. *radiata* was bred by Mr. Mossop in August 1836, from larvae found on Elder. From these, some specimens were sent to Mr. James C. Dale, father of Rev. C. D. Dale (Nat.1894, p.355). A pair (Taken by A.T.) at Theddlethorpe, 1895, were sent to Professor Carr, University College, Nottingham.
Two other specimens of Mr. Mossop's which were fasciated (var. *fasciata* – having a number of the dark markings joined together, side by side, to form a continuous flat plate) were described in the Nat. 1894, p.356.
The var. *eboraci* was taken at Mablethorpe in 1880, and noted in Nat. 1894, p. 361 (C. D. Ash).
A very fine ab. *zatima* was taken at Grimsby, 10th June 1962, (G.A.T.J.).
Ab. *zatima* has also resulted from an extensive breeding programme by R.E.M.P. (and others) from livestock originating from Gibraltar Point.
Other larvae found at Gibraltar Point by J. H. Duddington in 1958, were bred and selectively inbred over a ten year period.
This resulted in the gradual deepening of the wing ground colour to a rich orange shade in both the males and the females produced. Specimens were sent to the British Museum (Natural History) for identification, and it was decided that this was an, as yet, un-named colour variety.
L.F.P. Most low growing plants

195. *Cycnia mendica* Clerck　　　　　　　　**Muslin Ermine**
Widespread in its distribution across the county on heathland in May and June – but it has not been recorded too frequently.

Records include: Sleaford, 21st April 1961, (G.N.H.); near Holdingham, "larvae on coltsfoot", 30th July 1961, (G.N.H.); Manton and Scotton Commons, late May 1963, (J.H.D.); and Brigg "to M.V. light", May 1964, (A.H.N.).

　　　　　　　　　　　　　　　　L.F.P. Low growing plants

196. *Diacrisia sannio* L.　　　　　　　**Clouded Ermine**
　　　(russula L.)　　　　　　　　　　**(Clouded Buff)**
Strictly a heathland species which flies in June and early July. Lep. Lincs. 1 stated that it was not common, and only a few localities have been mentioned consistently in the records over the last fifty or more years. These localities are Linwood, Scotton, Laughton and Twigmoor. The most recent L.N.U. records are for: Linwood, 1971, (A.D.T.); Scotton Common, "males and females", 28th June 1978, (J.H.D.); and Laughton, "larvae found September 1981", (A.J.).

　　　　　　　　　　　L.F.P. Dandelion and other low plants

197. *Phragmatobia fuliginosa* L.　　　　　　　**Ruby Tiger**
Common in the county, being frequently recorded in a variety of woodland and heathland habitats in May and June, and sometimes in small numbers in the Autumn. Records include: North Somercotes, "larvae more abundant than usual", 1919, (Rev. S. Proudfoot); Sandilands, (Sutton-on-Sea), "larvae on ragwort", 3rd Sept. 1927, (A. Roebuck); Gibraltar Point, 1950, (L.N.U.); Woodhall Spa, 8th May 1954, (F.L.K.); Boston, "at light", 3rd August 1968, (A.E.S.), etc.

　　　　　　　　　　　L.F.P. Dock, plantains and
　　　　　　　　　　　　　　　other low plants

199. *Parasemia plantaginis* L.　　　　　　　**Wood Tiger**
Appears to be predominantly found in the North of the county in shrubby heathland and areas of light woodland – mainly in June.

The early records are from Holton le Moor, 12th June 1908, (E.P.H.B.); Market Rasen, 1943, (T.H.C.); Wrawby Moor, 1950, (M.P.G.) etc.,

More recently, the species has been found at Osgodby Moor, Elsham, Woodhall, Scotton Common, Laughton, Irby Dales, Epworth Turbary and Risby Warren.

　　　　　　　　　　　　　　　　　　L.F.P. Plantains

200. *Arctia caja* L. **Garden Tiger**
This species has been recorded all over the county (in July/ August) this century. Lep. Lincs. 1 claimed it was "very common", but it does not appear to be so numerous, or so widely distributed during the last decade.
More recent records include: Brauncewell, 1st August 1958, (G.N.H.); Saltfleetby, 18th July 1958, (G.A.T.J.); Great Hale, 24th July 1960, (G.N.H.); Gibraltar Point, "hundreds of larvae", 29th May 1967, (B.W.); Barrow Haven, "larvae", 20th Sept. 1969, (R. Goy); Messingham Sand Quarries, "larvae very common 1979" (J.H.D.).

L.F.P. Almost omnivorous – a wide variety of low plants

201. *Arctia villica* L. **Cream-spot Tiger**
According to South (1961), this species is found "most frequent in the south-west", and it also occurs in the eastern counties to Cambridge and Norfolk". It flies in May and June.
It has been recorded in one locality in Lincolnshire: near Scotton Common, 12th June 1901, (J. F. Musham).

HYPSINAE

203. *Panaxia dominula* L. **Scarlet Tiger**
A species found mainly in the south of the country – generally south of Oxford and Hertfordshire. The moth emerges in June, and "seems partial to marshy ground" (South 1961). L.N.U. records had it in the Market Rasen district from 1876-80 (F.A.L. in Lep. Lincs. 1).
It was recorded at Linwood, 16th August 1922, "larvae on grasses", (A. E. Musgrave). An entry in the records relating to Tealby, stated "Letter received 11th September 1959 from L. H. Newman, "Betsoms", Westerham, Kent, stating "I planted a colony in the gardens at White Cottage, Tealby, Lincoln, a small estate belonging to Mr. Geoffrey Harmsworth" (G.A.T.J.)".

L.F.P. Various "weeds" and low plants.

Super-family PSYCHOIDEA
LIMACODIDAE

HETEROGENEINAE

208. *Apoda avellana* L. **Festoon**
   (limacodes Hufn.)
Flies in Oak woodlands in June and July. It is only known to have occurred in Lincolnshire in 1913, when Mr. R. E. M. Pilcher discovered a pair in cop. in Southrey Wood.

L.F.P. Oak

ZYGAENIDAE
ZYGAENINAE

215. *Zygaena trifolii* Esp.   **Broad-bordered Five-spot Burnet (Five-spot Burnet)**

Mainly found in damp areas from June into July. Lep. Lincs. 1, described it as "local" – in the Owston Ferry district (A.R.), and Littlecotes, July 1906, (E.H.F.). In 1927, at Ancaster it was "very common in July in the tall grasses along the roadside", (A.E.M.). During the mid 1950s, there seemed to be a population explosion, and it became very common in a number of areas, e.g. Scunthorpe, "hundreds of cocoons on grasses on rough ground", July 1956, (J.H.D.). After 1960, the numbers rapidly declined again, as instanced by the record for Mablethorpe, "formerly common, only a few specimens since 1960", (T. R. New). The last L.N.U. records are for Ancaster, 15th June 1969, (A.D.T.); Messingham Sand Quarries "most years in the 1970s", (J.H.D.); Scrubbs Meadow, Bardney, 14th July 1979, (R.J.).

L.F.P. Trefoils and Clover

217. *Zygaena lonicerae* Scheven   **Narrow-bordered Five-spot Burnet**

South (1961) wrote that "the moth, out in late June and July, occurs in woods and plantations" and is "also found in meadows, and on rough waste ground, as well as in marshes and salterns". Lep. Lincs. 1 described it as "locally common", and gave Sutton-on-Sea, 1902, (F.S.A.), as one example.

There have only been a small number of L.N.U. records this century. These include Ancaster, 5th July 1969, (A.D.T. and R.E.M.P.); and Boston, 1970 and 1971, (A.E.S.).

L.F.P. Trefoils, Clover.

218. *Zygaena filipendulae* L.   **Narrow-bordered Six-spot Burnet (Six-spot Burnet)**

Generally more common than the previous burnet species – and has a preference for a drier type of habitat. It flies (South 1961) "on chalk downs and sand hills on the coast" (July/August).

There are records from: Donna Nook, 31st May 1906, "larvae", (L.N.U.); North Rauceby, 1952, (L.N.U.); Holme Plantation, July 1959, (J.H.D.); Gibraltar Point, June/July 1960, (G.N.H.); Ancaster, July 1963, "common on roadside", (G.A.T.J.); Tydd Gote, 1st August 1970, "common", (A.E.S.).

L.F.P. Clover, BIrds'foot Trefoil.

PROCRINAE

221. *Procris statices* L. **Common Forester**
This species was well distributed across the county on damp heathland (in July) up to the 1960s. After this date, it has appeared to be less plentiful. Typical localities include: Burwell Wood, 1947, (F.L.K.); Roughton Moor, "very common", 4th July 1959, (G.A.T.J. and G.N.H.); Scawby and Scotton, June 1960, (J.H.D.); and Kirkby Moor, 23rd July 1967, (J.B.).
L.F.P. Sorrel

SESIIDAE

SESIINAE

223. *Sesia apiformis* Clerck **Poplar Hornet Clearwing (Hornet Moth)**
Can be found where Poplars occur, from June into July. Lep. Lincs. 1 called it "rare". There are subsequently 6 records for: South Somercotes, June 1919, (Canon G. Houlden); Grantham, "a pair brought to the museum in June 1932," (A.E.M.); Boston, 10 July 1952, "a pair in cop.", (Miss B. Hopkins); Mablethorpe, one specimen 1957, (T. R. New); Woodhall Spa, 18th July 1958, (F.L.K.); and Scunthorpe, 11th July 1961, (D.A.E.S.).
L.F.P. Poplars

224. *Sphecia bembeciformis* Hubn. **Osier Hornet Clearwing**
(Crabroniformis Lew.) **(Lunar Hornet)**
On the wing in damp areas where the larval foodplant occurs, in late June through July. The only L.N.U. records are for: Yaddlethorpe, 15th June 1970, (Martin Simpson); Museum Nature Reserve, Scunthorpe, 14th July 1971, (R.K.); and Scrubbs Meadow, Bardney, 6th July 1974, (G. Posnett).
L.F.P. Sallow, Willow, Poplar (Feeding in the stems.)

229. *Aegeria tipuliformis* Clerck. **Currant Clearwing**
A species well distributed across the county in gardens in June and July. It can be a pest as instanced by the record from Mr. C. G. Else for Barton on Humber, where larvae "seriously damaged my Currant bushes", 1943. The following are records by Mr. A. Roebuck for: Lincoln, "larvae killed shoots of Blackcurrant", (L.N.U. Trans., Vol. 11, p.39); Kesteven, 1947, "occurs regularly doing damage"; Dowsby and Rippingdale area, 1951, "present in large numbers on Blackcurrant plantations, the larvae tunnel the stems and overwinter in them. In some cases tits have found these and pecked holes in the stems to remove the larvae".
L.F.P. In stems of Gooseberry and Currant bushes.

136 THE BUTTERFLIES AND LARGER MOTHS

232. *Aegeria vespiformis L.* **Yellow-legged Clearwing**
(cynipiformis Esp.)
Reported to be "local" in Lep. Lincs. 1, with typical records being for; "N. Brocklesby, larvae and pupae abundant in stumps of Oak trees cut down the previous year", (J.P. and G.W.M.); Pelham's Pillar Wood, "several larvae and pupae taken by Mr. John Porter of Hull, from Beech stumps", June 1911, (G.E.M.).
L.F.P. Feeds under the bark, or in the stumps of felled deciduous trees like Oak, Chestnut, Elm, Birch, Beech etc.,

233. *Aegeria myopaeformis Borkh.*
Small Red-belted Clearwing
There are only L.N.U. records for the period up to Lep. Lincs. 1 for: Market Rasen district, 1876-1880 (F.A.L.); Holbeach district, "fairly common", (L.M.C.).
L.F.P. In bark of Apple, Pear, Hawthorn

234. *Aegeria culiciformis L.* **Large Red-belted Clearwing**
Mainly out in May and June in Birch woods. There are L.N.U. records for: Gainsborough (F.M.B.), S. Skellingthorpe, 7.8.1902 and 2.5.1903 (J.F.M.), (both Lep. Lincs.1.). There is also one later mention for: Skellingthorpe, Big Wood, "empty pupae cases in Birch stumps at end of July", (Rev. F. L. Blathwayt).
L.F.P. Inner bark of Birch trees, and old stumps.

235. *Aegeria formicaeformis Esp.* **Red-tipped Clearwing**
Seen in marshy, wet areas in July and August. There is one early L.N.U. record with the species simply noted for "Lincolnshire, 1832", (James Rennie). Additionally there is one later mention for Scunthorpe, for a railway cutting near the steelworks, 12.6.1949, (J.H.D. reported in Entomologist, Vol. 82, p.117).
L.F.P. Osier

COSSIDAE

ZEUZERINAE

264. *Zeuzera pyrina L.* **Wood Leopard**
(The Leopard)
The larvae feed internally in the branches and twigs of various trees (e.g. Ash and fruit trees) and may take up to three years to mature. The adult moth emerges in July/August.
Lep. Lincs. 1 classed the species as "scarce", but at the present time, it is quite plentiful in some areas.

L.N.U. records include: Skellingthorpe, 1915, (Rev. Blathwayt and Rev. Proudfoot); Fiskerton and Barton on Humber, 1950, (M.P.G.); West Torrington, 5th August, 1955, (B.J.); Boston, July 1976, (A.E.S.); and Scotton Common, July 1979, (J.H.D.) etc.
L.F.P. Various trees; Ash, Maple, Sycamore etc.

COSSINAE

265. *Cossus cossus* L. **Goat**
(ligniperda F.)
Another larva which feeds internally, mostly in Willows, Elm and Ash trees, with the larvae feeding for several years before pupation, and with the possibility of considerable damage to the tree. The moth flies in July in the main. Generally it is well distributed across the county, and recorded from: Belton Park, 1932, (A.E.M.); Holton le Moor, 1950, (Miss J. Gibbons); Normanby, "larvae", 1951 and 1955, (M.P.G.); Sleaford, "bred from larvae", 19th September 1961, (G.N.H.).
L.F.P. Willows and various trees.

Super-family HEPIALOIDEA
HEPIALIDAE
HEPIALINAE

266. *Hepialus humuli* L. **Ghost Swift**
(Ghost Moth)
Abundant in meadows and grassy areas in the countryside and even in gardens, where the larvae feed underground on a variety of plants – and can cause considerable damage on occasion, to agricultural crops. The moth is out in June and July. Localities in the records: Kesteven, 1943, "larvae virtually destroyed a field of wheat", (A. Roebuck, L.N.U. Trans., Vol. 2, p.37); Sleaford, "females to light", 10th July 1958, (G.N.H.); Limber, "males to light", 1st July 1965, (G.A.T.J. and J.H.D.) etc.,
L.F.P. Roots of grasses and various plants such as Dandelion, Dead-nettle

267. *Hepialus sylvina* L. **Wood Swift**
(Orange Swift)
Quite widespread in its range, and fairly frequently recorded (July and August) on heathland, or in light woodland, where bracken occurs.

Typically recorded at Scawby Woods, "quite numerous" July 1956, (J.H.D.); Sleaford, 13th August 1963, (G.N.H.); and Wyberton, 9th August 1971, (A.E.S.).

L.F.P. Roots of Bracken, Dock etc.

268. *Hepialus fusconebulosa* Deg. **Map-winged Swift**
(*velleda Hubn.*)
Found in light woodland among bracken in June in the main, and into early July. Lep. Lincs. 1 classed it as "rare", and gave as its habitats Market Rasen and S. Wyberton (1880). At the present time it is certainly plentiful in suitable localities, and even occurs in gardens. Haunts include: Scawby Woodlands, 11th July 1951, (G. Else); Tumby Woods, 17th June 1961, (L.N.U. – A.D.T.); and Kirkby Moor, 1973, (R.E.M.P), etc.

L.F.P. Roots of Bracken

269. *Hepialus lupulina L.* **Common Swift**
Very abundant – can be found in large numbers all over the county in open countryside and in gardens (May and June), and comes readily to light at dusk. Records include: Sleaford, "bred from larvae found on Chrysanthemum roots", June 1956, (G.N.H.); Riby 30th June 1959, (G.A.T.J.); Scunthorpe Museum grounds, 1st June 1960, (D.A.E. Spalding); Stixwould Wood, "numerous", 8th June 1968, (A.D.T.).

L.F.P. Roots of Grasses and other plants

270. *Hepialus hecta L.* **Golden Swift**
Quite frequently recorded in areas where bracken occurs, and on the wing from June into July.
L.N.U. records include: Doddington, 12th July 1958, (L.N.U.); Linwood, 17th July 1959, (G.A.T.J.); Alkborough, June 1976, (J.H.D. – Museum Society Meeting).

L.F.P. Roots of Bracken

Super-family NOCTUOIDEA

NOCTUIDAE

AGROTINAE

272. *Euxoa cursoria Hufn.* **Coast Dart**
This very variable species is found on sand hills on the "east coast of England from Suffolk northwards" (South, 1961). The moth is on the wing from July into August. There are only two L.N.U.

records for: Mablethorpe, 1952 – 1965, "common", (T. R. New); and Gibraltar Point, 1969, and most years in the 1970s, (R.E.M.P.).

L.F.P. Various Grasses and plants such as Hound's-tongue

273. *Euxoa nigricans* L. **Garden Dart**
Common across the county – especially found in gardens on budleia bloom in July and August. Records included: Ruskington, "common" August 1956, (J. Hossack); Grimsby, "plentiful to M. V. light", 26th August 1961, (G.A.T.J.); Boston, 21st August 1972, (A.E.S.).

L.F.P. Clovers and Plantains

274. *Euxoa tritici* L. **White-line Dart**
The moth is out in July/August, in areas where sandy soils occur. Lep. Lincs. 1 gave the species as "frequent," and noted it occurring at the Humberstone foreshore – Cleethorpes, Sutton on Sea (at sugar), Wrawby Moor etc.,
There are a number of forms of the moth, with ab. *rhabdota* being rather larger than the normal *tritici* (found more along the coast), and this larger form being "more of an inland insect", (South 1961).
More recently there have been records from: Barton on Humber, 21st August 1956, (C. G. Else); Scunthorpe, "common", 19th August 1962, (J.H.D.); Kirkby Moor, 1972, (A.D.T.). Other areas mentioned include Elsham, Saltfleetby, Mablethorpe, Greetwell etc.

L.F.P. Chickweeds and Bedstraws

276. *Euxoa obelisca Schiff.* **Square-spot Dart**
South (1961) wrote that this moth was out "in August and September, in its special haunts", with the larvae feeding on "plants growing in rocky places by the sea or on hillsides".
Lep. Lincs. 2 claimed that the species existed in two areas in the vicinity of Lincoln about the year 1900, when "a few came to sugar" (W.D.C.) It has not been recorded since, and the early record cannot be verified.

L.F.P. Rock Rose, Lady's Bedstraw

277. *Agrotis segetum Schiff.* **Turnip Dart**
Noted as being abundant in Lep. Lincs. 2, it remains common across the county in June, with a smaller emergence in the Autumn.

One of the earliest L.N.U. records stated "1923-25, in the Holland division, larvae often causing serious losses in strawberry plantations". (H. W. Miles, L.N.U. Trans. Vol. 6, p.185). In 1935, "larvae abundant in Lincs. " (A. Roebuck, L.N.U. Trans. Vol. 9, p.48). More recent records covered Linwood, 15th September 1959, and Woodhall, 16th July 1965, (both G.A.T.J.), etc.,
L.F.P. Most low growing plants

278. *Agrotis vestigialis Hufn.* **Archer Dart**
(valligera Schiff.)
Essentially this is a coastal species, flying in July and August. Lep. Lincs. 2 said it was "fairly common on the coast, but occasionally met with inland". It can now, however, be found commonly inland on many sandy localities within the county, and it also visits flowers in gardens.

It was recorded: at North Somercotes, 1920, in the garden of Rev. S. Proudfoot; Sandilands, "at light", 1954, (G.A.T.J.); Black Walk Nook (Manton Common), July 1974 and 1976, (R.J.).
L.F.P. Bedstraws and various Grasses

280. *Agrotis clavis Hufn.* **Heart and Club**
(corticea Schiff.)
This is mainly a coastal species, on the wing in June/July. Lep. Lincs. 2 had it as "rather scarce", and it has not been frequently recorded this century.

It was found at Grasby Bottom, 15th June 1949, (G.A.T.J.); Barton on Humber, 21st August 1957, (C. G. Else); Elsham, June 1955, and Scotton Common, 4th July 1970. (both J.H.D.).
L.F.P. Various low plants, e.g. Goosefoot, Knotgrass.

281. *Agrotis deticulatus Haw.* **Light Feathered Rustic**
(cinerea auct.)
South (1961) wrote "The moth flies in May and June, and is usually found on hills and downs in chalk or limestone districts."

L.N.U. records contain two entries for the species. Firstly for "Ruskington, August 1956, a typical female was taken on buddleia – unusual in August in this district, – J. Hossack". The Lep. recording secretary noted, however, that this was an "unconfirmed report – G.A.T.J.". Secondly, it turned up at Holbeck, 2nd July 1975, (G.A.T.J.).
L.F.P. Wild Thyme

282. *Agrotis puta* Hubn. **Shuttle-shaped Dart**
(radius Haw.*)*
Claimed to be "partial to low lying marshy ground and meadows, and is widely distributed over the whole of the South of England, but is seemingly rare in the north" (South), the species is out in May/June and August.
It was recorded in Lincolnshire in 1921 by Rev. F. S. Alston (L.N.U. Trans. Vol. 5, p 154). In 1950 it was common on buddleia in the Scunthorpe area (J.H.D.) and it was last recorded at Gibraltar Point, 1975, (M.T.).
L.F.P. Dandelion, Dock, Knotgrass.

285. *Agrotis exclamationis* L. **Heart and Dart**
One of the most plentiful of the noctuid moths, abundant over the whole county, occurring in all types of habitat from late May into June/July, and up to the second generation in early September.
Larvae can occasionally be a nuisance, on certain agricultural crops (e.g. lettuce and root crops such as Turnips), as instanced by the record "1935, larvae abundant in Lincs. - A. Roebuck" (L.N.U. Trans. Vol. 9, p.48). Many localities are mentioned in the records - e.g. Osgodby Moor, Brocklesby, Saltfleetby, Epworth, Woodhall, Mablethorpe, Boston, etc.
L.F.P. Chickweed, Plantains etc.

286. *Agrotis ipsilon* Hufn. **Dark Dart**
(suffusa Schiff.) **(Dark Sword Grass)**
An indigenous species, with the population being supported by frequent immigrants. It is commonly found in the county - often seen on "sugar" in the Autumn.
Typical records include: Holme Plantation, 27th October 1954, (G.A.T.J.); Barton on Humber, 2nd October 1956, (C. G. Else); Sleaford, 29th October 1966, (Miss June Osgerby); and Gibraltar Point, September 1966, (A.D.T.).
L.F.P. Roots of Cabbage, Lettuce and many low plants

287. *Agrotis ripae* Hubn. **Sand Dart**
The initial entry for this species was by G.A.T.J. as L.N.U. lepidoptera records Secretary. It read: "The only knowledge I have of the occurrence of this species in the county is gleaned from Entomological works"; e.g. C. G. Barrett's "Lepidoptera of the British Isles" 1893/6 Lincolnshire coast, var. *grisea,* (see Tutt's ' British Noctuidae and their varieties, Vol. 2 p.67 and 71, 1892 edition, in which the author refers to specimens in his own collection which came from the Lincolnshire coast), Lep. Lincs. 2.

The only later entries state: 1952-65, Mablethorpe (T. R. New); and Gibraltar Point, 1969, (R.E.M.P.).

L.F.P. Saltwort and Sea Rocket etc.

289. *Lycophotia varia* Vill   **True Lovers' Knot**
(strigula Thunb.)
Found on moorland, or on regions where heather abounds. Not usually discovered in large numbers, but found widely on sandy *heathlands in July*/August.

Records include: Scotton Common, 1950, and Manton, 1955, (Both M.P.G.); Elsham, 1955 (J.H.D.); Twigmoor, Linwood and Laughton July/August 1957, 1958 and 1961 respectively (G.A.T.J.); Woodhall, 17th July 1964, (G.A.T.J); Kirkby Moor, 1966, (L.N.U.); and Linwood, 4th July 1981, S. Van den Bos.

L.F.P. Heather

290. *Actebia praecox* L.   **Portland Dart**
Lep. Lincs 2 said it was "scarce, but occasionally taken on the coast". It mainly inhabits coastal sandy areas – flying on duneland in July and August.

Early records had it at: Frodingham and Appleby, 28th July 1901, (H. H. Corbett – Naturalist, Sept. 1901); Scunthorpe, 28th August 1910, "one taken at rest on the sand by Mr. C. P. Arnold", (L.N.U.).

It was recorded at Woodhall Spa, "one taken at light", 6th August 1953, (F.L.K.). It now appears to have colonised the Scunthorpe District on sandy areas, as instanced by such records as: Greetwell, 16th August 1975, (J.H.D.); and Manton Common, July 1974 and 1976 (R.J.).

Other recently recorded areas are South Thoresby, 1971, and Kirkby Moor, 1973, (both R.E.M.P.)

L.F.P. Creeping Willow and various Grasses etc.,

292. *Peridroma porphyrea Schiff.*   **Pearly Underwing**
(saucia Hubn.)
A casual migrant in the county, not commonly recorded. The first L.N.U. recording after the Lep. Lincs. entries was from Grimsby, 14th July 1955, "at M.V.light", (G.A.T.J.). It was also at Barton on Humber, 28th September 1958, (C. G. Else); Mablethorpe, 1952-65, (T. R. New); Gibraltar Point, 1975, (M.T.).

L.F.P. Plantains and Dock

RECORDS OF LINCOLNSHIRE SPECIES

294. *Rhyacia simulans* Hufn. **Dotted Rustic**
  (pyrophila Schiff.)
This species flies in July, August and September. Its normal range extends from central England westwards. It was first recorded in the county on 6th August, 1976, at Muckton, when one came to light, (R.E.M.P.).
L.F.P. Dandelion, Groundsel and other low plants

295. *Spaelotis ravida* Schiff. **Stout Dart**
  (obscura Brahm.)
South (1961) referred to Mr. R. F. Bretherton's work on the history and distribution of this species in Britain (Entomologists Gazette, 1957, Vol. 8, 3-19; Vol. 8, 195-8). South noted "the chief preference of *ravida* seems to be for the edges of marshland, as in S. Essex, the Fens and most of its haunts in Lincolnshire and the Severn Valley".
Some years the species appears to be plentiful, and others absent. L.N.U. records had it common at Barton on Humber, 1944, (C. G. Else). It was recorded spasmodically up to the 1960s, since when it has been noted widespread across the county. Typical records include the following: Boston, "plenty", (1960); South Thoresby, "Hundred at M.V.light," (1971); South Thoresby, "Hundred plus at light", (1973); (all R.E.M.P.). It was at Goxhill, 8th August 1973, (C. Potts), and Brumby Common Nature Reserve, August 1976, (J.H.D.).
L.F.P. Dandelion and Docks

297. *Graphiphora augur* F. **Double Dart**
A woodland species – widespread and common in the county – on the wing in June and July. L.N.U. records include: Saltfleetby, 18th July 1958; Riby, 30th June 1959; Grimsby, 11th July 1962; and Limber, 1st July 1965, (all G.A.T.J.). etc.,
L.F.P. Hawthorn, Sallow, Birch and low plants.

298. *Diarsia brunnea* Schiff. **Purple Clay**
Lep. Lincs. 2 noted the status of this species as "frequent". Records for this century, however, have been mainly for small numbers in scattered localities, in June, July and August. Recorded at Manton, 1956, "larvae and adult moths", (J.H.D.); Linwood, 25th June 1959, (Rear Admiral A. D. Torless); Limber, 17th June 1964, (G.A.T.J.); Laughton, 30th June 1973, (J.H.D.).
L.F.P. Bramble, Sallow and various low plants

299. *Diarsia mendica* F. **Common Ingrailed Clay**
(festiva Schiff.)
A very common moth, variable in colouration and in size. Recorded (June and July) in woodlands all over Lincolnshire. e.g. Thunby, July 1922, (Rev. F. S. Alston); Ruskington, August 1956, (J. Hossack); Brocklesby, 24th June 1959, (G.A.T.J.); Woodhall, 17th July 1964, "numerous and variable", (G.A.T.J.); Humber Bank at Whitton, July 1976, (R.J.).
L.F.P. Primrose, Hawthorn, Sallow, Docks, etc.,

301. *Diarsia dahlii* Hubn. **Barred Chestnut Clay**
A heathland moth, out in August and September. Lep. Lincs. 2 had it as "scarce", and recorded the species in the Wrawby Moor area, "three at heather", 24th August 1908, (G.W.M.).
More recently it was found at Scotton Common in 1954, 1956, 1957 and 1970, (latter date being 30th August), (all J.H.D.). The only other record is for Laughton, 2nd September 1958, (G.A.T.J.).
L.F.P. Docks and Plantains

302. *Diarsia rubi* View **Small Square-spot**
(bella Borkh.)
Common in the county in a variety of habitats, in two generations (from May through to August/September). L.N.U. records include: North Somercotes, 15th May 1957, (C.G.E.); Grimsby, 3rd June 1959, and Holme, 20th July 1963, (both G.A.T.J.); Lincoln, 30th August 1968, and Saltfleetby, 13th September 1963, (both A.D.T.)
L.F.P. Dandelion and low plants

304. *Ochropleura plecta* L. **Flame Shoulder**
Common and widely distributed in all types of habitat, with two generations (June and September). Localities include: Haverholme, 1st June 1957, (L.N.U.); Doddington, 12th July 1958, (L.N.U.); Linwood, 25th June 1959, (A.D.T.); Sleaford, 6th June 1964, (G.N.H.); Kirkby Moor, 7th September 1967, (A.D.T.), etc.,
L.F.P. Various Low Plants

305. *Amathes agathina* Dup. **Heath Rustic**
At the time of Lep. Lincs. 2, this species was claimed to be "rare and local". It remains local today, but it can be found in reasonable numbers some years in the right heathland habitat. It was noted in Lincolnshire in 1921, (H.C.B., L.N.U.). Most recent records are those of J.H.D., from: Scotton Common, September

RECORDS OF LINCOLNSHIRE SPECIES 145

1956; Linwood Warren, "larvae", 13th October 1951; Messingham Moor, 1963; Laughton Common, September 1965; and Epworth Turbary, 21st August 1977.

L.F.P. Heather

309. *Amathes glareosa Esp.* **Autumnal Rustic**
Quite common on some of the heathland areas in the north of the county, though Lep. Lincs. 2 recorded it as "rare", in Scawby and the Middle Rasen area around 1877-1879 (R.T.C.). It flies in September mostly, and has recently been recorded at: Saltfleetby, 7th September 1965, (A.D.T.); Laughton, September 1956, and Greetwell, "plentiful", 6th September 1975, (both J.H.D.).

L.F.P. Heather and low plants

310. *Amathes castanea Esp.* **Grey Rustic**
(neglecta Hubn.) **(The Neglected)**
Found on heathery ground in August, there is a single L.N.U. record for Epworth Turbary, late August 1964, when a specimen came to a "Tilley" light, (J.H.D.).

L.F.P. Heather, Sallow.

311. *Amathes baja Schiff.* **Dotted Clay**
Common in wooded areas all over the county in July and August. Typically recorded: Swallow Vale, 2nd August 1955, (G.A.T.J.); Scunthorpe woodlands (Manton, Laughton, Twigmoor, Holme) in the 1960s, (J.H.D.); Kirkby Moor, 5th July 1964, "bred from larvae on elm", (G.N.H.); Wildsworth, 20th July 1974, (R.J.).

L.F.P. Hawthorn, Bramble, various low growing plants

313. *Amathes c-nigrum L.* **Setaceous Hebrew Character**
Commonly found in the county, mainly in the Autumn (Sept.), and sometimes in large numbers. Occasionally, it is discovered in June and July. More recent L.N.U. records include: Pelham Pillar Woods, 28th September 1959, (G.A.T.J.); Epworth Turbary, 25th June 1961, (A.E.S.); Sleaford, 16th October 1963, (G.N.H.); Mablethorpe 1952-65, (T. R. New); Gibraltar Point, 28th September 1958, (R. Goy), etc.,

L.F.P. Many low plants

315. *Amathes triangulum* Hufn.   **Double Square-spot**
Noted as "scarce" in Lep. Lincs. 2, but from the 1950s and 1960s to date, it has been very common in woodland areas in the north of the county. Recent records have included: Twigmoor, 30th June 1962, (J.H.D.); Limber, 17th June 1964, "plentiful to light", (G.A.T.J.); Woodhall, 17th June 1964, (L.N.U.); Sweeting Thorns, 24th June 1977, (R.J.).

L.F.P. Docks, and the foliage of various trees and low plants

316. *Amathes stigmatica* Hübn.   **Square-spotted Clay**
(rhomboidea Treits.*)*

A local species within the British Isles, "partial to woods", (South 1961). Among other regions, South mentions that the species is in Yorkshire. It flies in July and August.

In the "Lepidoptera of Yorkshire", (reprinted from the Naturalist, 1967-70), the entry for this species reads: "This species does not appear to occur in the county, and as other species are frequently confused with it, it seems possible that it never did," (p.13).

In 1976 and 1977, during a Humber Bank Survey it was reported that a single specimen occurred at Whitgift, 23rd July, and on Goole Moors, 28th July. (A. Grieve), but neither of these records can be substantiated.

L.F.P. Dandelion, Dock, Low plants

317. *Amathes sexstrigata* Haw.   **Six-striped Rustic**
(umbrosa Hubn.*)*

Common across the county in August, when the moth can be seen to be very fond of honey-dew and nectar bearing flowers. It is also attracted to light in numbers. Recorded: Barton on Humber, 29th August 1956, (C.G.E.); Near Byards Leap, 16th August 1959, (G.N.H.); Brigg, August 1964, (A.H.N.); Woodhall, 15th August 1966, (A.D.T.).

L.F.P. Docks, Plantain, Bramble etc.,

318. *Amathes zanthographa* Schiff.   **Square-spot Rustic**
This species is abundant all over the county. Colouration of the upper surface of the forewings can be very variable – through greys, reds, to dark shades approaching black. It flies in August/September, and is very partial to "sugar". L.N.U. records include: Chapel St. Leonards, 25th August 1960, (G.A.T.J.); Laughton, 26th August 1964, "plentiful at sugar", (A.D.T.); Boston, late August 1967, "very common", (J.B.); and Horncastle, 5th September 1968, (A.D.T.).

L.F.P. Grasses and low plants

319. *Axylia putris* L. **Flame Rustic**
(The Flame)
Lep. Lincs. 2 indicated that this species was "not common". From the 1950s, however, it has been commonly recorded all over the county in June/July – and especially at m.v. lights. An early record was for North Somercotes, 1920, in the garden of the Rev. S. Proudfoot. More recently it was "very common" in Grimsby, June to July 1956, (G.A.T.J.); also at Saltfleetby, 18th July 1958, (Rear Admiral Torless); Sleaford, June 1964, (G.N.H.); Boston, June, 1972, "common", (A.E.S.); Barrow Haven, 16th July 1977, (R.J.).

L.F.P. Hedge Bedstraw and other low plants

320. *Anaplectoides prasina* Schiff. **Green Arches**
(herbida Hubn.)
This is a woodland species which flies in June and July. Lep. Lincs. 2 classed it as "rare", and listed it at Skellingthorpe, 9th July 1895, (Canon Fowler). There are only a few L.N.U. records for this century for: Grimsby, 18th July 1956, "at light", and Brocklesby, 26th June 1957, (both G.A.T.J.).

L.F.P. Dock and various low plants

321. *Eurois occulta* L. **Great Brocaded Rustic**
(Great Brocade)
South (1961) wrote that "Scotland appears to be the British home of the species, and it is found in most woods throughout that country". It is also reported in South that two specimens have occurred in Lincolnshire – dated August 1896, and August 5th 1955.

The form found in Lincolnshire is the grey continental form which arrives here by migration in August and September. Lep. Lincs. 2 reported the moth at Market Rasen in 1896, and on 15th August 1897, (W.L.). It was also at North Somercotes in 1914, when "13 in splendid condition, besides several others much worn", were taken in his vicarage garden by Rev. S. Proudfoot. More recently it turned up at Scunthorpe on 7th August 1959, (J.H.D.). There are also a number of recordings by R.E.M.P. for Boston, "the grey form", July 27th to July 31st 1973; Kirkby Moor, 5 specimens, and S. Thoresby – 7 specimens, August 1976.

L.F.P. Dock, Plantain, Primrose, Dandelion etc.,

322. *Gypsitea leucographa Schiff.* **White-marked**
This specifies flies in the spring in Kent, Sussex, and various Southern counties. It is also well recorded from Bishop's Wood, near York, and occurs from Lancashire across to Durham.
  There is one L.N.U. record only; Sleaford, 9th Sept 1967, Miss Elaine Kay, (G.N.H.).
L.F.P. Bilberry, Sallows

323. *Cerastis rubricosa Schiff.* **Red Chestnut Rustic**
Frequently found in the county – flies in March and April, and is often seen visiting sallow blossom in wooded areas.
  An early record placed it at Doddington Woods, "common", (Rev. F. L. Blathwayt). It was recorded by G.N.H. at Ruskington, 20th April, 1956, and at Roughton Moor, 11th April, 1964, etc.
L.F.P. Many low plants

324. *Naenia typica L.* **Gothic Type**
(The Gothic)
A common moth, widespread across the county in June and July, and often found in gardens. In 1948 it was at Grimsby in Nun's Holme Gardens (G.A.T.J.). It was also recorded at: Barton on Humber, at sugar, 10th June 1959, (C.G.E.); Boston, "larvae common on saxifrage", March 1969, (A.E.S.).
L.F.P. Blackthorn, Hawthorn and many kinds of herbage.

NOCTUINAE

327. *Euschesis comes Hubn.* **Lesser Yellow-underwing**
  (orbona F.)
Common and widespread in all kinds of habitat in July, August and September. The species comes readily to light and is frequently found on buddleia blossom. It was recorded at Barton on Humber, "on sugar", 21st September 1959, (C. G. Else); Near Cranwell, 3rd September 1961, (G.N.H.); Saltfleetby, 13th September 1963, (A.D.T.); Boston, 24th August 1968, (A.E.S.)., etc.,
L.F.P. Grasses and a variety of low plants

328. *Euschesis orbona Hufn.* **Lunar Yellow-underwing**
  (ubsequa Hubn.)
Very rare in the county. Flies in July/August, and has been recorded by M. P. Gooseman at Bottesford in the 1950s. The only additional record is for Atkinson's Warren, Scunthorpe, 25th August 1971, (R.K.).
L.F.P. Grasses and other low plants

RECORDS OF LINCOLNSHIRE SPECIES 149

329. *Euschesis Janthina Schiff.* **Lesser-bordered Yellow-underwing (Lesser Broad-border).**
Fairly common – mainly along woodland margins, though it does occur in other habitats (July/August).
At Sleaford in 1956, it was found "in abnormal numbers", (A. Pilkington). It was also near Cowbit, June 1956, (E. J. Redshaw); and at Fishtoft, 20th July 1967, (J.B.) etc.
L.F.P. Many low plants – e.g. Primrose, Dock, Bramble, Cuckoo-pint, Dead-nettle etc.

330. *Euschesis interjecta Hubn.* **Least Yellow-underwing**
Lep. Lincs. 2 classed this species as scarce, but it seems to have been discovered in increasing numbers from the 1960s onwards. Most recent records are for: Alkborough, 5th July 1970, (M.P.G.); Holbeck, 12th July 1975, "plentiful", (G.A.T.J.); Scunthorpe, August 1976, "plentiful on buddleias", (J.H.D.).
L.F.P. Grasses, low plants such as Common Mallow

331. *Noctua pronunuba L.* **Common Yellow Underwing**
This species can be found from June through to September. It is abundant all over the county, with larvae occasionally being a pest in gardens (as reported by A. Roebuck in 1935 – "larvae abundant", in L.N.U. Trans. Vol. 9 p.48).
It was found at Sleaford, 1956, "in abnormal numbers," (G.N.H.); Easton Hall near Grantham, 6th July 1957, (L.N.U.); Grimsby, "very common during season in m.v. trap", 5th June 1957, (G.A.T.J.); Syston Park, 10th July 1965, (L.N.U.); and Boston, July and August 1967, (J.B.).
L.F.P. Widely polyphagous – many plants both wild and cultivated

332. *Lampra fimbriata Schreber* **Broad-bordered Yellow-underwing**
Found in woodland areas mainly, in July and August, and as described in Lep. Lincs. 2 – "well distributed over the county".
Lep. Lincs. listed Binbrook, Grimsby, Legbourne Wood, Wrawby Moor, Holbeach, Skellingthorpe etc. as localities for around 1905.
The species seems to have been on the increase in the north of the county (S. Humberside region) since the early 1960s. Recent L.N.U. records include: Brigg, "to m.v. lamp", August 1964, (A. H. Neale); Gibraltar Point, 9th August 1969, (A.D.T.); Boston, 28th July 1973, (A.E.S.); Scunthorpe railway embankment, 7th September 1978, (R.J. and J.H.D.). L.F.P. Birch, Sallow, Dock, Elder, Dead Nettle etc.

HELIOTHINAE

334. *Pyrrhia umbra Hufn.*   **Bordered Orange**
(marginata F.)   **(Bordered Sallow)**

Early L.N.U. records suggest that this was a "rare" species in the county. It was at Market Rasen, 1877-79, (G.W.M.); and Panton on "sugar", (G.H.R.). It was also discovered near Barton on Humber, when one moth "with crippled hind wing was taken at sugar on 21.6.1905". (Lep. Lincs. 2).
It was "bred in June 1957 from a larva discovered on antirrhinum flowers at Sleaford, and from pupae found" (G.N.H.).
L.F.P. Rest Harrow, Henbane etc.

335. *Heliothis viriplaca Hufn.*   **Marbled Clover**
(dipsacea L.)

According to South (1961), this species "occurs most commonly in the eastern counties, especially on the Breck area".
The only L.N.U. records are those mentioned in Lep. Lincs. 2 for: Manton Common (Dr. George), Woodhall Spa, "saw a pair taken by J.C.L. – G", 8-1901 (Nat. 1901, p. 365); S. Nocton, in clover field, 1893, E. Porter.
L.F.P. Many Low plants. Harrow, Campions, Clover etc.,

340. *Heliothis peltigera Schiff.*   **Dark Bordered Straw**
(Bordered Straw)

A migrant species, observed "more or less rarely in many English counties" (South 1961). It has only been recorded twice in Lincolnshire, according to L.N.U. records, the first record being for a single specimen on Manton Common, late June 1955, (J.H.D.); the second being for Muckton, 17th July 1975, (R.E.M.P.).
L.F.P. Low herbage, Rest Harrow, Clover, etc.,

341. *Heliothis armigera Hubn.*   **Scarce Bordered Straw**

A rare migrant to this country in the autumn. This species is very widely distributed throughout the world, only being missing from Arctic and Antarctic regions. Its larvae are capable of serious crop damage (e.g. in America on Cotton, Tobacco, Fruit, Tomatoes etc.).
A specimen was found by Mr. P. W. R. Walter on December 8th 1979 on the steelworks along Dawes Lane, Scunthorpe.
L.F.P. Does not normally breed in this country – but abroad eats many garden and wild flowering plants

RECORDS OF LINCOLNSHIRE SPECIES

ANARTINAE

342. *Anarta myrtilli* L. **Beautiful Yellow Underwing**
Quite a plentiful species on heathland – found from May to August. Localities mentioned in records include: Woodhall, Linwood, Scotton, Manton, Market Rasen, Roughton Moor, Epworth, Messingham Common etc. Some years the species is more common than others, and it appeared to thrive in 1978 in a number of areas in South Humberside – "large numbers at Scotton Common", (J.H.D.).
L.F.P. Heather

HADENINAE

345. *Mamestra brassicae* L. **Cabbage Dot**
A common moth across the county – out in June/July – occasionally being a pest on flowers and vegetables in gardens and in agricultural areas. Recorded from Sandilands, Grimsby, Ruskington, Sleaford, Lincoln, Scunthorpe etc.
L.F.P. Low herbage and leaves of some trees.

346. *Melanchra persicariae* L. **White Dot**
(The Dot)
Lep. Lincs. 2 implied that this species was "not common" up to the turn of the century. It is now much more numerous in the county, and can be particularly common in gardens where larvae feed on herbaceous plants. The moth flies in June and July. Typical records are for: Bottesford, "larvae very common", 1955, (M.P.G.); Sleaford, 14th July 1958, (G.N.H.); Limber, 15th July 1964, (J.H.D.); Barrow Haven, "larvae common", 20th September 1969, (R.G.).

347. *Polia hepatica* Clerck **Silvery Arches**
(tincta Brahm)
A species only recorded in Lep. Lincs. 2; 1840-50, N. Lincoln, (F.M.B.); S. Skellingthorpe, one, (C.P.A.).
L.F.P. Docks, Plantain, Knotgrass

348. *Polia nitens* Haw. **Pale Shining Arches**
(Advena Schiff.) **(Pale Shining Brown)**
Regarded as "Frequent" in Lep. Lincs. 2., it was recorded at Gainsborough, 1860, (F.M.B.); Linwood, (1857); Mablethorpe, (G.H.R.); Market Rasen, 1877-79, (F.A.L.); Panton, (G.H.R.); Theddlethorpe, 1904, (A. E. Gibbs, Ent. 1905, p.81); Haverholme Priory, "fairly common", (C.P.A.) etc.

152   THE BUTTERFLIES AND LARGER MOTHS

In the 1950s it was commonly found at light in Grimsby, (G.A.T.J.), and it was at Barton on Humber in 1956, (2nd July), and 1958, (11th July), (C. G. Else). The sole record for the 1960s is for Boston, 9th May 1962, (R.E.M.P.).

L.F.P. Knotgrass, Dandelion, Bilberry etc.

349. *Polia nebulosa* Hufn.   Grey Arches

A woodland species, frequently found on the wing in late June and July, and often seen at rest on tree trunks and telegraph poles. Also – a visitor to light and sugar.

It was recorded at Elsham, 7th July 1955, (G.A.T.J.); Twigmoor, 1st July 1962, (J.H.D.); Limber, 17th June 1964, "common to light", (G.A.T.J.); Laughton, 30th June 1973, (L.N.U.).

L.F.P. Docks

351. *Diataraxia oleracea* L.   Bright-line Brown-eye

This moth flies in June and July, and sometimes in the autumn. It is widespread and has been recorded from: Ruskington, August 1956, (J. Hossack); Twigmoor, 18th July 1957, (J.H.D. and G.A.T.J.); Saltfleetby, 18th July 1958, (G.A.T.J.); Sleaford, 28th July 1963, (G.N.H.); Mablethorpe, "very common", 1962-65, (T. R. New); Whitton, 16th July 1976, (R.J.).

L.F.P. Many Low plants.

353. *Ceramica pisi* L.   Broom Brocade

Commonly found in June and July in all kinds of habitat, with larvae plentiful in the autumn. Examples from the records include: Cleethorpes, 1951, (L.N.U.); Riby, 30th June 1959, (G.A.T.J.); Skegness, June 1965, (G.N.H.); Barrow Haven, September 1969, (R. Goy), etc.

L.F.P. Many herbaceous plants such as Broom, Bramble

354. *Hada nana* Hufn.   Light Shears
(dentina Esp.)   (The Shears)

This species appears to have been numerous in June at the time of Lep. Lincs. 2, with records for Alford, Ashby (Brigg), Gainsborough (at sugar), Market Rasen, Hartsholme, and Haverholme Priory (very common) between 1890 and 1901. Since that time there have been very few L.N.U. records – a late record being for Saltfleetby St. Clements, 7th June 1967, (A.D.T.). In May/June 1982, however, it came to light at Langholme, Twigmoor and Bagmoor (J.H.D. and R.J.).

L.F.P. Low plants – Dandelion, Knotgrass etc.

RECORDS OF LINCOLNSHIRE SPECIES 153

355. *Scotogramma trifolii Hufn.* **Small Nutmeg**
(chenopodii Schiff.)
According to Lep. Lincs. 2, this species was "rare" and only recorded from Cleethorpes and Gainsborough. Since 1948, however, it has been commonly found some years in a few localities, in two generations. It was recorded at Grimsby each year 1954-1959, (G.A.T.J.), and was plentiful at Scunthorpe, 1959-61, (J.H.D.). The latest L.N.U. records are for Holbeck, 21st June 1965, (G.A.T.J.), and Scunthorpe, 5th August 1976, "plentiful on buddleia", (J.H.D.).
L.F.P. Various Goosefoots, Beet crops and Onions

357. *Hadena w-latinum Hufn.* **Light Brocade**
(genistae Borkh.)
Rare early this century – only recorded at S. Allington and Haverholme Priory (June 1906) by P. Wynne and J.D.C. respectively. The species remains uncommon with only five subsequent L.N.U. records. The next sighting was for Scunthorpe, July 1952, "bred from larva", (J.H.D.). Other records are for Grimsby, 1956 and 1957, "single specimens to light", (G.A.T.J.), and Gibraltar Point, 1975, (M.T.).
L.F.P. Broom, Spotted Persicary and other low plants.

358. *Hadena suasa Schiff.* **Dog's Tooth**
(dissimilis knoch)
Found mainly in marshy places in the county, and flies in June/July. It was considered "local" early in the century, and localities in the records were Ashby (Brigg), Barton on Humber, Theddlethorpe, S. Boultham, and Hykeham. It remains local with records for Gibraltar Point, 6th July 1959, (Percy Cue); Boston, 11th August 1973, (A.E.S.); and East Halton Skitter, "plentiful", 18th June 1976, (S. Van den Bos).
L.F.P. Plantains, Dock etc.

359. *Hadena thalassina Hufn.* **Pale-shouldered Brocade**
Common from May into July in woodlands across the county. More recent records include: Brocklesby, 26th June 1957, (G.A.T.J.); Laughton, 9th June 1959, (M.P.G.); Brigg, May 1964, (A.H.N.); Holme Plantation, 25th May 1973, (J.H.D.).
L.F.P. Heather, Sallow, Willows etc

361. *Hadena bombycina Hufn.* **Glaucous Shears**
(glauca Hubn.).
According to South (1961), in England this species "occurs chiefly in hilly districts of the Northern counties from Staffordshire to Cumberland" in May and June. There are two L.N.U. records for it. It was "taken near Skegness in 1858 by Mr. G. Gascoyne of Newark" (Lep. Lincs. 2); and it was at Barton on Humber in 1939, (C. G. Else).
L.F.P. Heather, Sallow, Willows etc.

363. *Hadena bicolorata Hufn.* **Broad-barred White Gothic**
(serena Schiff.)
This moth is out from June to August, and can often be seen at rest on tree trunks or fences, or visiting flowers. It has been recorded down the length of the county from Scunthorpe and Grimsby in the North to Boston and Sleaford. Typical localities include: Grimsby 9th July 1956, (G.A.T.J.); Sleaford, 3rd June 1965, (G.N.H.); Lincoln 1971, (A.D.T.); Boston, June 1972, (A.E.S.); and Scunthorpe, June 1978, (J.H.D.).
L.F.P. Flowers and seeds of Hawk's-beard Sow-thistle

366. *Hadena conspersa Schiff.* **Common Marbled Coronet**
(nana Rott.)
Mainly a coastal species – out in June and July – but this species has spread inland in the county, and become more frequent.
The first L.N.U. record was for North Somercotes, 1917, (Rev. S. Proudfoot). This was followed by a sighting at Scawby Woods, 21st June 1951, (J.H.D.). It was at Manton (1958) and Kirton Lindsey Quarries, (1967) (Both J.H.D.), and in 1971 it was reported as "extending range and commoner in Lincs", (R.E.M.P.). It was found at Holbeck, 5th July 1975, (G.A.T.J.).
L.F.P. Campions – (especially Bladder Campion) on unripe seeds.

367. *Hadena compta Schiff.* **Varied Coronet**
Apart from early records in this country for odd migrant specimens in the 1800s, this species first occurred in numbers on the Kent coast in June 1948 and 1949. Since then, its range has been extended across the country, and it has reached Lincolnshire. It was first recorded at Alkborough on 21st July 1975, (M.P.G.); then at Muckton, 6th August 1975, (C. G. Wright). It was also noted at Boston, 26th June 1976, where it was "common", and again at Boston in June 1977, (Both A.E.S.).
L.F.P. Sweet-William

RECORDS OF LINCOLNSHIRE SPECIES

368. *Hadena bicruris* Hufn. **Lychnis Coronet**
(capsincola Hubn.)
Common in the county where Campions occur. It flies in June/July, with an occasional second generation in the autumn. Typically recorded at Laughton, 2nd September 1958, (G.A.T.J.); Tattershall, July 1962, (G.N.H.); Boston, 13th June 1972, (A.E.S.).
L.F.P. Unripe seeds of Campions etc.

370. *Hadena rivularis* F. **Campion Coronet**
(cucubali: Schiff.)
This species is out in June with a partial second generation in August. Lep. Lincs. 2 claimed that the status of the species was "not common", and the number of records this century would confirm this designation. Lep. Lincs. 2 gave it at Binbrook, Gainsborough, Lincoln and Waltham. More recently it was at Manton Common, June 1958, (J.H.D.); Brigg, May 1964, (A.H.N.); and Holbeck, 5th August 1975, (G.A.T.J.).
L.F.P. Leaves and unripe seeds of Campions.

371. *Hadena lepida* Esp. **Tawny Shears**
(carpophaga Borkh.)
This species flies in May, and sometimes later in the year in August and September. It was considered rare at the time of Lep. Lincs. 2, and was recorded from Binbrook, July 1906, (S.B.S.) and Gainsborough, (F.M.B.).
It was next recorded at Manton Common "larvae in Bladder Campion", and at Messingham Moor, (larvae again) in June 1965. It was also at the Kirton Lindsey Quarries in June 1967. (All records by J.H.D.).
L.F.P. Unripe seeds of Campions etc.

373. *Anepia irregularis* Hufn. **Vipers'-bugloss Gothic**
(echii Borkh.)
According to South (1961), this species was "first recorded in the country in July 1868 near Bury St. Edmonds resting on vipers' bugloss", as well as "in other parts of the Breck Sand district of Suffolk and Norfolk". South also mentioned a specimen presented to the Lincoln Museum.
In L.N.U. records this is confirmed in the following record: "N. East Ferry, one specimen bred from larvae taken on Vipers' Bugloss about 1896 by A.R., who kindly presented it to the County Museum at Lincoln". (Lep. Lincs. 2).
L.F.P. Spanish Catchfly (Unripe seeds).

374. *Heliophobus albicolon* Hubn.   **White Colon**
Found in England generally in coastal areas in June, and again in August. Lep. Lincs. 2 said it was "scarce, but less so on the coast" and gave habitats for it near Ashby (Brigg District), (R.T.C.); and it was at Theddlethorpe in 1904 (A. E. Gibbs – Entomologist 1905, p.81).
It was next found at Scunthorpe, July 1956, (J.H.D.); and at Gibraltar Point in 1969, (R.E.M.P.), and 1974, (M.T.). It was last recorded at Bagmoor, 4th June 1982 (R.J.).
                                                                     L.F.P. Low plants growing in sandy areas.

375. *Heliophobus reticulata* Vill.   **Bordered Gothic**
(saponariae Borkh.)
Predominantly a coastal species in the county – on the wing in June, July and into August. It was well recorded at the time of Lep. Lincs. 2, when it was claimed to be "frequent", and found at Barton on Humber, Binbrook, Near Croxby, Hammeringham, Market Rasen, Haverholme Priory and Skellingthorpe etc.
Since this period, the frequency seems to have declined, and there are fewer L.N.U. records. It was at N. Somercotes, 8th July 1919, (Rev. Proudfoot); Barton on Humber, 24th June 1952, (M.P.G.); Grimsby, 28th July 1952, (G.A.T.J.); Barton on Humber again, 6th August 1959, (C. G. Else); and Mablethorpe 1965, (T. R. New). More recently it was found at Washingborough on 4th July 1981, by Mr. & Mrs. A Binding.
                                                                          L.F.P. Knotgrass, etc.

376. *Tholera popularis* F.   **Feathered Gothic**
Frequently found in the county in grassy areas in August and September. As examples: Bottesford, 22nd August 1955, (M.P.G.); Laughton, 9th September 1959, (M.P.G. and J.H.D.); Benniworth, 3rd September 1960, (G.N.H.), etc.
                                                                 L.F.P. Mat-grass and other grasses.

377. *Tholera cespitis* Schiff.   **Hedge Gothic**
                                                                        (Hedge Rustic)
Lep. Lincs. 2 had the species as "scarce" – and recorded it from Near Market Rasen and Near Haverholme Priory (both 1895).
It was next recorded at R.A.F. Digby, 1956, (J. Hossack), and at Laughton, 1956, (J.H.D.). Since this period it has been found at Brigg, August 1964, (A.H.N.); and Scotton Common, May 1976, "larvae plentiful on hard grasses", (J.H.D.).
                                                       L.F.P. Mat-grass, Annual Meadow-grass etc.

RECORDS OF LINCOLNSHIRE SPECIES 157

378. *Cerapteryx graminis* L. **Antler**
Found commonly in July and August in grassy areas throughout the county. It was recorded at Nocton Wood, 1st August 1942, (L.N.U.); Twigmoor, July 1957, (G.A.T.J. and J.H.D.); Saltfleetby, 14th August 1967, (A.D.T.), etc.
L.F.P. A wide variety of grasses.

380. *Xylomyges conspicillaris* L. **Silver Cloud**
South (1961) wrote "The moth is out in April and May, but is very local in England". There is a single L.N.U. record for 1922, for "Stainton, in the spring", (Rev. F. S. Alston).
L.F.P. Trefoils and low plants.

ORTHOSIINAE

382. *Orthosia gothica* L. **Common Hebrew Character**
Common in all kinds of habitat in the spring (March/April/May) and can be often seen at sallow blossom. In the 1960s it was recorded at Sleaford, Grimsby, Roughton Moor, Ruskington, Bardney, Lincoln, Boston, Irnham, etc.
L.F.P. Many trees, shrubs and low plants.

383. *Orthosia miniosa Schiff.* **Blossom Underwing**
This species is out early in the year (March/April) in oak woodlands. L.N.U. records show its status as "Scarce" in the county. Lep. Lincs. 2 placed it in the Alford district 15th April 1891, (E.W.); and at Hartsholme Wood, "a few in April 1896", (W.L.). The only recent records are for Twigmoor, "larvae in June", 1948, (J.H.D.); and New Park Wood, Bardney, 14th May 1966, (G.N.H.).
L.F.P. Mainly low "scrub" oak.

384. *Orthosia cruda Schiff.* **Small Quaker**
(pulverulenta Esp.)
Common in wooded localities (March/April) – another regular visitor to sallow catkins and other blossom in the spring. Localities in the records include Bardney, Roughton Moor, Irnham, Holme, Laughton, Barton on Humber, Sleaford etc.
L.F.P. Oak, Hawthorn, Sallow and other trees.

385. *Orthosia stabilis Schiff.* **Common Quaker**
Flies in the spring (March/April) and has been commonly recorded across the county in such areas as Irby Dales, Linwood, Laughton, Snelland, Roughton Moor, Boston and Lincoln etc.
L.F.P. Trees such as Oak, Sallow, Elm etc.

386. *Orthosia populeti* F. **Lead-coloured Drab**
Classed as "scarce" in Lep. Lincs. 2, this species has only been found in a few localities since the early 1900s, but within these areas it has turned up in numbers (March/April). The first L.N.U. record was for Big Wood, Appleby, 4th June 1881, "larvae", (Mrs. Cross).
In the mid 1940s it was discovered at Sweeting Thorns (near Scunthorpe) as "larvae in June", and it was "quite common at light at Laughton in the mid 1950s", (both J.H.D.). Further south it was "quite plentiful at Boston" 1962, and at S. Thoresby, 1973, (both R.E.M.P.). The last record is for Doddington woodlands, 27th May 1981 (J.H.D.). when larvae were found.
L.F.P. Aspen and Poplars.

387. *Orthosia incerta* Hufn. **Clouded Drab**
(instabilis Schiff.)
Very common in wooded districts in March/April, and another visitor to sallow blossom. It has been extensively recorded in every area of Lincolnshire.
L.F.P. Various species of Trees – e.g. Birch, Sallow, Oak

388. *Orthosia munda Schiff.* **Twin-spot Quaker**
Flies in woodlands in March and April, and visits sallow blossom, but this species has not been commonly recorded in the county. There are L.N.U. records for: Boultham, "at sallows", 4th April 1907, (A.S. and C.P.A. in Lep. Lincs. 2); Knaith, 25th May 1957, "larvae", (L.N.U.); Roughton Moor, "to sugar", 11th April 1964, (L.N.U.); Scotton Common Nature Reserve, 10th April 1974, and April 1978, (R.J.).
L.F.P. Elm, Oak, Sallow and other trees.

389. *Orthosia advena Schiff.* **Northern Drab**
(opima Hubn.)
A species which flies in the spring (March/April). The only mention in L.N.U. records comes from R.E.M. Pilcher, who has found single specimens at S. Thoresby and at Boston. (No other data available).
L.F.P. Sallow, Willow, Birch.

390. *Orthosia gracilis* Schiff. **Powdered Quaker**
Found in all wooded localities, and commonly seen at blossom in the spring. Widely recorded from such areas as: Doddington Woods, "common beginning of April 1914", (Rev. Blathwayt); Grimsby "plentiful at light", in the 1950s, (G.A.T.J.); Bardney, 14th May 1966, (G.A.T.J.), etc.
L.F.P. Poplars, Willows, Bramble and a variety of low plants

391. *Panolis flammea* Schiff. **Pine Beau**
(piniperda Pan.*)* **(Pine Beauty)**
Visits sallow blossom near pine plantations in April. At the time of Lep. Lincs. 2 it was classed as "frequent". The species has not been reported from many localities since the early 1900s, though it ought to occur in the larger pine plantations. Records include: Linwood, "larvae", 1923, (A. E. Musgrave); Linwood, 1963 and 1965, (J.H.D.); Scotton Common, "a few each year in the 1970s" – a typical date being 17th April 1978, (R.J.).
L.F.P. Scots Pine.

LEUCANIINAE

392. *Meliana flammea* Curt. **Flame Wainscot**
This species is generally found in fen areas with "scattered and not dense reed beds" (South 1961), flying in May and June. There is a single county record for Kirkby Moor, where it came to light on 2nd June 1974, (R.E.M.P.).
L.F.P. Common Reed.

393. *Leucania pallens* L. **Common Wainscot**
Found in all grassy habitats across the county in two generations (June/July and August/September). It frequently visits flowers, and can be found in gardens. Localities mentioned in the records include: Riby, Ruskington, Limber, Brocklesby, Barton on Humber, Woodhall, Stamford etc.
L.F.P. Grasses such as Annual Meadow-grass, Couch Grass.

395. *Leucania impura* Hubn. **Smoky Wainscot**
Very common across the county from July into August. Recorded from Hendale Wood, Humberstone, Limber, Woodhall, Kirkby on Bain, Boston, Crowle Waste, and many other areas.
L.F.P. Grasses – e.g. Cock's-foot Grass, Tufted Sedge and Reeds.

396. *Leucania straminea* Treits. **Southern Wainscot**
This species was considered "rare" at the time of Lep. Lincs 2, "in suitable spots very local in Lincolnshire". (July and August). In 1956, it was found near Cowbit, "2 over reed beds" (E. J. Redshaw). In 1959 and 1960 it was described as "very local, a few in my garden to light" (near Boston) (R.E.M.P.). It was found at Amcotts in July 1964 and 1965, (J.H.D.), and last recorded at Kirkby Moor 1973, (R.E.M.P.).
L.F.P. Common Reed, Reed Grass.

397. *Leucania pudorina* Schiff. **Striped Wainscot**
       (impudens Hubn.)
A species found locally in marshy areas in July/August. Lep. Lincs. 2 gave only one example from Boultham (C.P.A.) – identified by Mr. E. A. Atmore. Since this period, there have been three other L.N.U. records: Epworth Turbary, July 1964; and Twigmoor, 29th June 1974, (both J.H.D.); with the last record for Epworth Turbary, 1977, (S. Van den Bos).
L.F.P. Common Reed.

398. *Leucania obsoleta* Hubn. **Obscure Wainscot**
South (1961) described this species as "very local, chiefly found among reeds in Norfolk, Huntingdonshire, Cambridge, Sussex and the Isle of Wight, and also occurs in marshy places along the banks of the Thames from Buckinghamshire to Essex and Kent." The moth flies in June and July. There was a single Lep. Lincs. 2 record – for Lincoln – when one example came to light (C.P.A. – identification confirmed by E. A. Atmore). It was next taken at Saltfleet Haven, "male taken from reed by Carey Rigall" – Rev. S. Proudfoot (G.E.M.). In 1959 and the early 1960s it was recorded near Boston "in my garden, very local in the county" (R.E.M.P.).

In 1964/65 (June) it was found at Amcotts (J.H.D.), extending its range into the (S. Humberside) north of the county. It was still at South Thoresby (Boston area) in 1965 and 1973 (R.E.M.P. and A.E.S. respectively) and was at Whitton on the Humber bank on 16th July 1976, (J.H.D. and R.J.).

The R.S.P.B. warden at Blacktoft Sands reported in 1976 that this was "the commonest wainscot to m.v. light trap", with a hundred plus specimens recorded over a short period of time.
L.F.P. Common Reed.

RECORDS OF LINCOLNSHIRE SPECIES 161

399. *Leucania litoralis* Curt. **Shore Wainscot**
Found in coastal regions on sandhills where Maram grass flourishes, in June and July. This was noted in Lep. Lincs. 2, when the species was described as "very local – taken near Skegness in 1858 by Mr. Gascoyne of Newark". (J.C.L.–C.). The only other mention of the species is for Gibraltar Point (South of Skegness), in 1969, (R.E.M.P.).
L.F.P. Maram Grass.

400. *Leucania comma* L. **Shoulder-striped Wainscot**
A moth found in grassy places in June and July, widely distributed and common across the county. It was at Somerby, 1950, (M.P.G.); Bleasby, 24th April 1955 and Scunthorpe 1962, (J.H.D.); and Holme Lane (near Scunthorpe), 24th June 1977, (R.J.).
L.F.P. Grasses – such as Cock's-foot.

404. *Leucania vitellina* Hubn. **Delicate Wainscot**
A migrant visitor, found mainly along coastal areas in August and September. A single specimen turned up to light at Gibraltar Point in 1978, (M.T.).
L.F.P. Grasses.

407. *Leucania lythargyria* Esp. **Clay Wainscot**
Common in most grassy localities in the vicinity of woodlands from June into July and August. Recorded widely from places such as Grimsby, Martin (near Horncastle), Linwood, Messingham Moor, Roughton Moor, Boston, etc.,
L.F.P. Grasses – such as Meadow-grass, and low plants such as dandelion.

408. *Leucania conigera* Schiff. **Brown-line Wainscot**
(Brown-line Bright-eye)
Found rather locally in damp places and in coastal districts in June, July and August, Lep. Lincs. 2 described it as "local" from Barton on Humber, Cleethorpes, Owston Ferry, Cowbit, Haverholme Priory, and Holbeach. More recently it was at Grimsby, 24th July 1954, and "common at light", (G.Á.T.J.); Scunthorpe, 31st July 1959, (J.H.D.); Limber, 17th June 1964, (G.A.T.J.); and Whitton, 16th July 1976, (R.J.).
L.F.P Couch Grass

NONAGRIINAE

411. *Rhizedra lutosa* Hubn. **Large Wainscot**
(crassicornis Haw.)
A species which occurs across the county (September and October) where dykes, ditches and marshy areas occur – and it is particularly plentiful in the S. Humberside region. It was recorded at S. Ferriby, 29th October 1951, (M.P.G.); Grimsby, when it was noted a number of times in the 1950s (G.A.T.J.). In the 1960s it was recorded at Scunthorpe, Boston, Epworth Turbary, and Gibraltar Point, with ab. *rufescens* Tutt. (a reddish shade) being taken at Scunthorpe (J.H.D.).
L.F.P. The root of reed (Phragmites).

413. *Arenostola pygmina* Haw. **Small Wainscot**
(fluva Hubn.)
A rather local species on wet grassy heathlands, flying in August, September and occasionally early October. It has been typically recorded from Freshney Bog (L.N.U. Trans. Vol. 6, p.17, G.A. Grierson); Legsby, 1954, (R. C. Cornwallis); Epworth Turbary and Scotton Common, September 1964, (J.H.Ð.); and Austacre Wood, 1971, (A.D.T.).
L.F.P. Various Sedges and Grasses.

414. *Arenostola extrema* Hubn. **Concolorous Wainscot**
(concolor Guen.)
South (1961), noted how this species was first discovered in Yaxley Fen and other Fens in Huntingdonshire and Cambridgeshire in the mid 1800s. It subsequently disappeared to turn up again in Hunts. and Northants in clearings in large woodlands where the moth was found flying at dusk in mid June and July. It was also noted how this species has "a dawn flight also".
There is a single L.N.U. record for 1974, when R.E.M.P. discovered a breeding colony in the Bardney Forest area.
L.F.P. Bulrush and some Grasses.

415. *Arenostola fluxa* Hubn. **Mere Wainscot**
(hellmanni Ev.)
Quite a rare species in the county at the time of Lep. Lincs 2, and remains relatively so today. It is a wetland species and is on the wing from late July into early September. It was first recorded near Market Rasen, 1877-79, "a few each year". (W. L. Panton). Other Lep. Lincs. 2 records were for Lincoln Fen and Skellingthorpe. It was also recorded at Skellingthorpe in 1915, "six at moth trap",

RECORDS OF LINCOLNSHIRE SPECIES 163

(Rev. F. L. Blathwayt and Rev. S. Proudfoot) and again on 12th August 1962, (A.D.T.). There are two records for Bardney Forest, 1973 and 1974, (R.E.M.P.).
L.F.P. Various reeds (inside the stems.)

417. *Arenostola elymi Treits.* **Lyme-grass Wainscot**
Common along the coastline of the county in June and July mainly. Typical records cover the Cleethorpes, Mablethorpe, Skegness and Gibraltar Point regions.
L.F.P. Lyme-grass.

419. *Arenostola phragmitidis Hubn.* **Fen Wainscot**
Lep. Lincs 2 classed this species as local, but it appears now to be well distributed in marshy areas and dykes in the county. It can be seen in August and September. Early records placed it in the Brigg district, Market Rasen, Owston Ferry, Sutton on Sea, Haverholme Priory and Wyberton. More recently there are records for Amcotts, 1962 and 1964 (July and September); and Holme Plantation, September 1962, (both J.H.D.); Fishtoft, 4th August 1967, (J.B.); and Gibraltar Point, August 1969, (R.E.M.P.).
L.F.P. Stems of Reeds.

421. *Nonagria algae Esp.* **Reed Wainscot**
    (cannae Ochs.) **(Rush Wainscot)**
A fen species normally found in counties to the South of Lincolnshire in August and September. It has been recorded once in the county at Burton Pits, near Lincoln, when it came to light in 1974, (M.T. confirmed by R.E.M.P.).
L.F.P. Bulrush and Reed-mace.

422. *Nonagria sparganii Esp.* **Webb's Wainscot**
The first British specimens of this moth were discovered in the counties along the south coast of the British Isles in the late 1800s. It flies in August and September. In 1974 one was taken at S. Thoresby, (R.E.M.P.), and this is the sole L.N.U. record.
L.F.P. Yellow Flag, Bur-reed, Reed-mace.

423. *Nonagria typhae Thunb.* **Bulrush Wainscot**
    (arundinis F.)
Lep. Lincs. 2 reported that "This species has only been recorded for a few localities, but doubtless it is fairly common where the bulrush grows freely". This suspicion has been confirmed over the years, as it is now known that this species has colonised most ponds and areas where bulrush occurs.

In the 1800s it was found at Lincoln Fen (24.8.1898, J.F.M.); and Market Rasen (1876-1880, F.A.L.). More recent localities include Saltfleetby, Systen Park, and the Burton Pits Nature Reserve, etc.

L.F.P. Reed-mace.

424. *Nonagria geminipuncta Haw.* **Twin-spot Wainscot**
Found in August in marshy areas more to the South of the British Isles than the county of Lincolnshire. All L.N.U. records date from after 1956, and for areas in the south of the county.
The first mention of the species reads: "Boston 1959, fourteen specimens in my garden the third year in succession – R.E.M.P.". In 1962 and 1964, 2 and 1 specimens were recorded in this S. Thoresby locality, and it was seen subsequently in 1973 and 1974 (all R.E.M.P.). The last record was for Boston, 28th August 1977, (A.E.S.).

L.F.P. Reed

425. *Nonagria dissoluta Treits.* **Brown-veined Wainscot**
This species occurs in wet areas in July and August. Lep. Lincs. 2 classed it as "scarce and local" from Ashby (Brigg), Gainsborough, Lincoln and Saltfleetby. Additional records for this century include Boston, "common 1959"; Gibraltar Point, 1969; S. Thoresby, 1973-74 (all R.E.M.P.). It was recorded also at Alkborough, 5th July 1970, (J.H.D.); and Tetney Blow Wells, August 1976, (S. Van den Bos.).

L.F.P. Common Reed.

427. *Coenobia rufa Haw.* **Rufous Wainscot**
(despecta Treits.) **(small Rufous)**
A local species in marshy localities in July and August. It was first noted in L.N.U. records at Twigmoor, 27th August 1959, (Rear Admiral Torless), when it came to light. It was also taken at Amcotts, July 1964, (J.H.D.).

L.F.P. Stems of Rushes.

428. *Chilodes maritima Tausch.* **Silky Wainscot**
(ulvae Hubn.)
Mainly found to the south of Lincolnshire (June to early August) in fen areas. It was first recorded for the L.N.U. at Boston, 14th July 1964, (R.E.M.P.), when "amongst a number of type specimens, a rather rare *bimaculata* form" occurred. It was also noted that "the last locality known for this moth was ploughed up a few years ago, I wonder where it came from?"
It next came to notice at Twigmoor, 29th June 1974, and at Manton, (Black Walk Nook), 7th July 1974, (Both R.J.). Again, among type specimens, at the latter locality ab. *bipunctata* occurred.

L.F.P. Common Reed.

CARADRININAE

429. Meristis trigrammica Hufn.   **Treble-line**
(trilinea Schiff.)
A species seen in woodland in June and July. Lep. Lincs. 2 classed the species as "common" and the records indicate that it must have been widespread across the county divisions. Since that time, however, it has only turned up in small numbers. The woodlands mentioned in the records are Skellingthorpe, Elsham, Laughton, and Limber with the last records being for Twigmoor, 17th June 1977, (R.J. and J.H.D.), and Holme Plantation, 19th June 1981 (J.H.D.).
                                                               L.F.P. Plantains and other low plants.

430. Caradrina morpheus Hufn.   **Mottled Rustic**
Found frequently in the county in a variety of habitats. Early records placed the species near Market Rasen and S. Boultham in the late 1800s. More up to date it has occurred at Grimsby, Sleaford, Twigmoor and Barton etc.
                                                                        L.F.P. Many low plants.

431. Caradrina alsines Brahm.   **Uncertain**
According to early records, this species was widespread over the county in such localities as Scunthorpe, Barton on Humber, Binbrook, Market Rasen (Legsby Wood), Sutton on Sea, Theddlethorpe and Haverholme Priory. It was claimed that the species was "frequent but less common than the mottled rustic".
    There are recent records for Lincoln, 1st July 1968, (A.D.T.); Boston, 16th July 1968, (A.E.S.); and Scunthorpe in the 1970s, (J.H.D.).
                                                        L.F.P. Chickweed, Docks, etc.

432. Caradrina blanda Schiff.   **Smooth Rustic**
(Taraxaci Hubn.)   **(The Rustic)**
Common in June, July and August in a variety of habitats. It was considered "frequent" early this century at such places as Binbrook, Gainsborough, Market Rasen, Skegness, Haverholme Priory, Lincoln Fen and Wyberton. There are later records for Brocklesby, 24th June 1959, (G.A.T.J.); Woodhall Spa, 16th July 1965, (G.N.H.); Gibraltar Point, September 1969, (R.G.).
                                                                  L.F.P. Low Plants.

435. *Caradrina clavipalpis* Scop.　　**Pale Mottled Willow**
　　　(quadripunctata F.)
A species which has been recorded each month from June to October, and is widely distributed across the county. Up to date records include Scunthorpe, 21st October 1959, (J.H.D.); Sleaford, 23rd July 1963, (G.N.H.); Boston, September 1967, (J.B.), etc.

　　　　　　　　　　L.F.P. Grasses, seeds such as Peas and Corn, Wheat and Grain – can be an agricultural pest.

436. *Laphygma exigua* Hubn.　　**Small Mottled Willow**
A rare migrant species – it appears in the south of England most years, and sometimes turns up in numbers. It was found in Lincs. on 9.9.1932, (R.E.M.P.). There is a further L.N.U. record for Mablethorpe – for one specimen for each of the years 1958, 1963 and 1964, (T. R. New), and a record for Boston, 9.9.1954, (R.E.M.P.).

　　　　　　　　　　L.F.P. Low plants

APAMEINAE

438. *Dypterygia scabriuscula* L.　　**Bird's-wing**
　　　(pinastri L.)
A woodland species found locally in the county. The moth flies in June and July, with an occasional second generation in the autumn.

First L.N.U. records are for the middle 1950s when it was plentiful at Laughton Common on "sugar" in July, (G. Hyde). Subsequently, it has been seen at Laughton, Wildsworth, and the neighbouring Scotton Common Reserve on a number of occasions up to 1978, (J.H.D. and R.J.).

　　　　　　　　　　L.F.P. Sorrel, Docks, Knotgrass.

441. *Apamea lithoxylaea* Schiff.　　**Common Light Arches**
　　　　　　　　　　　　　　　　　　　　(Light Arches)
Generally distributed over the county in June, July and August but it may not be as plentiful now as it was at the time of Lep. Lincs. 2 – when it was considered to be "very common".

It was recorded at Sandilands, 13th August 1954, (G.A.T.J.); Barton on Humber, 3rd July 1956, (C. G. Else); Manton Common, July 1960, (J.H.D.); Sleaford, 8th August 1962, (G.N.H.); Boston, "at light", and June and July 1968, (A.E.S.), etc.

　　　　　　　　　　L.F.P. Stems of grasses, e.g. Meadow-grass.

442. *Apamea sublustris Esp.* **Reddish Light Arches**
Flies in June and July, mainly in Limestone localities, and "chiefly on the coast" (South: 1961).
Lep. Lincs. 2 classed it as "not common", and after these early records (for Barton on Humber, Market Rasen, Skegness, Theddlethorpe, S. Boultham, Haverholme Priory and Wyberton), there are only 2 L.N.U. records. These are for North Somercotes, 1914, "in vicarage gardens", (Rev. S. Proudfoot) and "Lincolnshire", Rev. F. S. Alston, (L.N.U. Trans. Vol. 5, p.154).
L.F.P. Stems and roots of various grasses. (According to P. B. M. Allan (1949) "Foodplant unknown up to 1948").

444. *Apamea monoglypha Hufn.* **Dark Arches**
(polyodon L.)
Found in abundance all over Lincolnshire between June and August, and is subject to considerable colour variation, from pale, to melanic looking specimens.
L.F.P. Stems of grasses.

447. *Apamea crenata Hufn.* **Cloud-bordered Brindle**
(rurea F.)
Quite well distributed over the county and on the wing in June and July. The typical form is a greyish white colour, but the species is extremely variable, and attractive reddish brown forms are commonly found in Lincolnshire.
Typical records include: Nocton Wood, 10th June 1909, (L.N.U.); Haverholme, 1st June 1957, (L.N.U.); Saltfleetby St. Clements, 7th June 1967, (A.D.T.); and Boston, 26th June 1972, (A.E.S.).
L.F.P. Grasses, Primrose, Cowslip.

448. *Apamea sordens Hufn.* **Rustic Shoulder-Knot**
(basilinea Schiff.)
Very common and widespread in May and June. This species has been known to be an agricultural pest. L.N.U. records confirm this with the following entries: Humberston, 25th August 1927, "larvae feeding on the developing grain in wheat ears", (A. Roebuck); Lincolnshire, 1942, "a little damage caused on wheat by larvae", (A. Roebuck, L.N.U. Trans. Vol 10, p.135).
L.F.P. Grasses, various low plants, grain etc.

168 THE BUTTERFLIES AND LARGER MOTHS

449. *Apamea unanimis* Hubn. **Small Clouded Brindle**
Local in the county in moist areas in June. Places mentioned in Lep. Lincs. 2 include: Althorpe, Barton on Humber, Binbrook, Lincoln, Panton, Theddlethorpe and Haverholme Priory.
It appeared at Barton on Humber again on 10th June 1959, (C.G.E.). It was additionally at: Boston, 21st July 1968, (A.E.S.); Snakeholme Wood, 1972, (A.D.T.); Gibraltar Point, 1973, (M.C.T.); and Dunholme, June 1979, (K.S.).
L.F.P. Grasses in marshy, fen areas.

450. *Apamea pabulatricula* Brahm **Union Rustic**
(connexa Borkh.)
South (1961) described this species as "very local in the British Isles", and apart from certain districts in Scotland, found only in woods in "South Yorkshire, as near Rotherham, Sheffield, Barnsley and also in Lincolnshire".
Lep. Lincs. 2 called it "scarce", and noted it had been at Skellingthorpe, and Legsby, 27th July 1901, (C.P.A.).
It was at Skellingthorpe in 1914/15, at sugar, when taken by the Revs. Blathwayt and Proudfoot. It was last recorded "near Lincoln, August 1916, when 60 were found during the first fortnight of August".
L.F.P. Grasses (?)

451. *Apamea oblonga* Haw. **Crescent Striped**
(abjecta Hubn.)
Found along the coast between June and August, with the larvae feeding on grasses growing in salt marshes, and along tidal rivers.
It was at North Somercotes in 1914/17/20, (Rev. Proudfoot). More recent records are for Kirton Marsh, 16th August 1951, (L.N.U.); Holbeck, 17th June 1975, (G.A.T.J.); East Halton Skitter, 16th July 1976, (S. Van den Bos).
L.F.P. Sea Meadow-grass etc.

452. *Apamea infesta* Ochs. **Large Nutmeg**
(sordida Borkh.)
At the time of Lep. Lincs 2 this species was classed as "frequent", (from June into July) with records from Barton on Humber, Binbrook, Cleethorpes, Mablethorpe, Legsby Wood, Haverholme Priory and Skellingthorpe. Since the early 1900s, however, sightings have been much less plentiful. It was at Grimsby, late June 1956/57, (G.A.T.J.); Sleaford, 26th June 1958, (G.N.H.); and other single specimens have been found in the 1970s.
L.F.P. Grasses.

453. *Apamea furva Schiff.* **Confused Brindle**
(The Confused)
L.N.U. records for this species commence "Mr. Burton informs me that he has 5 separate notes of the occurence of this insect at Gainsborough in 1860. I understand that *FURVA* is chiefly a coast species, and is pretty well confined to the N. and W. of England, so its occurence inland in Lincolnshire, though a long time ago, is most interesting".
There are other early records for Gainsborough in 1860, and the Lincoln district in 1852 (F.M.B.). The only recent note of the species was for Gibraltar Point, 18th June 1966, (B.W.).
L.F.P. Grasses – e.g. Meadow-grass.

454. *Apamea remissa Hubn.* **Dusky Brocade**
(obscura Haw.)
A common moth, seen in woodlands across the county in June and July. At Scunthorpe 17th June 1960, the variety *remissa* was taken, (J.H.D.). It was found at Limber, 17th June 1964, (G.A.T.J.); Gibraltar Point bird observatory, 14th to 18th June 1966, (B.W.); and Barrow Haven, 16th July 1977, (R.J.).
L.F.P. Grasses such as
Reed-grass, Couch-grass

455. *Apamea scolopacina Esp.* **Slender Brindle**
The species was classed as "scarce" early this century, with records for Gainsborough, Market Rasen and Skellingthorpe mentioned in Lep. Lincs. 2, (dated July/August). There are only two further L.N.U. records, reading: N. Lincoln, 1916, "twenty during the first fortnight in August", (Rev. Proudfoot); and S. Thoresby, 1978, (R.E.M.P.).
L.F.P. Grasses, e.g.
Couch-grass, Quaking-grass.

456. *Apamea secalis L.* **Common Rustic**
(didyma Esp.)
Abundant in July and August in the county. In "Lincolnshire, 1944, scattered attacks were made on roots of cereals – A. Roebuck, L.N.U. Trans. Vol. 2 p. 83".
There are later records for Ruskington, July 1956, (J. H. Hossack); Barton on Humber, 15th August 1958, (C.G.E.); Sleaford, 4th August 1961, (G.N.H.); Boston, "common July and August 1967", (J.B.); Lincoln, July 1969, (A.D.T.), etc.
L.F.P. Grasses, e.g.
Cock'sfoot and various
Fescues.

457. *Apamea ophiogramma* Esp. **Double-lobed**
A moth usually found in marshy areas in August. Lep. Lincs. 2 considered it "rare", and the only early localities in the records are Cowbit (M. Hufton), and Haverholme Priory. "Two specimens in 1906 on flowers of sedges", (J.D.C.).
More recent records are for the 1970s, for the Humber bank area at Barrow, 8th August 1973, (C. Potts); and Alkborough (date unknown – M.P.G.); and 1980s for Fiskerton, (M.T.); and Dunholme, July 1980, (K.S.).

L.F.P. Reed Grass and Canary Grass.

458. *Apamea ypsillon* Schiff. **Dismal Brindle**
(fissipuncta Haw.*)* **(Dingy Shears)**
Found locally in the county in July and August, where Poplars and Willows are growing.
Early records placed it at Alford, 5th August 1890, (E.W.); Haverholme, "fairly common" 8th July 1901, (J.D.C.); and Stamford, "larvae", 29th June 1905, (L.N.U. meeting). It may well be more widespread than the recent L.N.U. records indicate, as since the 1950s it has only been recorded in the Grimsby, (22nd July 1956 – G.A.T.J.), Scunthorpe (July 1964 – J.H.D.), and Fiskerton (1980 – M.T.) areas.

L.F.P. Willow and Poplars.

461. *Eremobia ochroleuca* Schiff. **Dusky Sallow Rustic**
This species used to occur only in the South of the county on areas of chalk grassland (mainly around Ancaster). It could be seen during the day sitting on Thistle and Knapweed heads. Since the 1950s it has spread all over the county, and has reached the Humber. Most recent records include: Honington (near Grantham), 12th September 1965, Kirton Lindsey quarries, August 1966, (J.H.D.); Ancaster, 20th August 1967, (A.D.T.); Goxhill, 8th August 1973, (C. Potts);
It has also turned up in the "peat" areas of the Isle of Axholme in the 1970s (Epworth Turbary, J.H.D.).

L.F.P. Grasses – Cock's foot and Couch Grass.

462. *Procus strigilis* Clerck. **Marbled Minor**
Abundant and widespread in June and July, and variable in colour from greyish shades through to black. It has been widely recorded from such localities as Sleaford, Ruskington, Twigmoor, Saltfleetby, Brocklesby, Barton on Humber, Mablethorpe and Boston.

L.F.P. Cook's foot and Couch Grass.

463. *Procus latruncula Schiff.* **Tawny Minor**
Flies in June and July. It is reported by R. E. M. Pilcher to be common and widespread towards the south of the county. During the Humber Bank survey in 1976, it was recorded at Whitgift, Fockerby and Blacktoft Sands between 13th June and 6th July, "in good numbers", (A. Grieve).
L.F.P. Cock's foot Grass.

464. *Procus versicolor Borkh.* **Rufous Minor**
According to South (1961), "The moth is on the wing in June and July, but so far the larva has not been noted in Britain. It is presumed to feed on grasses". He also wrote (pp 287 and 288): "It was recorded from the Forest of Dean in 1938 and appears to be widely distributed, and has been reported from Lincolnshire..." There are, however, no L.N.U. records for this species and the authors have no knowledge of Lincolnshire lepidoperists finding it.
L.F.P. See above.

465. *Procus fasciuncula Haw.* **Middle-barred Minor**
Plentiful across the county – often seen flying at dusk in June and July in grassy areas. Typically recorded from Grimsby, Scawby, Anderby Creek, Woodhall Spa, Scunthorpe, etc.
L.F.P. Tufted Hair-grass.

466. *Procus literosa Haw.* **Rosy Minor**
Mainly a coastal species – quite frequently found flying in July and August. It does, however, breed inland in the more "sandy" areas.
It was at N. Somercotes in 1920, (Proudfoot); Scunthorpe, 17th August 1962, (J.H.D.); Frieston, "abundant along the seashore, and along the river bank at Fishtoft near Boston", July and August 1967, (J.B.); Brigg, July and August 1978, (A. Goodall).
L.F.P. Stems of various Grasses.

467. *Procus furuncula Schiff.* **Cloaked Minor**
(bicoloria Vill.)
Another coastal species which can occur on sandy areas inland in July and August. It is very variable indeed in its colouration.
Localities in the records include: "Doddington Road and Six Fields, very abundant at dusk 1923", (A.E.M.); Scotter Common 14th August 1954, (G.A.T.J.); Scunthorpe, "hundreds flying over rough ground", 26th July 1960, (J.H.D.); and Dunholme, 18th August 1978, (K. Saville).
L.F.P. Grasses.

468. *Phothedes captiuncula Treits.* **Least Minor**
(*expolita Staint.*)
According to South (1961), "This species was first discovered in Britain in a locality near Darlington in 1854. It is now obtained in several places in that county, and in Northumberland. Also found in North Lancashire, Westmorland, and once in Yorkshire". There is a single L.N.U. record for the species for Barton on Humber, 16th August 1958, (C. G. Else).

L.F.P. Glaucous Sedge
(in the stem).

469. *Luperina testacea Schiff.* **Flounced Rustic**
Early this century this species was found to be "frequent" in the county in August and September. It has a variable wing colouration, and has always been noted in the records for being commonly attracted to m.v. light and to "sugar" in the Autumn. It continues to be found commonly in such localities as Swanpool and Boultham, Humberston, Scunthorpe, Chapel St. Leonards, Saltfleetby, Laughton, Boston etc.

L.F.P. Grass roots.

472. *Euplexia lucipara L.* **Small Angle-shades**
Common in June, July and occasionally in August across Lincolnshire. The larvae can be a pest in greenhouses, feeding on ferns and other greenhouse plants. It has been recorded from: Grimsby, Revesby, Aswarby Thorns, Linwood, Tetney Blow Wells, and Boston, among many other localities.

L.F.P. Ferns and Willow-
herb etc.

473. *Phlogophora meticulosa L.* **Large Angle-shades**
This moth is common in every area of Lincolnshire. It occurs in the spring, can be found through to the summer, and even into the autumn months. During this later period numbers of the species are increased by migration from the continent – as indicated by the L.N.U. record reading: "South Thoresby, 14th October 1977, three thousand nine hundred and twenty nine in three traps – R.E.M.P.".

Larvae can be found feeding at any time of year – even in the middle of a mild winter, and they can sometimes do considerable damage to herbaceous plants in gardens, and to Chrysanthemums and Geraniums under glass. The moth is not variable, but there are two colour forms of the larvae – brown and green.

L.F.P. Widely polyphagous on
both wild and cultivated
plants.

476. *Thalpophila matura* Hufn. **Straw Underwing**
(cytherea F.)
Fairly common in the county in July and August in sandy districts.
L.N.U. records include: Swanpool and Boultham district, 1924, (A.E.M.); Manton Common, 28th July 1955, (M.P.G.); Scawby, 18th July 1957, (J.H.D.); Rimac, 13th August 1962; (G.A.T.J.); Roughton Moor, 17th July 1964, (G.N.H.); etc. Additionally, where the blown sand extends to the Humber, the species was found at Whitton, 16th July 1976, (R.J.).

L.F.P. Grasses.

AMPHIPYRINAE

478. *Petilampa minima* Haw. **Small Dotted Buff**
(arcuosa Haw.)
Well distributed over the county, and has been particularly well recorded in the more sandy areas in June, July and August.

Localities mentioned in the records include: Twigmoor, 18th July 1957; Chapel St. Leonards, 20th August 1960; Laughton, 30th June 1973; and Manton Common, 9th July 1974, (all J.H.D.).

L.F.P. Tufted Hair-grass.

480. *Hydrillula palustris* Hubn. **Marsh Buff**
(Marsh Moth)
In the 1800s, as written by South, this species was recorded mainly as individual specimens in a number of scattered localities. There is a single L.N.U. Lep. Lincs. 2 entry, reading "One example of this rarity has been recorded – N. Lincolnshire coast sandhills, a male at light, 21.6.1902, J.F.M. and C.P.A.".

Subsequently, all L.N.U. records stem from R.E.M.P. and are as follows: South Thoresby, one at light 13th June 1970; South Thoresby and another locality, 1971; South Thoresby (5th site) 1973; Gibraltar Point, 3 at light, 1975.

It appears, therefore, that in this region of Lincolnshire, thriving colonies now exist.

L.F.P. Meadow-sweet and other low plants.

481. *Celaena haworthii* Curt. **Haworth's Crescent**
(Haworth's Minor)
In early records this moth was considered "very rare", and it is noted "I have only two records of this insect's occurrence in Lincolnshire; both may be regarded as authentic. I have seen Mr. Reynold's specimens". It was found "N. Barton on Humber, taken in the garden of the late Mr. William Gray, Cliff House, in August

1878 (Prof. R. Meldola, F.R.S.) and at East Ferry District (A.R.)".

More recently it was discovered at Rush Bottom Close in 1975, (R.E.M.P.), and a strong colony was found at Crowle Waste in August 1978, (R. Key).

L.F.P. Cotton Grass.

482. *Celaena leucostigma* Hubn. **Brown Crescent**
(fibrosa Hubn.) **(The Crescent)**

A species found in marshy areas – on the wing in July, August and September. Early this century it was thought to be rare, and at Haverholme Priory (J.D.C.). It was next at Barton on Humber, 4th September 1956, and again in 1957, (C. G. Else). It was then at Saltfleetby, 13th September 1963, (A.D.T.); the Boston area, 1965, (R.E.M.P.); Alkborough, 22nd July 1970, (J.H.D.); and Whitton, 16th July 1977, (R.J.).

L.F.P. Sedges and Yellow Flag.

484. *Hydraecia oculea* L. **Common Ear.**
(nictitans Borkh.) **(The Ear Moth)**

Generally distributed across the county – in the main in wet areas – and flies in July, August and September.

It has been recorded from Crowle Waste, N. Somercotes (a black form taken by Rev. Proudfoot), Sandilands, Barton on Humber, Laughton, Gibraltar Point, Epworth Turbary and Lincoln.

L.F.P. Meadow-grasses.

485. *Hydraecia paludis* Tutt. **Saltern Ear**

This species is slightly larger than the preceding one, and is more ochreous in colour. According to South, it is found "on the wing" in August and September, and occurs all round the coast of England on both sand dunes and salt marshes". It was discovered at Gibraltar Point on 16th August 1959 and odd specimens have been found in the Scunthorpe area (recorded by J.H.D.).

L.F.P. Common Grass

486. *Hydraecia lucens* Freyer **Large Ear**

This species is out in August and September. It is known to occur in a number of localities in Yorkshire (The Lepidoptera of Yorkshire, Yorkshire Naturalists' Union, 1970, p.18).

One specimen was reputedly found at Whitgift during the Humber Bank Survey on 9th August 1976, (A. Grieve).

Elsewhere it was recorded at South Thoresby, 24th August 1968, and at Boston, 11th August 1964, (both R.E.M.P.).

L.F.P. Grasses.

488. *Gortyna micacea* Esp. **Rosy Ear**
(Rosy Rustic)
Widely distributed in Lincolnshire, quite common in the autumn, and partial to visiting Ragwort flowers. Larvae are occasionally an agricultural pest. Recorded at Grimsby, Sleaford, Humberston, Pelham's Pillar Woods, Gibraltar Point, Heckington, Epworth Turbary, Boston and Barrow Haven etc.
L.F.P. Docks, Plantains, etc.

489. *Gortyna petasitis* Doubl. **Butterbur Ear**
Early in the century it was considered rare in the Gainsborough district, where it was "bred from Butterbur (Petasites), (F.M.B.)" as recorded in Lep. Lincs. 2.
More recently it was taken at Laughton Common, 26th August 1964, (A.D.T.). There are almost annual records for South Thoresby, (R.E.M.P.), between 1969 and 1978. Some of these indicate the species occurring in numbers – e.g. 1971, "2 a penny", and 1977, "19 in trap". It may well occur in other localities where the foodplant is widespread.
L.F.P. Butterbur.

490. *Gortyna flavago* Schiff. **Orange Ear**
(ochracea Hubn.*)* **(Frosted Orange)**
Appears to be widespread over Lincolnshire, occurring between August and October. It was recorded at Skellingthorpe, 1942, (F.L.K.); Candlesby, 10th September 1955, (B.J.); Limber, "plentiful" 28th September 1956, (G.A.T.J. and J.H.D.); Martin, 8th September 1956, "larvae on burdock", (G.N.H.); Boston, late September 1967, (J.B.); and Gibraltar Point, 28th September 1969, (R.G.), etc.
L.F.P. Larvae feed internally on stems of Burdock, Thistle, Mullein, Figwort, Potato etc.

493. *Cosmia pyralina* Schiff. **Lunar-spotted Pinion**
The first L.N.U. record for this species was for 1921, (Rev. F. S. Alston, L.N.U. Trans. Vol.5, p.154) and the only other record available is for South Thoresby, 1978, (R.E.M.P.).
L.F.P. Oak, Elm etc.

494. *Cosmia affinis* L. **Lesser-spotted Pinion**
At the turn of the century this species was considered to be scarce. There are only a handful of subsequent records. These are for: Barton on Humber, 1945, (C. G. Else); Grimsby 1954 and 1955, "to m.v.lamp", (G.A.T.J); Martin, 9th August 1956, "from larvae on elm leaves", (G.N.H.); South Thoresby, 1970, (R.E.M.P.); and Boston, 23rd August 1972, (A.E.S.). L.F.P. Elm.

495. *Cosmia diffinis* L. **White-spotted Pinion**
Uncommon early this century, and only few L.N.U. records available. These are for: Lincoln, Market Rasen 1895 (W.L.), and Haverholme Priory (Lep. Lincs. 2); Legsby, 29th May 1940, (F.L.K.); Boston, "single specimen 1964", (R.E.M.P.).
L.F.P. Common Elm.

496. *Cosmia trapezina* L. **Dun-bar**
A common woodland species which flies in June, July and August. The species is most variable – from the type form being pale buff, to a rich red colouration. It has been recorded from Holton-le-Moor, Nocton Wood, Appleby, Sleaford, Kelby, Bourne, Manton, Linwood and Syston Park etc.
L.F.P. Foliage of many forest trees – the larvae being notoriously cannibalistic.

497. *Enargia paleacea* Esp. **Angle-striped**
(*fulvago* Hubn.) **(Angle-striped Sallow)**
Before the turn of the century this species was very local, with its main habitat being on the Lincs./Notts border near Gainsborough. Up to 1955 there are only two records for Skellingthorpe (1914 and 1916, Rev. Proudfoot and Rev. Blathwayt). From 1955, the species appears to have extended its range across the county, and it has been recorded each year – from localities as far apart as Barrow (on the Humber) and Boston.
 Typical recordings are for: Laughton (the main area), 11th September 1963, (J.H.D.); Kirkby Moor, 1970, "good numbers", (R.E.M.P.); and Holme Plantation, 26th August 1976, (J.H.D.), etc.
L.F.P. Birch and Aspen.

499. *Zenobia retusa* L. **Double Kidney**
This species was only recorded in one locality – Skellingthorpe, up to 4.8.1901, (J.F.M.), in Lep. Lincs. 2.
 There is a single additional record for the Boston area, 1965, (R.E.M.P.).
L.F.P. Sallows, Willows.

500. *Zenobia subtusa* Schiff. **Olive Kidney**
(The Olive)
Appeared scarce at the time of Lep. Lincs. 2 and there are few further records. It was at Laughton, "larvae plentiful" 1918, (Rev. Proudfoot); Grimsby, 30th August 1948, (G.A.T.J.); Scunthorpe,

"plentiful to light", August 1955, (J.H.D.); Dunsby Wood, 7th July 1959, (J.H.D.); Boston, "single specimen 1964", (R.E.M.P.); and South Thoresby.

L.F.P. Black Poplar and Aspen.

501. *Panemeria tenebrata Scop.* **Small Yellow Underwing**
(arbuti F.)
This species flies "on sunny days in May and early June, and is more or less common in grass-bordered lanes, hay meadows etc." (South 1961). It has, however, been less frequently found in Lincolnshire since the time of Lep. Lincs. 2, which is to be expected with the shrinking of this type of habitat.
There are recent records for Twyford Forest, June 1936, (A.E.M.); Barton on Humber, 1957, (C. G. Else); Horncastle, 3rd June 1968, (A.D.T.); and Isle of Axholme, 1975.

L.F.P. Chickweeds.

502. *Amphipyra pyramidea L.* **Common Copper Underwing**
Uncommon in the last century, but now quite widespread in the county in August and September.
Typical records for the 1950s and 1960s include Doddington Woods, 1953, (G. Houlden); Skellingthorpe, 1956, (J.H.D.); Skellingthorpe, "larvae 12th March 1957", (Canon G. Houlden); Laughton 1959, 1962, 1965, (J.H.D. and A.D.T.), etc.

L.F.P. Oak, Birch and other shrubs.

502A. *Amphipyra berbera Rungs* **Drab Copper Underwing**
In the 1970s it was recognised that a proportion of the Copper Underwings found in the county were a separate species, named as above.
A large percentage proved to be *A.berbera* in some localities.
It has been recorded from Burton Pits, 1975, (M.T.), and South Thoresby, 1977 and 1978, (R.E.M.P.).

L.F.P. Trees and Shrubs.

503. *Amphipyra tragopoginis Clerck* **Mouse**
Common in the county from July through into the autumn. Localities in the records include Appleby, Chapel St. Leonards, Epworth, Saltfleetby, Laughton, Boston, Mablethorpe etc.

L.F.P. Hawthorn, Willows, Honeysuckle.

504. *Rusina tenebrosa* Hubn. **Brown Feathered**
(umbratica auct.) **(Brown Rustic)**
A very common woodland species in June. Found at Grasby Woods, Scawby Woods, Ruskington, Brocklesby, Linwood, Broughton, and numerous other localities.
L.F.P. Many herbaceous plants.

505. *Mormo maura* L. **Old-lady**
A species frequently found at the time of Lep. Lincs. 2, with a considerable number of areas throughout Lincolnshire mentioned in L.N.U. records.
Since this period it has only turned up spasmodically. There are records from Barton on Humber, 1945 and 1948, (C. G. Else and M.P.G. respectively). It was at Barton again on 17th August 1957, (C. G. Else); at Haxey, 1975, (Mr. Snow); and South Thoresby, 1977, (R.E.M.P.).
L.F.P. Hawthorn, Birch and low plants.

APATELINAE

506. *Cryphia perla* Schiff. **Marbled Beau**
Earlier this century, this species was thought to be widely distributed. Recent records would suggest that it may be less plentiful today. It flies in July and August. Most L.N.U. records are for the 1940s and 1950s for: Barton on Humber, "common 1944", (C. G. Else); Grimsby, 11th July 1955, "m.v.light", (G.A.T.J.); Sleaford, 10th July 1958, (G.N.H.); and Scunthorpe, July 1959, "found in numbers", (J.H.D.).
L.F.P. Lichen (Growing on old walls.

511. *Moma alpium* Osbeck **Scarce Merveille-du-jour**
(orion Esp.)
There is a single L.N.U. record for this species from Lep. Lincs. 2. It reads "Near Sleaford, one example at sugar, (J.D.C.), Mr. Coward assures me that there is no doubt of the identity of this insect". There is no possibility of verifying this old record.
L.F.P. Oak.

512. *Apatele leporina* L. **Miller**
At the time of Lep. Lincs. 2 this species was described as "scarce". From the 1940s it has been found in increasing numbers on bushy heathlands where birch scrub occurs. (Mainly in June and July).
The more recent records include: Doddington, 1946, (H.S.);

RECORDS OF LINCOLNSHIRE SPECIES 179

Manton Common, 25th June 1957, (M.P.G.); Somersby Parish, 25th June 1957, C. G. Else); Burton Gravel Pits Nature Reserve, 1971, (A.D.T.); Brigg, July and August 1978, (A. & R. Goodall).

L.F.P. Birch, Oak, Poplar, Sallow.

513. *Apatele aceris* L. **Sycamore Dagger**
L.N.U. records for this species read: "Scarce, S. Haverholme Priory, not common, (J.D.C.)", (Lep. Lines. 2). "Spalding – one larva (Springfields) 20th September 1970, (A.E.S.)". It has also been found at Boston in recent years, (R.E.M.P.).

L.F.P. Sycamore, and a number of other trees.

514. *Apatele megacephala* Schiff. **Poplar Dagger**
(Poplar Grey)
Well distributed over the county where Poplars occur. It flies mainly in June and July – sometimes with a second generation in August. Occasional melanic specimens have been captured. Localities mentioned in the records include: Grimsby, Broughton Woods, Laughton, Sleaford, Martin/Timberland, Woodhall, Boston etc.

L.F.P. Poplar and Willow.

515. *Apatele alni* L. **Alder Dagger**
(Alder Moth)
Thought, at the time of Lep. Lincs. 2, to be "scarce but well distributed – with the records chiefly referring to the discovery of larvae of this species". Since the more frequent use of mercury vapour lamps, it has been found to be more widely distributed (in June). Examples of records are: Irby Dale, 1913, "larvae", (F. W. Sowerby); Laughton, 19th June 1959, (G. Hyde); Scawby Woods, 1962, "3 larvae", (J.H.D.); South Lincs, "6 specimens 1974", (R.E.M.P.); Scawby Woods, 11th June 1974, (R.J.); Dunholme, June 1980, (K. Savage).

L.F.P. Alder etc.

517. *Apatele tridens* Schiff. **Dark Dagger**
Early this century it was known that "this species is probably overlooked by reason of its similarity to the following species (Grey Dagger), but it appears to be well distributed". The same may be said today, as it is very difficult to separate these species without an examination of the structure of their genitalia, or without having had them in the larval stage. L.N.U. records for

this species are, therefore, mainly for larvae, as identification of this stage can be positive.
Such larval records include: Lincolnshire, 1921, (H. C. Bee, L.N.U. Trans. Vol. 5, p.154); Scawby, 10th September 1955, "larvae", (M.P.G.); Grainthorpe, 12th September, 1959, "one larva taken on wild rose", (G.A.T.J.); Boston, July and August 1967, "several larvae", (J.B.).
The adult moth is on the wing in June.

L.F.P. Hawthorn, Sloe, Birch, Sallow etc.

518. *Apatele psi L.* **Grey Dagger**

Common in May, June and July, across the county. It has been recorded from: Holton Le Moor, Morton Carrs, Grimsthorpe Park, Grantham, Grimsby, Brocklesby, Saltfleetby, Woodhall, Boston and Lincoln.

L.F.P. Hawthorn, Blackthorn etc.

520. *Apatele menyanthidis View.* **Light Knot-grass Dagger**

Lep. Lincs. 2 classed the species as "scarce and local", from Near Broughton Woods, 1895, (A.E.H.); and Scotton Common, (F.M.B.).
All subsequent records stem from J.H.D. and are for Epworth Turbary 1962 and 1965, and Scotton Common Nature Reserve, June 1969, 1970, 1972, 1973. The species flies in late May and June.

L.F.P. Birch, Sallow, Creeping Willow etc.

523. *Apatele rumicis L.* **Dusky Knot-grass Dagger**
(The Knot grass)

Common in Lincolnshire in June and July in all kinds of habitat – with occasional specimens found in August and September.
Typically recorded from: Holton le Moor, 12th June 1908, (L.N.U. – F. P. H. Birtwhistle); Littlecoates, 1914, (F.W.S.); Gibraltar Point, 1950, (L.N.U.); Scunthorpe, June 1959, (J.H.D.); and Barrow Haven, 20th September 1969, (R.G.).

L.F.P. Dock, Sorrel, Plantains.

524. *Craniophora ligustri Schiff.* **Crown**
(The Coronet)

Scarce in the county throughout this century. In the early 1900s it was near Gainsborough, Market Rasen, Woodhall and Haverholme Priory. All recent records have been submitted by R.E.M.P. and are for South Thoresby, 1970, 1973, and 1978. The moth flies in June and July.

L.F.P. Ash and Privet.

RECORDS OF LINCOLNSHIRE SPECIES

525. *Simyra venosa* Borkh. **Powdered Dagger**
(albovenosa auct.) **(Reed Dagger)**
The only L.N.U. record is from Lep. Lincs. 2 for "Near Lincoln, 1840 and 1850 (F.M.B.)". The divisions starred in conjunction with this record are Martin (Blankney), and Marton. According to South, the moth is out in June and it occurs mainly in fenland areas.

L.F.P. Reed, Meadow-grass.

CUCULLIINAE

527. *Cucullia umbratica* L. **Common Shark**
(The Shark)
Well distributed and frequently found in the county. It visits Honeysuckle, Valerian, Sweet William in gardens in June and July. It was found at: Hartsholme, 1925, (A.E.M.); Grimsby, 2nd July 1954, "fairly plentiful during the season", (G.A.T.J.); Scunthorpe, 15th June 1973, (J.H.D.); and Holbeck, 18th June 1975, (G.A.T.J.), etc.

L.F.P. Sow-thistles, Hawksbeard, Lettuce etc.

528. *Cucullis asteris* Schiff. **Starwort Shark**
A coastal species – the larvae feeding on Sea Aster growing in salt marsh areas. The moth emerges in July mainly.
There are L.N.U. records for: Freiston, 1946, "larvae found on foreshore, the purple variety, finished feeding on the flowers of Golden-rod", (J. H. White); Kirton Marsh (near Boston), 1951, (L.N.U. Miss B. Hopkins); Saltfleetby, July 1969, "larvae plentiful", (J.H.D.); and Gibraltar Point, 9th August 1969, (A. D. Townsend).

L.F.P. Sea Aster (Starwort) flowers and Golden Rod.

529. *Cucullia chamomillae* Schiff. **Chamomile Shark**
Considered "very rare" at the time of Lep. Lincs. 2, and recorded only from the Holbeach district, and Wyberton, April 1897, (J.C.). The species has been found in small numbers, subsequently, in April/May.
Records include: Barton on Humber, 3rd May 1957, (C. G. Else); Sleaford, 1st May 1964, (G.N.H.); Scunthorpe, April 1977, (P. Walter); Brumby Common Reserve, "one larva", July 1977, (J.H.D.).

L.F.P. Chamomile and Mayweeds.

531. *Cucullia absinthii* L. **Pale Wormwood Shark**
(Wormwood Shark)
The first L.N.U. record for this species was for Grimsby, 25th July 1955, (G.A.T.J.), when one "turned up to light". This was followed by one at Scunthorpe, 1964, (J. Trinder). It has been found commonly in the Scunthorpe area by J.H.D. in the 1970s, (1970, '72, '73, '75, '76, '77). The larvae have been found feeding on Wormwood, and a few on Mugwort, (with as many as 60 larvae counted on one large plant!) and the moths have been observed visiting Valerian flowers and Campions.
L.F.P. Wormwood, Mugwort.

533. *Cucullia verbasci* L. **Mullein Shark**
Larvae are commonly found in damp areas on Figworts, and often turn up in gardens on Mullein and Buddleia. The moth flies in May.
Records include: Appleby, 19th June 1917, (L.N.U.); Denton district, 1924, (H. L. Bond); Eaton and Reepham, 1926, (A.E.M.); Scunthorpe, Sweeting Thorns, 1969, (R.J.); Billinghay, April 1961, (G.N.H.); Glentworth and Dimston Heath, 1972, (A.D.T.); and Susworth, 14th June 1975, (L.N.U. meeting).
L.F.P. Mulleins and Figworts.

XYLENINAE

537. *Lithomoia solidaginis* Hubn. **Bilberry Brind**
(Golden-rod Brindle)
There is a single L.N.U. entry for this species for Little Cawthorpe, 22nd September, 1973, (R.E.M.P.).
L.F.P. Bilberry, Heather, Sallow etc.

538. *Lithophane semibrunnea* Haw. **Tawny Pinion**
A local species – on the wing late in the year (September to November) and sometimes seen the next spring at Sallow blossom after hibernation.
The only L.N.U. records came from R. E. M. Pilcher, and are for Boston, September 12th-18th 1944, and Tytton Hall Park, Boston, September 7th 1948. L.F.P. Ash, Privet, Plum.

543. *Lithophane ornitopus* Hufn. **Grey Shoulder-knot**
(rhizolitha F.)
Lep. Lincs. 2 labelled this species as "rare", with two records for Tealby (1877-79) and Horkstow (1892). The only additional L.N.U. records are for Barton on Humber, June 1957 and 1959, (C. G. Else) – "on sugar". L.F.P. Oak.

544. *Xylena exsoleta* L.   Cloudy Sword-grass
(Sword-grass)

Considered common early this century in a number of divisions within the county. Since this time, there have been few records, and the species has only been found locally, in the autumn. There are records for: Barton on Humber, 1944/5, (C. G. Else); Barton on Humber, 16th July 1951, (M.P.G.); Laughton Common, October 1959, "light and sugar", (M.P.G. and J.H.D.); and Wildsworth, (Laughton), September 1972, (R.J.).

L.F.P. Many low growing plants.

545. *Xylena vetusta* Hubn.   Red Sword-grass

A rare species in Lincolnshire. There are records for "1892 Market Rasen and South Hartsholme Wood, (W.D.C.); and 1891, Haverholme Priory, 9th October." (Lep. Lincs. 2). There is a single further entry for Barton on Humber, 1945, (C. G. Else).

L.F.P. Low Plants

546. *Xylocampa areola* Esp.   Grey Early
(Lithorhiza Borkh.)   (Early Grey)

Not considered common early this century, but now known to be quite well distributed – a frequently found species. There are records for Thonack Park, 9th April 1952, (M.P.G.); Laughton Common, 16th March 1961, (A.D.T.); Kirkby Underwood, 15th April 1963, (G.N.H.); Roughton Moor, 11th April 1965, (G.A.T.J.); and Scotton Common, 25th April 1970, (R.J.).

L.F.P. Honeysuckle.

DASYPOLIINAE

550. *Brachionycha sphinx* Hufn.   Common Sprawler
(cassinia Schiff.)

A local species – flies in woodlands very late in the year (November). Because of this time of emergence, the species has been mainly found in the larval stage, but Lep. Lincs. 2. contains a record for: Haverholme Priory, 1902, "15 males at light in five minutes", (J.D.C.).

More recent records for larvae include: Barton on Humber, 1946, (C. G. Else); Skellingthorpe and Kexby, "larvae found in May 1955", (M.P.G.); Aswarby Thorns, near Sleaford, 7th June 1958, (G.N.H., L.N.U. meeting); and Bardney Forest, 1974, (M.T.).

L.F.P. Foliage of various trees.

552. *Bombycia viminalis* F. **Minor Shoulder-knot**
Considered "locally common" at the time of Lep. Lincs. 2, but there are very few records for the last seventy years. It occurs in woodlands and along forest rides in June and July.
It was at: Holton le Moor, 12th June 1908, "larvae", (F.P.H. Birtwhistle, L.N.U.); Brumby Wood, 21st June 1959, (J.H.D.); Holme, 20th July 1963, (G.A.T.J.); and Lincoln July 1969, (A.D.T.).
There are more recent records for South Thoresby in the 1970s, (R.E.M.P.).

L.F.P. Sallows, Willow, Aspen etc.

553. *Aporophyla lutulenta Schiff.* **Deep Brown Rustic**
(Deep Brown Dart)
Mainly a coastal species out in August and September. It was considered "local and scarce" early this century, and there are very few subsequent records.
Early records had it at Ashby (Brigg) district, (R.T.C.), and Market Rasen, 1877/79 (F.A.L.). It was next found at Laughton, "several in 1964" (J.H.D.); Roughton Moor, 1973, (M.C.T.); and Messingham Sand Quarries, 18th September 1976, "a strong colony", (R.J.).

L.F.P. Grasses and low plants.

555. *Aporophyla lunula Stroem* **Black Rustic**
(nigra Haw.)
Lep. Lincs. 2 classed this species as "local and scarce", and South (1961) mentioned it as being "local and rare" in Lincolnshire. It is, according to South, "Chiefly a Northern species, but it occurs in some of the Southern Counties".
It was first found at Cowbit (Chas. M. Hufton) and at Hartsholme, "scarce", (W.D.C. in Lep. Lincs. 2). The only other record is for Laughton, "one to light", 9th September 1959, (G.A.T.J.).

L.F.P. Hair-grass, Dock, Sorrel.

557. *Allophyes oxyacanthae* L. **Green-brindled Crescent**
A common moth, partial to sugar in the autumn (September and October). It was found at: Scunthorpe and Laughton, 1955, when it was described as "plentiful with 50 the var.capucina", (J.H.D.); Pelham's Pillar Woodland, 28th September 1959, (G.A.T.J.); near Kelby, October 1961, and 1963, (G.N.H.); and Boston, 30th September 1968, (A.E.S.).

L.F.P. Hawthorn, Blackthorn etc.

559. *Griposia aprilina* L. **Common Merveille-du-Jour**
Perhaps less frequently found today, as its habitat (oak woodlands) has been reduced in size this century in the county. It was common and abundant in a number of localities at the time of Lep. Lincs. 2. Since then, its main locality has been Laughton, and there are L.N.U. records (in the autumn) for 1956, '58, '59, '60, '63, '64, '69, '72, '73 (by J.H.D., G.A.T.J., A.D.T., R.J.) for this region. It was also at Linwood, on 12th September 1959, (J.H.D.).
L.F.P. Oak.

562. *Eumichtis adusta* Esp. **Dark Brocade**
Found in a variety of habitats – woodland, heathland etc. – in late May and June. It was particularly common at Grimsby, being recorded almost annually between 1949 and 1959 (G.A.T.J.). It was at Scunthorpe, June 1959, (J.H.D.), and was "common" at Hendale Wood, 20th June 1975, (R.J.).
L.F.P. Grasses and various low plants.

563. *Eumichtis lichenea* Hubn. **Feathered Ranuncule
(Feathered Ranunculus)**
A coastal species, recorded in Lincolnshire in sand dune areas in September/October. It was thought to be "local" early this century, and there are few further records – all for the Gibraltar Point area – for 19th September 1957, 1969, 1972, 1978 (records by R.E.M.P., J.H.D., A.D.T., and T. R. New). In 1980, R.E.M.P. found the species again at Gibraltar Point, and recorded that it was "common all along the coast".
L.F.P. Low plants like Hound's-tongue, Ragwort, Dock etc.

564. *Parastichtis suspecta* Hubn. **Suspected**
A very local species in wooded areas in July and August. At the time of Lep. Lincs. 2 it was found at Ashby, Market Rasen (common some years) and Skellingthorpe. Later records include Lincoln, 1916, (Proudfoot); Woodhall Spa, 16th July 1965, (G.A.T.J.); and Riseholme Wood, 18th July 1967, (A.D.T.).
L.F.P. Birch and Sallow.

565. *Dryobotodes eremita* F. **Brindled Green Mottle**
(protea Schiff.) **(Brindled Green)**
Another Oak woodland species, with the moth out in the autumn. It was more frequently found in the early part of this century and in the 1800s. Localities mentioned in early L.N.U. records are: Ailby, Binbrook, Brocklesby, Gainsborough, Lincoln, Market Rasen, Panton, Saxby, Wrawby Moor, Haverholme and Skellingthorpe. Apart from single records in the 1950s for Holme

Plantation and Barton on Humber, all other specimens have been found at Laughton (1955, '57, '58, '59, '60, '62 etc.), and Scawby in the region of the Ermine Street in the 1960s (J.H.D.).
L.F.P. Oak.

567. *Dasypolia templi Thunb.* **Brindled Ochre**
According to South (1961) this species "frequents rocky places on the coast and on hills", and its range is along the south and west coasts, and from Yorkshire northward. In Lincolnshire, however, it has been recorded in inland localities – in sandy areas. The moth is out in October. There are early records for Barton on Humber, 10th October 1912/13, (G. W. Mason). The next L.N.U. record is also for Barton, but for 18th October 1952, (M.P.G.). It was at Scunthorpe in October 1956, '57 and '59 (J.H.D.). It was also reported from Mablethorpe, from 1957 to 1965, (T. R. New).
L.F.P. Cow Parsnip (internal feeder).

568. *Antitype flavicincta Schiff.* **Large Ranuncule (Large Ranunculus)**
A species which was considered particularly common at the time of Lep. Lincs. 2 – and which was recorded from a large number of the county divisions. It can be found in a number of types of habitat in September. It was at Alford, September 1946, (F. L. Kirk); Woodhall Spa, 20th September 1965, (A.D.T.); Boston, 20th September 1970, (A.E.S.); and South Thoresby, 1970, (R.E.M.P.), etc.
L.F.P. Dock, Groundsel, low plants.

569. *Antitype chi L.* **Grey Chi**
There are very few records for this species this century. It was "scarce" at the time of Lep. Lincs. 2 and found only in the East Ferry district (E. Reynolds) and at Risby Warren, 21st September 1907, (G.W.M.).
The only subsequent reliable records read "Lincolnshire, 1921, H. C. Bee, L.N.U. Trans. Vol.5, p.154), and Scunthorpe area, September 1948, (J.H.D.).
L.F.P. Dock, Dandelion etc.

571. *Eupsilia transversa Hufn.* **Satellite**
(satellitia L.)
A woodland species which flies in the autumn. It was quite commonly found at sugar patches in the 1950s. It was frequently recorded at Laughton, e.g. 9th October 1955, (M.P.G. and J.H.D.). It was at Holme Plantation, 27th October 1954, (G.A.T.J.); Skellingthorpe, 5th October 1957, (L.N.U.); and Barton on Humber, 1958, (C. G. Else) etc.
L.F.P. Oak, Beech, Birch and other forest trees.

572. *Jodia croceago* Schiff. **Orange Upperwing**
A species which flies in September and October in Oak woodlands, and hibernating specimens can become active the following March. South noted "The species does not appear to have been noticed in the eastern or northern counties of England". There is one L.N.U. record reading: "Mr. H. C. Bee records hibernating examples at Sallow bloom on 28th March 1918. (G. W. Mason)".

L.F.P. Oak.

574. *Omphaloscelis lunosa* Haw. **Lunar Underwing**
Quite frequently discovered in the autumn in sandy areas – and partial to light and sugar.
Records include: Barton, 1950, (M.P.G.); Humberstone, 18th September 1957, "plentiful in m.v. trap", (G.A.T.J.); South Ferriby, 11th September 1957, (J.H.D.); Gibraltar Point 1972, (A.D.T.); and Whitton, 24th September 1976, (R.J.).

L.F.P. Meadow Grass etc.

575. *Agrochola lota* Clerck. **Red-line Quaker**
A species seen in damp places in September and October, where the foodplant occurs. It has been recorded from Martin, Heckington, Scotton Common, Epworth Turbary, Sleaford, Boston and Burton Gravel Pits etc.

L.F.P. Sallow and Willow.

576. *Agrochola macilenta* Hubn. **Yellow-line Quaker**
South wrote that this species is "local or scarce in the Midlands. Barrett said it is abundant in Lincolnshire, but I have records from few localities". The most recent sightings have been made at Skellingthorpe, Roughton Moor, Roxton Wood, Messingham and Laughton. Numbers recorded at this latter locality peaked dramatically in the 1950s. The population increased inexplicably, to enable up to 75 specimens to be counted at one sugar patch, and numbers beyond counting to be met with during the course of one evening. During this population explosion, examples of many different forms of the species were observed. At the present time (1979) it is rarely met with.

L.F.P. Heather, Oak Poplar etc.

577. *Agrochola circellaris* Hufn. **Brick**
 (*ferruginea* Esp.)
A common moth in September and October, in all wooded localities – partial to sugar. It has been found found at: Bottesford, Holme, Laughton, Pelham's Pillar Woods, Roxton Wood, Morkery Wood (near Castle Bytham), etc.

L.F.P. Ash, Elm, Willow.

THE BUTTERFLIES AND LARGER MOTHS

578. *Agrochola lychnidis Schiff.*   **Beaded Chestnut**
(pistacina F.)
A very plentiful species, flying between September and November, and exhibiting a number of forms. Places mentioned in L.N.U. records include: Grimsby, Holme Plantation, Linwood, Limber, Sleaford, Epworth, Wilsford, Boston and Scawby etc.
L.F.P. Buttercup, Dock, Dandelion.

579. *Anchoscelis helvola L.*   **Flounced Chestnut**
(rufina L.)
This is a woodland species, flying in the autumn, (Sept/Oct). Lep. Lincs. 2 recorded that it was "not common except in one or two localities" – these being Hartsholme and Skellingthorpe. It was found in the Limber Woods on 28th September 1959, (G.A.T.J.); Laughton, 13th September 1966, and Linwood, 1961 to 1963, "a very good period for this species", (both J.H.D.); and Holbeck, 18th September 1975, (G.A.T.J.).
L.F.P. Oak, Elm and other trees.

580. *Anchoscelis litura L.*   **Brown-spot Chestnut**
(Brown-spot Pinion)
Quite a common species in the autumn, particularly liking sugar and light. It was recorded at: Sleaford, Laughton, Humberstone, Ruskington, Mablethorpe, Horncastle, Boston, Gibraltar Point etc.
L.F.P. Bramble, Rose, Sallow etc.

581. *Atethmia xerampelina Esp.*   **Centre-barred Sallow**
Found quite commonly in August and September where Ash is plentiful. Lep. Lincs. 2 listed a considerable number of localities for the species, and commented as to its status that it was "Frequent, probably common, now that the habits of the larvae are more widely known". More recent records include: Candlesby, 10th September 1955, (Bruce James); Grimsby and Laughton, 1955 and 1958 respectively, (G.A.T.J.); and Goxhill, 4th September 1973, (C. Potts).
L.F.P. Ash.

582. *Tiliacea citrago L.*   **Orange Sallow**
A local moth, found where limes occur, in August and September. This century larvae were found at Skellingthorpe, August 1958, (J.H.D.). The moth was recorded at Scotgrove Wood, 1971, (A.D.T), and at Twigmoor, "3 larvae found", 19th May 1977, (J.H.D.).
L.F.P. Lime.

583. *Tiliacea aurago* Schiff. **Barred Sallow**
Very rare in the County – confined mainly to Beechwoods in chalk/limestone areas where it flies from September into October. There are only two reliable records for: Ruskington, September 1956, "one only", (J. Hossack); and South Thoresby, 1970, (R.E.M.P.).

L.F.P. Beech, Maple, Sycamore.

584. *Citria lutea* Stroem **Pink-barred Sallow**
 (flavago F.)
A fairly common species in the Autumn, particularly where sallows thrive in damp places.
Records for the post-war period include: Skellingthorpe, September 1948, (H.M.S.); Sleaford, 29th September 1957, (G.N.H.); Epworth Turbary, 15th September 1962, and Woodhall Spa, 20th September 1965, (both A.D.T.), etc.

L.F.P. Sallow catkins in the early stages, and later on a variety of low plants.

585. *Cirrhia icteritia* Hufn. **Common Sallow**
 (fulvago L.)
Plentiful where Sallow grows, flying from August through to October. Localities mentioned in the records include: Susworth, Ruskington, Boston, Humberstone, Barton on Humber, Sleaford, Laughton, Linwood, Epworth, Woodhall, Lincoln, etc.

L.F.P. Sallow – and in later stages low plants.

586. *Cirrhia gilvago* Schiff. **Dusky-lemon Sallow**
In the main this species is found in the Southern Counties of England, but South wrote (1962) "The earliest known British specimens of the typical form were captured in the neighbourhood of Doncaster about 1842-3", and at the time of Lep. Lincs. 2, it was well known in Lincolnshire. It was recorded from: Near Appleby, "larvae from elms", 20th May 1882, (Mrs. Cross); Binbrook, 1906, "in fair numbers", (S.B.S); Bourne in Lincolnshire, (Stainton's Manual, Vol. 1, p.253); Haverholme Priory, "very common", (J.D.C.). More recently it has been seen at Grimsby, Laughton, Scawby, (bred from seeds on Wych Elms) and Sleaford, etc.

L.F.P. Wych Elm, Common Elm.

590. *Conistra vaccinii* L. **Common Chestnut**
A common and very variable moth, found in wooded areas in the autumn and early winter – and often seen at sugar. It also appears, after hibernation with the sallow blossoms in the spring. Typically recorded at: Holme Plantation, Laughton, Linwood, Epworth, Roughton Moor, Woodhall Spa, Bardney etc.
L.F.P. Elm, Oak and other forest trees, followed by low growing plants when larvae are full-sized.

591. *Conistra ligula* Esp. **Dark Chestnut**
(Spadicea Staint.)
Considered to be common at the time of Lep. Lincs. 2, but today it appears less common than the previous species. The moth flies in October and November, and according to South "it pairs in the Autumn, and lays its eggs at the end of December or early January".
It was at Boston, 20th October 1956, (C.G.E.); Sleaford, 27th October 1965, (G.N.H.); Boston, "to light", 12th December 1967, (A.E.S.); Horncastle, 25th December 1972, (A.D.T.).
L.F.P. Hawthorn, Oak, Sallow, and later on low plants

HYLOPHILIDAE

WESTERMANNIINAE

592. *Bena prasinana* L. **Green Silver-lines**
(fagana F.)
Lep. Lincs. 1 had this species as "common", but perhaps today it is a little less frequently found. It is met with in small numbers in wooded localities in June and July. The more recent records are for: Scawby, 23rd May 1955; Laughton, May 1956; Holme Plantation, 1974; Twigmoor, 29th June 1974 (all J.H.D. and R.J.). It was also at Kirkby Moor, June 1968 (A.E.S.); and Roxton Wood, near Keelby, 21st-28th May 1967, (Miss Hillary Lubin). etc. etc.
L.F.P. Oak, Beech, Birch.

593. *Pseudoips bicolorana* Fuessl. **Scarce Silver-lines**
(quercana Schiff.)
Lep. Lincs 1 noted this species to be rare. It is now found in most of the Oak woods in small numbers. It was seen at Spanby, 5th July 1941, (L.N.U. – F.L.K.); Manton Common, "larvae very plentiful May 1953", (J.H.D.); Woodhall Spa, "2 to light, late

arrivals – approximately 12.30 a.m.", 17th July 1974, (G.A.T.J. – G.N.H.); Kirkby Moor, 19th – 28th May 1968, "larvae fairly common on Oak", (A.E.S.); Skellingthorpe, July 1973 and 1974 (R.J.); Norton Disney, 4th October 1975, "3 larvae" (M.P.G.), etc.

L.F.P. Mainly Oak.

594. *Earias clorana* L. **Cream-bordered Green (Cream-bordered Green Pea)**
A local moth, found in damp places around Osiers and Sallows in June – with only a few L.N.U. records. It was at Scunthorpe, June 1953, (J.H.D.); Aswarby Thorns, 7th June 1958, (G.A.T.J.); Laughton, 19th June 1959, (Mr. G. Hyde); Boston, 10th June 1973, (A.E.S.); and South Thoresby, 1978, (R.E.M.P.).
L.F.P. Willows and Sallows.

NYCTEOLINAE

595. *Nycteola revayana* Scop. **Large Marbled Tort.**
(undulana Hubn.)
An oak feeder, South (1961) wrote "This is a most variable species . . . The moth seems to be out from August to April . . . in the later months of the year it seems to hide in Yews and Hollies . . It may be seen regaling itself on overripe blackberries, or on the ivy blossom, and it is not an infrequent visitor to the sugar patch".
The entry in the L.N.U. records reads "Rare, N. Ashby, (Brigg) District, R.T.C. S. Haverholme Priory, generally common, J.D.C.". Lep. Lincs. 1. 1952, Manton 13th July, (2 larvae, 1 bred) (J.H.D.). 1978, Bardney Forest, (R.E.M.P.).

L.F.P. Oak

PLUSIIDAE

EUSTROTIINAE

603. *Lithacodia fasciana* L. **White-spot Marbled (Marbled White-spot)**
A local moth, occasionally seen in June and July at rest on tree trunks (Larch and Pine). Lep. Lincs. 2 reported that it was local early this century, with early records placing the species at Linwood and Market Rasen. It still occurs in these two localities as instanced by: Market Rasen, 7th July 1951, (C. G. Else); Linwood, 16th June 1958, (G.A.T.J.). It was also found at Troy Wood, Tumby, near Louth, 17th June 1961, (L.N.U.); and Bardney, 6th July 1974, (J.H.D.). It was last recorded at Linwood on 4th June 1981 (J.H.D.).

L.F.P. Purple Moor Grass.

606. *Eustrotia uncula Clerck* **Silver Hook**
(uncana L.)
Like the previous species, also fond of marshy areas, and seen in June and July. Following the drainage of many inland areas in the county, it has seemed to become more scarce in the recent part of this century. In the 1950s it appeared quite commonly in colonies in damp areas such as Manton Common, Epworth, Scotton, Linwood etc. The most recent records have been from Gibraltar Point, 1969, Crowle and Manton, 1970, (all R.E.M.P.); Scawby Woods, 6th June 1976, (R.J.).

L.F.P. Sedges.

CATOCALINAE

608. *Catocala fraxini L.* **Clifden Nonpareil**
A migrant species which flies in the autumn (September) and it has been known to breed in this country during certain periods (e.g. in Kent and Norfolk). There are L.N.U. records for: Hogsthorpe, September 1875, (R. Garfit, in Nat. 1887, p.69, and Nat. 1890, p.150); Great Coates, 2nd September 1976, "found in grasses", (G.A.T.J.); Gibraltar Point, 1976, (R.E.M.P.) and again in 1976, (A.D.T.).

L.F.P. Poplar, Aspen.

610. *Catocala nupta L.* **Red Underwing**
Occurs in areas where poplars grow, and flies in August and September. It was thought to be local at the time of Lep. Lincs. 2, with only two recordings from Tattershall and Haverholme.

From 1930 to the early 1940s it was recorded annually from mainly the south of the county in places like Tallington near Stamford, Tattershall, Bucknall, Washingborough, Holywell, Threekingham (South of Sleaford), Boston, Swineshead, etc. During the 1940s it extended its range to reach the Humber – turning up at Barton on Humber, Grimsby, Humberstone, etc. While it has remained fairly widespread in the South of the County, it seems to have become less frequent in the Humberside area, some years not being seen at all. It was recorded at Lea, Gainsborough, 9th September 1977 (F. Brasier). At the time of writing (1979) a number of sightings are being made. Two were found in the Berkeley area of Scunthorpe at the end of August, (R.J.); one was seen in Brumby in early October (J.H.D.) and two came to sugar at Laughton, on 1.10.79 (R.J. and A.J.).

L.F.P. Poplar and Willow.

613A. *Clytie illunaris* Hubn. **Trent Double-stripe**
In June 1964 a greenish noctuid larva was found by Mr. D. S. Brown feeding at night on the foliage of Horse Radish beside the River Trent at Amcotts, Nr. Scunthorpe, S. Humberside. It pupated, the pupa being dark brown and wrinkled rather resembling that of the spectacle moth *(Unca triplasia L.)*. A female moth emerged on the 25th August 1964 and was fawn, marked with darker brown having a wing span of one and six-tenths of an inch. It was identified by Mr. P. R. Seymour of the British Museum as *Clytie illunaris* Hubn., an inhabitant of Southern Europe and Western Asia. It was recorded at Rothampstead by Mr. R. A. French. This moth was new to Britain and was named the Trent Double Stripe by I. R. P. Heslop.

L.F.P. Wild Horse Radish?

615. *Euclidimera mi* Clerck **Mother Shipton**
A species which flies commonly in the sunshine in grassy places, in May and June. It has been recorded from: Stainton Wood, Little Ponton, the Valley of Ancaster, Grimsthorpe Park, Willingham Forest, Aswarby Thorns, Manton, Risby Warren, Gibraltar Point etc.

L.F.P. Clovers.

616. *Ectypa glyphica L.* **Burnet Companion**
Not considered to be too common early this century, and the same may be said today. The moth flies over meadows and in woodland clearings in June and July. It was at Holywell, 17th July 1955, (M.P.G.); Aswarby Thorns, 7th June 1958, (L.N.U.); Near Creeton Lime Quarry, 8th June 1963, (G.N.H.); Scotgrove and Austacre Wood, June 1968, (A.D.T.); Bardney, 6th July 1974, (J.H.D.).

L.F.P. Clover, Trefoils, Black Medick.

PANTHEINAE

617. *Colocasia coryli L.* **Nut-tree Tuffet**
A species which flies in May and June, and again in the autumn, but mainly in the Southern counties of Berks, Bucks, and Devon.

It was described as "rare" at the time of Lep. Lincs. 2, but was claimed to be near Gainsborough (F.M.B.). Also, it was at Newball, with "one or two larvae, about September 1894, taken by the late Rev. C. Wilkinson, and identified by G.H.R. 6.6.1886 (J.F.M.)."

There is one further unconfirmed mention for Mablethorpe in the records for the period 1952-65 (T. R. New).

L.F.P. Beech, Birch, Hazel.

619. *Episema caeruleocephala* L. **Figure of Eight**
A common species, frequently found in the larval form in May and June along hedgerows. The moth is out in October and November.
It has been widely recorded from such sites as Legsby, 1954, "unusually common this autumn", (R. Cornwallis); Bottesford Moor, 20th September 1959, (J.H.D.); near Anwick, September 1964, (G.N.H.); Brigg, 17th November 1964, (A. H. Neale); and Brattenby, 6th June 1969, "larvae", (A.D.T.), etc.
L.F.P. Hawthorn, Sloe etc.

PLUSIINAE

621. *Polychrisia moneta* F. **Silver Eight**
(Golden Plusia)
According to South, "The British history of the species dates back to only 1890. In that year on July 2nd, the late Mr. Christy, of Watergate, Emsworth, found a specimen in his illuminated moth trap".
The first L.N.U. record dates from 14th July 1914, Louth, when a specimen was taken on Delphinium (J. Larder). It was next at N. Somercotes 1918, "occuring more or less commonly". It is now common in gardens all over Lincolnshire. More recently it was at Grimsby, 11th July 1961, (G.A.T.J.); Saltfleet, 16th July 1963, (G.N.H.); Boston, 28th July 1970, (A.E.S.); Scunthorpe, 20th June 1973, (J.H.D.); and throughout the 1970s, larvae at Scawby have destroyed the flowers on several Delphinium beds. (R.J.).
L.F.P. Delphinium and Monkshood.

623. *Plusia chrysitis* L. **Common Burnished Brass**
A very common species in the area, double brooded in good years, with the moth out between June and September. It is found along hedgerows and verges, and in gardens, where it is partial to visiting flowers.
It was found typically at such localities as Hendale Wood, Lincoln, Belton Park, Humberstone, Easton Hall near Grantham, Grimsby, Brocklesby, Keddington, Scunthorpe, Saltfleet, Woodhall, Limber, Boston, etc.
L.F.P. Stinging Nettle and other low plants.

626. *Plusia bractea* Schiff. **Gold Spangle**
Mainly a northern species in the British Isles, and not often met with in the county. It is out in June and July, and visits flowers.
In Lep. Lincs. 2. only one specimen was known, taken in the

garden of Mr. Reynolds at Owston Ferry. More recently it was at S. Thoresby, "one specimen 1969", (R.E.M.P.), and Lincoln, 1973, (M.C.T.).

L.F.P. Dandelion, Dead Nettle, low plants.

627. *Plusia festucae* L. **Common Gold Spot**
This moth is out in June and July (and sometimes in August/September in a second generation), in damp areas where larvae are found on various sedges, and inside Yellow-flag. Early this century it was considered scarce. Now it is found widely in suitable habitats. It was recorded at Somercotes in 1914 and 1918 (Rev. S. Proudfoot and Canon G. Houlden respectively). It was found quite plentifully at Barton on Humber in the 1950s (M.P.G.), and at Scunthorpe and Alkborough, 1976, when it was "common", (J.H.D.) etc.

L.F.P. Sedges, Reeds, Yellow-flag.

627A. *Plusia gracilis* Lempke **Lempke's Gold-spot**
There is a single entry in L.N.U. records for this species, claiming that it turned up at Mablethorpe, 1964, as "one specimen, only recently separated from *P. festucae*". (T R. New).

L.F.P Reeds and Sedges etc.

630. *Plusia jota* L. **Plain Golden Y**
Very common in Lincolnshire from the end of June into July. Recorded from Skellingthorpe, Barton on Humber, Grimsby, Swallow Vale, Brocklesby, Scawby, Woodhall, Denton Reservoir, Boston – and many other localities.

L.F.P. Honeysuckle, Hawthorn, Nettle etc.

631. *Plusia pulchrina* Haw. **Beautiful Golden Y.**
This species was described as common in Lep. Lincs. 2, but it only appears to have been found in one's or two's earlier this century. It seemed to remain fairly scarce up to the start of the 1970s. Over the last few years, however, the numbers seen have increased to equal, and occasionally exceed those of the Plain Golden Y, when they are flying together in June and July.
It was at Grimsby, 1953, "2 to light", (G.A.T.J.); Linwood, 16th June 1955, "one to light", (G.A.T.J.); Scawby, 1st July 1967, (J.H.D.); Twyford Forest, 21st June 1969, (R.G.); Hendale Wood, 20th June 1975, (R.J.), etc.

L.F.P. Nettles, Groundsel etc.

632. *Plusia ni Hubn.* **Silver V**
(brassicae Ril.) **(ni moth)**
A very rare migrant to the British Isles. There is a record for Boston, 20th August 1965, "in my garden at mercury vapour lamp", (R.E.M.P.). In September 1982 a second specimen was found by Mr. Pilcher. The species was also seen at M.V. light at Epworth Turbary on September 25th, 1982 (R.J.).
L.F.P Cruciferae.

635. *Plusia gamma L.* **Common Silver Y**
A really abundant species some years – and very widely noted in the county. It is a migrant species, often seen in the early summer, with a second generation hatching in early autumn. There is an interesting L.N.U. record reading "Aug/Sept 1936, Theddlethorpe, Mablethorpe. On Aug 23rd and 24th I spent some time watching groups of these moths coming in off the sea, they were flying in small groups, and oddly close to the sand and against a fresh breeze from off the land. Later enormous numbers were seen all along the coast by several observers, continuing into September" (F. T. Baker, L.N.U. Trans. Vol.9, p.119).
L.F.P. Most low plants.

636. *Plusia interrogationis L.* **Scarce Silver Y**
Lep. Lincs. 2 claimed this species was found "occasionally on the hills in Lincolnshire: Near Louth, one example V. Crow, and Grantham, 23.7.1873 – Isaac Robinson". The only other records are for: Mablethorpe, 1952-65 rare, (T. N. New); S. Thoresby, 16th August, 1977, "Scandinavian Form", (R.E.M.P.).
L.F.P. Heather and Bilberry.

638. *Unca triplasia L.* **Dark Spectacle**
A rather local species, mainly turning up as single recorded specimens in gardens, visiting flowers at dusk.
It was in the Denton district in 1924 (L. H. Bond); Barton on Humber, 2nd July 1957, (C.G.E.); Boston, late June 1967, (J.B.); Scunthorpe, 17th June 1973, (J.H.D.), etc.
L.F.P. Nettle and Hops.

639. *Unca tripartitia Hufn.* **Light Spectacle**
(urticae Hubn.)
More plentiful than the previous species –comes to light as well as to garden flowers from May through to August in a protracted emergence. Found at Hendale Wood, Bottesford, Ruskington, Sleaford, Saltfleetby, Brocklesby, Scunthorpe, Limber, Boston, etc.
L.F.P. Stinging Nettle.

OPHIDERINAE

641. *Acontia luctuosa* Schiff. **Four-spot**
Flies in clover-fields, on chalky slopes etc. in the sunshine in May/June, and again in August/September.
There is a single county record for Tallington, near Stamford, 1934, "a specimen taken at light", (Miss Stow, checked by A.E.M.).
L.F.P. Small Bindweed.

644. *Lygephila pastinum* Treits. **Plain Blackneck**
(The Blackneck)
A species considered "local" early this century – and the same obtains today. At the time of Lep. Lincs. 2 it was at Alford, Well Vale, 14th June 1889), (E.W.); and Pelham's Pillar Wood, "a few each year," (T.W.M.).
It was found at Old Mere Road near Saltby, 8th July 1967, (J.B. and G.N.H.), and was at Crowle and Manton 1970, and Kirkby Moor, Goulceby, Marton, Crowle and Anton's Gowt (R.E.M.P.) etc.
L.F.P. Vetch.

648. *Rivula sericealis* Scop. **Straw Point**
(Straw Dot)
Found in damp marshy areas – mainly in July. Lep. Lincs. 2 called it "rare" at Haverholme Priory, with "4 on wing in garden" (J.D.C.). It was at Lincoln in 1901, (Rev. W. Beecher).
There is another record for Woodhall Spa – "to light, 17th/18th July 1965" (J.B., G.N.H. and G.A.T.J.), and a number were recorded at South Thoresby, 1979, (R.E.M.P.).
L.F.P. Grasses.

649. *Phytometra viridaria* Clerck. **Small Purple Bars**
(aenea Hubn.*)* **(Small Purple Barred)**
A day flying moth – over heathland in sunshine between June and early August. It was thought "local" at the time of Lep. Lincs. 2 at Gainsborough, Linwood, Market Rasen and Pelham's Pillar Woods. It was later recorded from Lynwode Warren and woods, 5th June 1911, (J.W.M., L.N.U.); Holywell, 3rd June 1915, (L.N.U.); Manton Common, 17th August 1957, and Scotton Common, 30th July 1973, (Both J.H.D.).
L.F.P. Milkwort, Lousewort.

GONOPTERINAE

651. *Scoliopteryx libatrix* L. **Herald**
A common moth. It flies in the late autumn, hibernates (often found in numbers in roofs, forestry huts, belfries etc.) and reappears in the spring, to be seen at sallow blossom.
It is very widely recorded from such localities as Lincoln, Laughton Common, Humberstone, Sleaford, Barton on Humber, Scotton Common, Great Hale, New Park Wood – Bardney, Mablethorpe etc.
L.F.P. Sallows and Willows.

HYPENINAE

652. *Bomolocha fontis* Thunb. **Beautiful Snout**
(crassalis Treits.*)*
A local heathland species, flying in June and July. There is a single L.N.U. record for Skellingthorpe, 1950, (T. C. Taylor).
L.F.P. Bilberry.

653. *Hypena proboscidalis* L. **Common Snout**
Common and plentiful in June, July and August, where nettles grow in quantity. It has been found at Grimsthorpe Park, Easton Wood (Near Grantham – L.N.U. meeting), Saltfleetby, Swallow Vale, Twigmoor, Linwood, Syston Park etc.
L.F.P. Stinging Nettle.

656. *Hypena rostralis* L. **Buttoned Snout**
Occurs in the south of England in August and September (where hops are grown), and reappears in the spring after hibernation.
The only L.N.U. records stem from the time of Lep. Lincs. 2 and read: "local. N. West Ashby, 1901, F.S.A; S. Allington, one 31.3.1904, disturbed when pruning a plum tree, P. Wynne; Haverholme Priory, very common, J.D.C."
L.F.P. Hops.

658. *Shrankia costaestrigalis* Steph. **Pionion-streaked Snout**
According to South, "this species is partial to moist localities, and its favourite haunts are fens, mosses, or marshy heaths, and the outskirts of damp woods". It flies between June and October.
The only L.N.U. records are from R. E. M. Pilcher, who took the species at Moor Farm, near Woodhall, and Bardney Forest in 1973.
L.F.P. Uncertain in the wild-
known to eat Thyme flowers in
captivity.

661. *Zanclognatha tarsipennalis* Treits. **Brown Fanfoot**
(The Fanfoot)
Quite frequently found in lanes and along hedgerows in June and July – but it is not so plentiful in numbers as the following species.
Records include: Grimsby, 5th July 1958, (G.A.T.J.); Amcotts, July 1962, (J.H.D.); Sleaford, 17th August 1965, (G.N.H.); and Austacre Wood, 1971, (A.D.T.), etc.

L.F.P. Ivy, Knotgrass.

662. *Zanclognatha nemoralis* F. **Small Fanfoot**
(grisealis Schiff.)
Lep. Lincs. 2 described the species as "frequent and generally common". It can still be disturbed in wooded localities in reasonable numbers today, in the months of June and July.
It has been recorded at: Hendale Wood, 16th July 1951, (G.A.T.J.); Welton le Marsh, 7th June 1952, (L.N.U.); Irby Dales, 25th June 1955, (L.N.U.); Brocklesby, 24th June 1959 and Woodhall Spa, 17th July 1964, (Both G.A.T.J.); and Riseholme, 25th June 1968, (A.D.T.), etc.

L.F.P. Wild Raspberry and Oak.

663. *Zanclognatha cribrumalis* Hubn. **Dotted Fanfoot**
(cribralis Hubn.)
Found in fens and marshy areas in June and July – and only mentioned once in the records for "Skegness, 16th July 1879, (G.T.P.)".

L.F.P. Marsh Rushes and Grasses.

665. *Herminia barbalis* Clerck **Common Fanfoot**
A local woodland species, flying in June and July.
Early this century it was recorded from such woodlands as Linwood, Legsby and Newball, Haverholme Priory and Skellingthorpe.
There are only three records for more recently: Grimsby, 9th July 1957, and Woodhall Spa, 16th July 1965, (Both G.A.T.J.); and Tumby Wood near Louth, 17th June 1961, (L.N.U. meeting).

L.F.P. Dead leaves of Oak and Birch.

666. *Laspeyria flexula* Schiff. **Beautiful Hook-Wing**
The species flies in June, July and August, with the larvae feeding on lichens growing on various trees.
South wrote that it is more widely distributed over the southern counties, and in the east to Norfolk. According to R.E.M. Pilcher, it can be "frequent" along the east coast of the county, where larvae feed on lichens growing on *H. rhamnoides*. Other

records are from: Manton, 28th July 1955, (Bruce James.); Linwood, 25th June 1959, (G.A.T.J.); Woodhall Spa, 17th July 1964, (G.N.H. and G.A.T.J.); New Park Wood, Bardney, 11th August 1979, (J.H.D.); and Linwood again, 4th June 1981, (J.H.D.).

L.F.P. Star Lichen, Wall Lichens.

Super-family GEOMETROIDEA

GEOMETRIDAE

ARCHIEARINAE

667. *Archiearis parthenias* L. **Common Orange-underwing**
Flies in the daytime on sunny days in March and April, around birch trees on heathland.
L.N.U. records include: Raventhorpe, 1950, (M.P.G.); Holme, 6th April 1950, (C.G.E.); Epworth, 7th April 1952, (M.P.G.); Scotton Common, April 1956, (J.H.D.); Grasby Bottom, 13th April 1958 (G.A.T.J.); Laughton Common, 16th April 1976, (J.H.D.); and Kirkby Moor, 2nd April 1978, "very common locally", (A.E.S.)

L.F.P. Birch.

668. *Archiearis notha* Hubn.
Light Orange-underwing
A species which flies in the sunshine among aspens in the spring. It was recorded in this manner in Linwood, 17th April 1911, (by F.P.H.B., F.W.S., and G.W.M.). It was rediscovered in Kesteven Forest, where it appeared quite common, by Mr. A. E. Smith on 26th and 27th March 1981.

L.F.P. Aspen.

OENOCHROMINAE

669. *Alsophila aescularia* Schiff. **March Usher**
(The March Moth)
Found commonly in woodland localities in the spring. Places mentioned in L.N.U. records include: Grimsby, Knaith, Culverthorpe near Ancaster, Linwood, Roughton Moor, Bardney, Wilsford Heath, Boston etc.

L.F.P. Many deciduous trees and shrubs.

GEOMETRINAE

671. *Pseudoterpna pruinata* Hufn. **Greater Grass Emerald**
(cytisaria Schiff.)
Frequently seen on heath and moorland in June and July. For example, there are a number of records by J.H.D. for Greetwell 1953; Scotton Common, 19th July 1959; Epworth and Messingham 1961; Holme Plantation, 25th July 1966; Scotton, July 1969 and 1977.
L.F.P. Furze, Common Broom.

672. *Geometra papilionaria* L. **Large Emerald**
Common and well distributed on Birch covered heathland in July and early August. There are numerous L.N.U. records for such localities as: Crosby Woods, Twigmoor, Woodhall Spa, Epworth, Skellingthorpe, Laughton, Holme, Wyberton etc.
L.F.P. Birch, Beech, Hazel etc.

673. *Comibaena pustulata* Hufn. **Blotched Emerald**
(bajularia Schiff.)
A species of oak woodlands – flying in June and July. It is not a common moth in the county, but it can be found in reasonable numbers in locally restricted areas.
It was at Skellingthorpe, 1914, (Rev. Blathwayt); Scawby Woods, 1955, (J.H.D.); Linwood, 1960, (A.D.T. and G.A.T.J.); Moor Farm and Kirkby Moor, 1974, (M.T.), etc.
L.F.P. Oak.

674. *Hemithea aestivaria* Hubn. **Common Emerald**
(strigata Mull.)
Flies in June and July along hedgerows and in woodlands. It was considered locally common early this century, but today it appears to be even more restricted, and only found in small numbers.
Records from the 1950s onwards include: Scunthorpe, 1951, (J.H.D.); Grimsby, 24th July 1954, (G.A.T.J.); Sleaford, 1957, "to light", (G.N.H.); Boston, July 1967, (J.B.); Boston, 2nd July 1968, (A.E.S.); Ermine Street, Twigmoor, 2nd July and 5th July 1973, (R.J.) etc.
L.F.P. Mugwort, Docks and other low plants.

675. *Chlorissa viridata* L. **Small Grass Emerald**
A species found in open woodland in May and June. The first L.N.U. record was by M.P.G. who caught a female at Skellingthorpe in June 1955. Ova were laid, and the larvae fed on Hawthorn.
Two moths emerged in 1979, from larvae discovered at Woodhall (in 1978) by R.E.M.P.
L.F.P. Heather, Furze and Bramble.

679. *Hemistola immaculata* Thunb. **Lesser Emerald**
(chrysoprasaria Esp.*)* **(Small Emerald)**
This is most frequently found in the southern areas of England, where it flies over chalk lands in June, July and August. There are three L.N.U. records for the county, for: Market Rasen, 7th July 1951, and Barton on Humber, 27th July 1951, (Both C.G.E.); and Tumby, Troy Wood, 17th June 1961, (L.N.U.).
L.F.P. Traveller's Joy.

680. *Jodis lactearia* L. **Little Emerald**
Locally, this species can be common in woodland habitats in May and June. It can be occasionally seen in numbers. It was at: Skellingthorpe, 26th May 1957, (J.H.D.); Linwood, 6th June 1958, "to light", (G.A.T.J.); Tumby Woods, 17th June 1961, (L.N.U. meeting); Newball Wood, June 1969, and Burton Gravel Pits, 1971, (Both A.D.T.); and Holme Plantation, 15th June 1976, (J.H.D.).
L.F.P. Birch, Oak, Blackthorn, Hawthorn, etc.,

STERRHINAE

681. *Calothysanis amata* L. **Large Blood-vein**
(The Blood-vein)
Another species which likes damp areas. The moth is on the wing in May and June, with an occasional second generation in August and September.
It was found at Grainsby Park, 24.7.57, (G.A.T.J.); Stubton Hall, June 1965, (L.N.U.); Scunthorpe, Laughton and Manton, "common from May 1960", (J.H.D.); Riseholme Wood, 18th August 1967, (A.D.T.); and Normanby le Wold, June 1976, (J.H.D.).
L.F.P. Dock, Sorrel, Knotgrass etc.

682. *Cosymbia albipunctata Hufn.* **Birch Mocha**
(pendularia auct.)
"A local species" (Lep. Lincs. 3) in Birch wooded areas. It can be found in May/June and again in July/August. It was "common" at Market Rasen early this century, and it has subsequently been recorded at Linwood on numerous occasions up to the 1960s (records from J.H.D., G.A.T.J., M.P.G., A.D.T., etc.) Other localities in the records include: Newball, Doddington, Skellingthorpe, Hendale Wood, and Stainton Plantation.
It was last recorded at Linwood on 4th June 1981 (J.H.D.).
L.F.P. Birch.

684. *Cosymbia annulata Schulze* **Maple Mocha**
(omicronaria Schiff.) **(The Mocha)**
A moth found in May/June, and August/September, along lanes or in woodland where Maple is plentiful. It is very rare in Lincolnshire, with a single entry in the records reading: "Rare, 1840-50, N. Gainsborough; Lincoln, (F.M.B.); S. Wyberton, rare, (J.C.L-C.)" (Lep. Lincs. 2).
There are indications in the records that the species was found in four of the Natural History Divisions in Lincs. at this time.
L.F.P. Maple.

687. *Cosymbia punctaria L.* **Maiden's Blush**
A local species, with a number of records for the period up to Lep. Lincs. 3, but with only a few mentions in more recent years.
It is flying in late May and June, with possibly a second generation in August and September.
Early recordings were made at: Legsby, Newball, Panton, Maltby Wood, Wrawby Moor and Hartsholme etc. More recently it turned up at Scawby, 1956, (J.H.D.); Knaith Park, 25th May 1957, (L.N.U.); Kirkby, near Woodhall Spa, 14th June 1962, (J.B.); and Kirkby Moor, 1972, (A.D.T.).
L.F.P. Oak and Birch.

688. *Cosymbia linearia Hubn.* **Clay Triple-lines**
(trilinearia Borkh.)
Found locally in Beech woods (May/June and August/September), but as noted in Lep. Lincs 3, "it can be common where it occurs".
Early this century it was located at Brocklesby, Limber and Pelham's Pillar Woods. The only subsequent reliable records are all for Hendale Wood; 1956 and 1957, (bred 20th May J.H.D.) and 29th June 1973, (R.J.).
L.F.P. Beech

689. *Scopula ternata* Schrank **Smoky Wave**
(fumata Steph.)
A species found mainly on moors, and damp areas, in June and July. It was reported by R.E.M.P. (in 1979) as having occurred in the Bardney and Woodhall areas earlier this century, though there are no confirmed L.N.U. records.
L.F.P. Sallow and low plants.

692. *Scopula promutata* Guen. **Mullein Wave**
(marginepunctata Auct.)
Essentially this is a coastal species – or found inland on sandy areas, where it flies in June or July.
L.N.U. records are very sparse, reading: "only one record, 1840-50, N. Lincoln F.M.B.". (Lep. Lincs. 3); and "Grainsby Park, 24th July 1951, (G. A. T. Jeffs)".
L.F.P. Low plants like Mugwort, Cinquefoil, Vetch etc.

694. *Scopula imitaria* Hubn. **Small Blood-vein Wave**
Not an uncommon species in some districts – flying mainly in July, with a few appearing in the autumn. It was at Sleaford, 24th July 1956, (G.N.H.); Grimsby, 7th July 1961, (G.A.T.J.); Boston, July 1968, "very common", (A.E.S.); Whitton, 16th July 1976, and Barrow Haven, 16th July 1977, (Both R.J. and A.J.).
L.F.P. Dock, Dandelion and low plants.

695. *Scopula emutaria* Hubn. **Rosy Wave**
A very scarce species on marshland in coastal areas, where it flies in July and early August. Its stronghold in Lincolnshire is at Gibraltar Point, where it is found most years.
In Lep. Lincs. 3 it was recorded at Sutton on Sea, 25.7.1909, (F. G. Whittle), and Cowbit, (Chas. M. Hufton).
In the 1950s, when the sea broke through to flood the Gibraltar Point marshes, numbers declined. It appeared again on 6th July 1959, however, "to light", (G.A.T.J.). It was seen again in 1969, (R.E.M.P.), and in 1970, when R. E. M. Pilcher reported that the species was "back to normal numbers".
L.F.P. Feeds on low plants in captivity. "Foodplant in the wild uncertain" (P. B. M. Allen, 1949).

698. *Scopula immutata* L.   **Lesser Cream Wave**
Found in boggy, marshy areas during June and July. In Lep. Lincs. 3 it was considered "Rare, probably often overlooked as it is very similar to *A.remutaria*". It was recorded from Sutton on Sea (F. G. Whittle) 25.7.1909. It was also at Theddlethorpe in 1904, (A. E. Gibbs – written up in Ent. 1905, p.81) etc.
Since the 1950s it has been found at Skellingthorpe 26th May 1957, (J.H.D.); and Saltfleetby, 18th July 1958, "common at m.v. lamp", (G.A.T.J.).
On 14th August 1965, it was discovered at Rimac, with "specimens quite fresh – I think the larvae feed on the Great Valerian or Meadow Sweet, growing in the fresh water marsh", (J.H.D.).
There are further records for Saltfleetby in 1965, (L.N.U.), and for the Burton Gravel Pits, 1973, (M.C.T.).
L.F.P. Wild Valerian, Meadow Sweet.

699. *Scopula lactata* Haw.   **Greater Cream Wave**
(remutaria Hubn.)
A common woodland species – flies among bracken in May and June. Typical records are for Twigmoor Woods, 5th June 1955, (G.A.T.J.); Scawby, 14th June 1955, (B.J.); Broughton Woods, June 1976, (R.J. and J.H.D.), etc.
L.F.P. Bedstraw, Dock, Sallow etc.

702. *Sterrha interjectaria* Boisd.   **Dwarf Cream Wave**
(fuscovenosa auct.)
More commonly found in the south of the country, with the species becoming more or less rare as one moves to the north. It flies in the summer. An entry in L.N.U. records refers to R.E.M.P. who has stated that the species is common around Gibraltar Point. (No dates available). It was recently found at Washingborough on 4th June 1981, by Mr. & Mrs. A. Binding.
L.F.P. Dandelion and low plants.

704. *Sterrha dilutaria* Hubn.   **Silky Wave**
(holosericata Dup.)
According to South this is a very local and uncommon species in the British Isles. The only L.N.U. records for it date from Lep. Lincs. 3, when it was claimed that it had been recorded from Legsby, Mablethorpe, Skegness, Theddlethorpe, Little Bytham etc. in the late 1800s and early 1900s.
L.F.P. Rock Rose.

707. *Sterrha dimidiata Hufn.* **Single-dotted Wave**
A fairly common and widespread species in July and August – though it is a small and insignificant moth and often possibly overlooked.
Seen at Saltfleetby, 18th July 1958, (G.A.T.J.); Sleaford, 5th August 1960, (G.N.H.); Manton Common, 3rd July 1960, and Laughton, 20th July 1974, (Both J.H.D.).
L.F.P. Burnet Saxifrage, Hedge Bedstraw etc.,

710. *Sterrha seriata Schrank* **Small Dusty Wave**
(virgularia Hubn.)
Quite common – found occasionally in gardens in June/July, and with a second generation in September. The most recent records are for: Sleaford, 2nd and 9th September 1959, (G.N.H.); Amcotts, July 1963, (J.H.D.); Syston Park, 10th July 1965, (L.N.U.); and Boston, late June 1967, (J.B.).
L.F.P. Ivy.

711. *Sterrha subsericeata Haw.* **Satin Wave**
Mainly a species which flies over heathy ground in June and July. According to R.E.M.P. it was taken at Moor Farm, (Kirkby Moor) on 6th August 1972. This is the sole L.N.U. record.
L.F.P. Dandelion and low plants.

712. *Sterrha sylvestraria Hubn.* **Ringed Wave**
(staminata Treits.) **(Dotted Border Wave)**
According to South (1961), this "moth occurs in July and August on bush sprinkled heaths, or heathy ground". There is a single known locality in Lincolnshire, near Scotton, where it has been seen between 1958 and 1978 in July by J.H.D.
L.F.P. Foodplant unknown in the wild – though it is said to take Bramble, Knotgrass and Wild Thyme in confinement.

716. *Sterrha straminata Borkh.* **Plain Wave**
(inornata Haw.)
A woodland species – on the wing in July. There is a single unconfirmed L.N.U. record for 1848, N. Gainsborough, Lincoln, F.M.B. (Lep. Lincs.3.)
L.F.P. A wide variety of low plants.

717. *Sterrha aversata* L. **Riband Wave**
Very common in the county in June and July in all kinds of habitat. It can be found commonly as the type form *(aversata)*, and with a small percentage being the darker banded form *(remutata)*. Widely recorded from such places as Grimsby, Laughton, Sleaford, Gibraltar Point, Woodhall, Syston Park, Boston and Lincoln etc.
L.F.P. Dandelion, Dock and other low plants.

719. *Sterrha biselata* Hufn. **Small Fan-footed Wave**
  (bisetata Rott.)
Common around wooded areas in July and August. Later L.N.U. records include: Manton Common, July 1957, (J.H.D.); Linwood, 17th July 1959, (G.A.T.J.); Sleaford, 10th August 1962, (G.N.H.); Saltfleetby, 14th August 1965, (L.N.U.); and New Park Wood, Bardney, July 1979. (J.H.D.).
L.F.P. Withered leaves (e.g. Dandelion and Bramble) and low plants.

720. *Sterrha emarginata* L. **Small Scallop Wave**
A species which frequents damp areas such as fens, and moist parts of woodlands. It flies mainly in July at dusk. It was considered scarce at the time of Lep. Lincs. 3. but it is now found in suitable habitats in numbers. Such localities include: Scawby, 14th July 1951, (C.G.E.); Scawby and Manton, "common July 1959", (J.H.D.); Woodhall Spa, July 1965, (G.A.T.J. and G.N.H.): Market Rasen, July 1968, (A.D.T.); and Laughton, 30th June 1973, (R.J.).
L.F.P. Bedstraws, Bindweeds, Broom.

721. *Rhodometra sacraria* L. **Vestal**
A casual migrant to the British Isles in the autumn. A specimen was recorded by R. E. M. Pilcher at Gibraltar Point in 1979.
L.F.P. Low plants like Knotgrass, Dock etc.,

LARENTIINAE

723. *Xanthorhoe quadrifasiata* Clerck
  (quadrifasciaria L.) **Large Twin-spot Carpet**
Generally this is a local species found mainly in the Southern half of England. At the time of Lep. Lincs. 3, only single specimens were recorded from such places as Lincoln, Panton, Croxby and West Ashby.

Today, however, it is found in all types of habitat, and is particularly plentiful in South Humberside. Records from the 1950s to the 1970s include Hendale, Scunthorpe, Limber, Broughton, Normanby Park, Laughton and Twigmoor, (all J.H.D.); and Sweeting Thorns, 24th June 1977, (R.J.).

L.F.P. Bramble, White Deadnettle and other low plants.

725. *Xanthorhoe ferrugata* Clerck **Dark Twin-spot Carpet** *(unidentaria Haw.)* **(Dark-barred Twin-spot Carpet)**
Very common and widespread in Lincolnshire from May to August in two generations. Typical L.N.U. records are for: Barton on Humber, 16th August 1959, (C.G.E.); Sutton on Sea, 8th August 1964, (A.D.T.); Twigmoor, 23rd May 1965, (G.A.T.J.); Boston, August 1967, (J.B.); Swallow Vale, 23rd June 1979, (R.J.).

L.F.P. Bedstraws, Groundsel, low plants etc.

726. *Xanthorhoe spadicearia* Schiff. **Red Twin-spot carpet** *(ferrugata staud. non Clerck)*
"Not so common as the previous species" (Lep. Lincs. 3 – and the same could be said today). There are two generations – May/June and July/August.
L.N.U. records include: Sleaford, 19th July 1959, (G.N.H.); Limber, 5th June 1962, (G.A.T.J. and J.H.D.); Sutton on Sea, 8th August 1964, (A.D.T.); and Saltfleetby, 14th August 1965, (L.N.U.), etc.

L.F.P. Low plants.

728. *Xanthorhoe designata* Hufn. **Flame Carpet** *(propugnata Schiff.)*
This species is more "plentiful in the southern half of England" (South 1961) in moist woodland areas between May and August. There is a single mention of it in the L.N.U. records – for the Brumby Woods near Scunthorpe. This specimen is awaiting confirmation.

L.F.P. Bedstraws and low plants.

729. *Xanthorhoe montanata* Schiff. **Silver-ground Carpet**
Found abundantly all over Lincolnshire – in wooded and damp areas particularly – in June and July. There are numerous records from such places as: Welton Le Marsh, Moreton Carrs, Pickworth, Telford, Irby Dales, Stubton Hall, Easton Hall, Stourton Park, Gibraltar Point, Epworth Turbary, Denton Reservoir, Syston Park, Saltfleetby, St. Clements, Horncastle, Boston etc.,

L.F.P. Docks and a wide variety of low plants.

730. *Xanthorhoe fluctuata* L. **Garden Carpet**
Abundant – found in all types of habitat over the county, from May to September in two generations. Places mentioned in the records include: Holton le Moor, Pickworth, Grimsby, Scunthorpe, Scawby, Boston, Lincoln, Woodhall etc.
L.F.P. Various Cruciferae.

731. *Nycterosea obstipata* F. **Narrow-barred Carpet**
(fluviata Hubn.) **(The Gem)**
A small, migratory species, seen in the late spring and early summer, and again in the autumn in August, September and October.
There is one record for Lep. Lincs. 3 for: Skegness, (A. H. Waters, Nat. Chron., 1895, p.128).
It was later at Grimsby, 25th August 1961 "female taken in bathroom", (G.A.T.J.); Boston, August 1964, Gibraltar Point, 16th September 1969, 19th October 1969, and 22nd August 1977, (All R.E.M.P.).
L.F.P. Low plants like Knotgrass, Groundsel etc.

733. *Colostygia pectinataria* Knoch **Spring Green Carpet**
(viridaria F.) **(Green Carpet)**
A fairly widespread species in Lincolnshire, flying in June and July. There are records for: Grimsthorpe Park, 1950, (L.N.U.); Hendale Wood, 18th June 1955, (M.P.G.); Stubton Hall, 1955, (L.N.U.); Riby, 30th June 1959, (G.A.T.J.); Muckton Wood, 14th July 1962, and Roughton Moor, 1st July 1968, (Both A.D.T.); Manton and Laughton, July 1974, (R.J. and J.H.D.). etc.
L.F.P. Bedstraws, Sorrel and low plants.

734. *Colostygia salicata* Hubn. **Striped Twin-spot Carpet**
This species is "chiefly an inhabitant of the northern counties" (South 1961) – flying in "May and June, and in some localities again in August and September". The entry in L.N.U. records reads: "Lincoln District 1840-50, (F.M.B.)" and "Welton Wood near Alford, 22nd May 1971", (L.N.U. Meeting).
L.F.P. Bedstraws.

735. *Colostygia multistrigaria* Haw. **Grey Mottled Carpet**
(The Mottled Grey)
This moth is out in March and April in damp areas. It was thought to be "local and scarce" early this century, though it now seems to be a little more common in the South Humberside area.
There are records for Market Rasen, Wrawby Moor, Scotton and Laughton Commons, and Woodhall Spa.
L.F.P. Bedstraws.

THE BUTTERFLIES AND LARGER MOTHS

736. *Colostygia didymata* L. **Small Twin-spot Carpet**
Common in May, June and July – and particularly plentiful in woodland areas to the west of Scunthorpe, where it has been seen in large numbers by J.H.D. Other localities in L.N.U. records include Grimsby, Riby and Laughton etc.
L.F.P. Wood Anemone, Campion, Sorrel etc.

737. *Pareulype berberata* Schiff. **Barberry Carpet**
Out in May/June, and again in August, and according to South "seems to be confined to the eastern counties".
There are two L.N.U. records for: East Ferry District, (A. Reynolds, Lep. Lincs. 3.); and North Somercotes, 22nd May 1918, (Rev. S. Proudfoot).
L.F.P. Barberry.

738. *Earophila badiata* Schiff. **Shoulder-striped Carpet (The Shoulder Stripe)**
Flies quite commonly in the spring where wild Rose flourishes. Recorded from Somerby, 1950, (M.P.G.); Bleasby, 24th April 1955, (B.J.); Scunthorpe, April 1962, (J.H.D.); Roughton Moor, 11th April 1965, (G.A.T.J.); Scotton Common, 4th May 1969, and 2nd May 1970, (Both R.J.).
L.F.P. Wild Rose.

739. *Anticlea derivata* Schiff. **Streamer Carpet**
(nigrofasciaria auct.) **(The Streamer)**
Quite common in the spring where wild Rose and Honeysuckle grow. Recorded from: Lincoln, Cowbit, Grimsby, Linwood and Scotton (The latter for May 1969 and 1970 – R.J. and J.H.D.).
L.F.P. Wild Rose and Honeysuckle flowers.

740. *Mesoleuca albicillata* L. **Beautiful Carpet**
Locally common in woodlands in June and July. There are early records for: Binbrook, Croxby, Haugham Wood, Maltby Wood, Market Rasen, Pillar Woods, Tothill, Haverholme Priory, Little Bytham etc.
More recent sites cover: Skellingthorpe, 1950, (T. C. Taylor); Grasby Bottom, 10th July 1953, (G.A.T.J.; Doddington, 25th June 1965, (B.J.); Laughton and Limber, 1970, (J.H.D.); Brumby Nature Reserve, June 1977, (J.H.D.); and Swallow Vale, 23rd June 1979, (R.J.).
L.F.P. Bramble etc.,

743. *Perizoma sagittata* F. **Marsh Carpet**
This local species occurs in marshy areas from the end of June into July. It was first recorded as a British species in 1848, with a specimen being obtained near Peterborough.
It was considered "rare and local" at the time of Lep. Lincs. 3, having occurred at South Billinghay (E. R. Walker), and in 1880 at Boultham Fen, when two were taken by E. Mead. It was later bred from 3 doxen wild larvae taken at Haverholme Priory in 1906 (J.D.C.), where it was also found "on the wing". More recently, larvae have been found on the banks of the River Idle, in Nottinghamshire, just over the Lincolnshire border. Possibly, if a thorough search of the food plant was made in the adjoining part of Lincolnshire, the species could be discovered in the county again.
L.F.P. Meadow-rue.

746. *Perizoma affinitata* Steph. **Large Rivulet**
(The Rivulet)
South (1961) wrote that this species was "Widely distributed" in June and July. It was noted as "rare" in this county in Lep. Lincs. 3, when it had been found near Gainsborough (F.M.B.) and Maltby Wood (B.T. Crow).
There are only 2 further records for: S. Thoresby, 1978, (R.E.M.P.); and Swallow Vale, 25th June 1979, (R.J. and J.H.D.).
L.F.P. Campions.

747. *Perizoma alchemillata* L. **Small Rivulet**
(rivulata Schiff.)
More frequently seen than the previous species – flying in June and July. It was taken at Skellingthorpe in 1915, (Rev. Blathwayt and Rev. Proudfoot) and in the Isle of Axholme 1955, and Scunthorpe, 3rd July 1965, (Both J.H.D.). In the 1960s and 1970s it has been found at Limber, Laughton, South Thoresby, and the Messingham Sand Quarries.
L.F.P. Hemp Nettle.

748. *Perizoma flavofasciata* Thunb. **Sandy Carpet**
(decolorata Hubn.)
Individuals of this species appear over a protracted period of emergence from June, through July and August. The L.N.U. records indicate that it has been discovered more frequently in the North of the County (including South Humberside.) Places mentioned in the records are: Grasby Bottom, Scotton Common, Irby Dales, Hendale Wood, Brocklesby, Laughton, Scunthorpe, Grimsby, Messingham etc.
L.F.P. Campions (Red, White and Bladder.)

749. *Perizoma albulata* Schiff. **Grass Rivulet**
Found in grassy meadows and along lanes in May and June. There are a considerable number of records from the period covered by Lep. Lincs. 3, when the species was considered to be fairly common. Since the early 1900s, however, there are only a handful of sightings for: Nocton Wood, 10th June 1909, (L.N.U.); Mablethorpe, 21st May 1961, (G.N.H.); Linwood, 15th June 1963, (J.H.D.); Gibraltar Point, 1973, and Moor Closes and Ancaster, 1975, (Both M.C.T.); and Ancaster again, June 25th 1979, (R.E.M.P.), when the species was "swarming – flying in the sunshine".
L.F.P. Yellow Rattle.

750. *Perizoma bifaciata* Haw. **Barred Rivulet**
 (unifasciata Haw.*)*
This moth flies in July and August, and is reputed to come to light and visit flowers (South 1961).
It was first recorded in the L.N.U. records in 1975, when R.E.M.P. found it at S. Thoresby. It turned up again there in 1978.
L.F.P. Red Bartsia, Eyebright.

751. *Perizoma minorata* Treits. **Heath Rivulet**
A local species, mainly found in the northern counties of the British Isles in July and August.
The only record of the species comes from Mr. Norman Holland of Sleaford, who recorded one specimen at m.v. light in Hatton Wood near Wragby on 13th July 1968.
L.F.P. Eyebright seeds

758. *Euphyia bilineata* L. **Yellow Shell**
Very common and widespread in the county, taken in all kinds of habitat in June, July and August. Typical localities in the records are: Welland Washes, Ancaster, Woodhall, Gibraltar Point, Woolsthorpe-by-Belvoir, Kirkby Moor, Sibsey etc.
L.F.P. Docks, Grasses, low plants.

759. *Melanthia procellata* Schiff. **Pretty Chalk Carpet**
Mainly a species of the southern counties of England, along hedgerows where the food plant thrives. The only L.N.U. county records are for Boston. One came to light on 31st July 1969 (A.E.S.), and it has been seen in this region by R.E.M.P. on a number of occasions since the 1940s.
L.F.P. Traveller's Joy.

RECORDS OF LINCOLNSHIRE SPECIES 213

760. *Mesotype virgata* Hufn. **Oblique-striped**
(lineolata Schiff.)
A coastal species, occurring in areas of sand hills in May/June and again in July/August.
The only entry in the L.N.U. records reads: "a specimen taken by W. Lewington of Market Rasen, on the coast, I believe at Mablethorpe. He showed it to me" (R.T.C.) (Lep. Lincs. 3).
L.F.P. Bedstraws.

761. *Lyncometra ocellata* L. **Purple Bar Carpet**
Appears in June and July, and occasionally in August and September, in small numbers throughout the county. It flies during the day time, and visits light at night.
Records include: Scotton Common, 14th August 1954, (G.A.T.J.); Linwood Warren, June 1958, (J.H.D.); Elsham, 13th August 1958, (L.N.U.); Kirkby Moor, July 24th 1966, (G.N.U.); Scotgrove Wood, 1971, (A.D.T.); and Swallow Vale, 25th June 1979, (R.J.).
L.F.P. Bedstraws.

762. *Lampropteryx suffumata Schiff.* **Water Carpet**
Common in April and May in some localities in the county.
It was at: Irby Dales, 9th May 1955, (L.N.U.); Kirkby Moor, 2nd May 1964, (G.N.H.); New Park Wood, 14th May 1966, (A.D.T.); Holme Plantation, Bottesford, 25th May 1973, (J.H.D.); Scawby Woods, 22nd May 1976, (R.J.); and Scotton Common, May 1978, (S. Van den Bos).

764. *Electrophaes corylata Thunb.* **Broken-barred Carpet**
Frequently seen resting on tree trunks in woodland in May and June.
Records include: Holton-le-Moor, 12th June 1908, (L.N.U.); Skellingthorpe, 2nd June 1952, (T. C. Taylor); Laughton and Twigmoor, 11th June 1959, (J.H.D.); Tumby Wood, 17th June 1961, (L.N.U.); and Holme, 10th May 1975, (R.J.).
L.F.P. Birch, Oak and the foliage of other trees.

765. *Ecliptopera silaceata Schiff.* **Small Phoenix**
A fairly common species where beds of Willow Herb occur – seen in May/June with an occasional second generation in August/September. There are records for: Limber, 10th May 1948, (G.A.T.J.); Bourne Wood, 16th June 1965, (G.N.H.); Scotton Common, 31st August 1957, (J.H.D.); and Linwood Warren, 15th September 1963, (A.D.T.), etc.
L.F.P. Willow Herbs.

214   THE BUTTERFLIES AND LARGER MOTHS

767. *Lygris prunata* L. **Large Phoenix**
(The Phoenix)
This species flies in gardens and along hedgerows in July and August, where the larval food plant occurs. It was considered "scarce" at the time of Lep. Lincs. 3 and was mentioned as being found at: Gainsborough, Market Rasen, Owston Ferry, Somersby, West Ashby, and Haltham.
  Since the early 1900s there have been only 3 reliable L.N.U. records: Grimsby, 27th July 1962, (G.A.T.J.); Riseholme Wood, 19th August, 1967, (A.D.T.); and Swallow Vale, 23rd June 1979, (R.J.).
L.F.P. Foliage of Currant bushes and Gooseberry.

768. *Lygris testata* L. **Common Chevron**
(The Chevron)
Found on heathland in August and September. Lep. Lincs. 3 called the species "common" but with the area of heathland in the county diminishing, it now appears to be rather local.
  It was recorded at Crowle, 28th August 1966, (L.N.U.); Scotton Common, 1st September, 1951, (C.G.E.); Twigmoor, 22nd August 1959, (Rear Admiral Torless); Greetwell, 16th August 1975, (J.H.D.).
L.F.P. Willow, Sallow, Aspen, Birch etc.

769. *Lygris populata* L. **Northern Spinach**
A scarce species in Lincolnshire. It flies typically in July and August in ancient woodlands, and particularly in damp areas.
  There were 3 records for before the turn of the century, and more recently sightings at: Legsby, 1954, (R. Cornwallis); Riseholme Wood, 1967, (A.D.T.); Scotgrove Wood, 21st July 1968, (A.D.T.).
L.F.P. Bilberry, Sallows and Goat Willow.

770. *Lygris mellinata* F. **Currant Spinach**
(associata Borkh.) **(Spinach)**
Quite frequently found in gardens (July/August) where the larval foodplant occurs. There were numerous records for the Grimsby district in the 1950s, (G.A.T.J.); with others for Scunthorpe, August 1958, (J.H.D.); Sleaford, 4th July 1963, (G.N.H.); Boston, June-July 1967, (J.B.); Twigmoor, 22nd July 1977, (R.J.).
L.F.P. Red and Black Currant.

771. *Lygris pyraliata* Schiff. **Barred Straw Chevron**
 (dotata L.)
Found along hedgerows and in woodland from the end of June into July. Up to date records include: Grainsby Park, 24th July 1951, (G.A.T.J.); Brocklesby, 24th June 1959, "common at light", (G.A.T.J.); Laughton, 20th July 1974, (J.H.D.), etc.
L.F.P. Bedstraws, Goosegrass etc.

772. *Cidaria fulvata* Forst. **Barred Yellow**
A fairly widespread species in July – found across Lincolnshire in gardens and along hedgerows.
Examples of records are: Grainsby Park, 14th July 1951, (G.A.T.J.); Chapel St. Leonards, 25th July 1962, (G.N.H.); Woodhall, 17th July 1964, (G.A.T.J.) and Whitton, 16th July 1976, (R.J.).
L.F.P. Rose.

773. *Plemyria rubiginata* Schiff. **Blue-bordered Carpet**
 (bicolorata Hufn.)
Not a very common species in the county, though it can be locally plentiful in orchards or other suitable habitats. It flies in July and August. At the time of Lep. Lincs. 3, it was recorded from: Gainsborough, Wrawby Moor, Haverholme Priory and Wyberton. More recently it was at: Amcotts, 23rd July 1963; "flying in numbers well before dusk", (J.H.D.); Woodhall Spa, 17th July, 1964, "to light", (G.A.T.J.); Boston, 1st August 1970, (A.E.S.), etc.
L.F.P. Alder, Birch, Plum, Apple etc.

774. *Chloroclysta siterata* Hufn. **Red-green Carpet**
 (psittacata Hubn)
A woodland species – seen in the autumn, and in the following spring after hibernation. There was a single record for Lincoln in Lep. Lincs. 3, in the 1840s, (F.M.B.). There are later records for: Grimsby, 24th July 1948, (G.A.T.J.); Woodhall, 20th September 1966, and Scotgrove Wood, 27th July 1969, (Both A.D.T.).
L.F.P. Deciduous trees like Oak, Birch, Ash etc.

775. *Chloroclysta miata* L. **Autumn Green Carpet**
More common than the previous species, and on the wing during the same period (i.e. autumn, and during the following spring, after hibernation) in wooded areas. It was "fairly common" early this century – e.g. it was "South of Lincoln, 13th October 1892, (J.F.M.); and in "Lincolnshire 1921, (Rev. F. S. Alston – L.N.U. Trans. Vol.5., p.154)".

The only more recent records are for a "larvae, record from Laughton, and emerged September 1958"; and for Manton, June 1976, (Both J.H.D.).

L.F.P. Oak, Alder, Birch etc.

776. *Dysstroma truncata* Hufn.   **Common Marbled Carpet**
(russata Borkh.)

An abundant species across Lincolnshire. It flies in May and June, with a second generation in the autumn, and can be found particularly along lanes and in woodland. It shows very variable colouration, and a large proportion of dark melanic looking specimens occur. Localities mentioned in the records include: Grimsby, Sleaford, Linwood, Laughton, Barton on Humber, Alkborough, Saltfleetby, Mablethorpe, Ruslingthorpe, etc.

L.F.P. Birch, Hawthorn, Sallow, etc.

778. *Dysstroma Citrata* L.   **Dark Marbled Carpet**
(immanata Haw.)

Flies over moorland and in wooded areas in July and August. Since Lep. Lincs. 3 there have been few L.N.U. records. These are for: Linwood, 16th September 1963, (A.D.T.); Limber, 17th July 1964, (G.A.T.J.); and Mablethorpe, 1965, (T. R. New). Possibly this species has been overlooked due to its close resemblance to the dark form of the previous species *(truncata)*.

L.F.P. Wild Strawberry, Sallow, etc.

779. *Thera obeliscata* Hubn.   **Grey Pine Carpet**

Common in Fir woods in May, June and July, with a second generation in the autumn. Records include: Holton Le Moor, 12th June 1908, (L.N.U. meeting); Pelham's Pillar Woods, 28th September 1959, (P. Cue); Linwood Warren, 15th September 1963, *obliterata* (A.D.T.); Pickards Plantation, near Market Rasen, 13th July 1963, (G.N.H.) etc.

L.F.P. Pine and Spruce.

781. *Thera cognata* Thunb.   **Chestnut-coloured Carpet**
(simulata Hubn.)

This species is found in July and August in highland localities in northern England, Wales, Scotland and Ireland, where the food plant Juniper occurs.

It was reputed to have been found in Lincolnshire in the mid 1800s (Lep. Lincs. 3) in such localities as Gainsborough, Lincoln, and Scunthorpe (Sweeting Thorns.) This mountain species bears a close resemblance to the Grey Pine Carpet *(T. obeliscata* Hubn.) and in the absence of specimens to examine, it must be considered likely that a wrong identification was made last century.

L.F.P. Juniper.

RECORDS OF LINCOLNSHIRE SPECIES 217

782. *Thera firmata* Hubn. **Reddish Pine Carpet**
(The Pine Carpet)
This species is out in June and July, and again in the autumn (September and October), in Pine woodland. It's quite scarce in the county. In Lep. Lincs. 3 it was recorded: "N. Ashby, Brigg (R.T.C.); Linwood, 1907, (S.B.S.); Market Rasen, one bred 8.9.04, (G.E.M.); S. Hartsholme (L.N.U. meeting – verified A.T.)" etc.
The only subsequent L.N.U. records are for Linwood, 15th June 1963, (J.H.D.) Roughton Moor, 29th June 1969, (A.D.T.); Scotton Common, 17th July 1976, (J.H.D.); and Muckton, 12th September 1979, (R.E.M.P.).

L.F.P. Scots Pine.

783. *Thera Juniperata* L. **Juniper Carpet**
This species is out in October and November, and in South (1961) it is said that "Berkshire, Kent, Surrey and Sussex appear to be the only English counties in which it is established, though it has been recorded in other counties in the south and midlands".
There is a single L.N.U. record for Boston, October 1979, (A.E.S.).

L.F.P. Juniper

784. *Hydiomena furcata* Thunb. **July Highflyer**
(sordidata F.)
Flies commonly in July and August, with a number of named colour forms being noted in the county.
Recent records include: Bourne Woods, 26th July 1956, (G.N.H.); Easton Hall, 6th July 1957, (L.N.U.); Grimsby, 12th July 1958, (G.A.T.J.); Woodhall, 16th July 1965, "very common", (G.A.T.J.); Tetney Blow Wells, 25th June 1967, (G.N.H.), etc.

L.F.P. Sallows, Willows etc.

785. *Hydriomena coerulata* F. **May Highflyer**
(impluviata Schiff.)
Can be found in numbers in damp areas in May and June, where the larval foodplant Alder occurs – but in the 1800s the species was considered scarce.
It was found at Tumby, near Louth, spring 1922, (Rev. F. S. Alston); Broughton Woods, 14th June 1974, (J.H.D. and R. J.); S. Thoresby, Spring 1978, (R.E.M.P.), etc.

L.F.P. Alder

786. *Hydiomena ruberata Freyer* **Ruddy Highflyer**
A local species – found in April and May in damp heathland areas among Sallows and Willows. In Lep. Lincs. 3 it was noted from: Althorpe, Ashby, Market Rasen, Scotton Common, Haverholme and Hartsholme. The only further records are for Scunthorpe, May 1956, and Scotton, 2nd May 1970, (Both J.H.D.).
L.F.P. Willows and Sallows.

787. *Philereme vetulata Schiff.* **Brown Scallop**
(rhamnata Schiff.) **(Dark Umber)**
Scarce in Lincolnshire in June and July, (larvae Buckthorn feeders). There are only a few L.N.U. records. In the 1800s it was found at Skegness, (A. H. Waters – Nat. Chron. 1895, p.129); S. Allington, (P. Wynne); Cowbit, (Chas. M. Hufton); Hartsholme and Haverholme Priory, (J.F.M. and J.D.C. respectively).
There is a further record for Skellingthorpe, 1915, (Rev. F. L. Blathwayt and Rev. S. Proudfoot), one for Burton Pits, 1977, (M.T.) and one for Gibraltar Point, 5th August 1979, (R.E.M.P.).
L.F.P. Buckthorn.

790. *Rheumaptera cervinalis Scop.* **Scarce Tissue**
(certata Hubn.)
This species flies in April, May and June. It is only noted once in L.N.U. records (Lep. Lincs. 3) as having occurred "N. Gainsborough, Lincoln, 1840-50, (F.M.B.)" and at "Market Rasen (W. L., Nat. 1898 p.50)". It is impossible to verify these old records, and there have been no observations of it this century.
L.F.P. Barberry and Berberis.

791. *Rheumaptera undulata L.* **Shell Scallop**
(The Scallop Shell)
Quite a rare species in woods and marshy areas (June and July) where Sallows thrive. Last century it was found at Burwell Wood, Linwood, Newball Wood and Hartsholme. There are four further records for: Scotton, "larvae found" 1971, (J.H.D.); Scunthorpe area, 1972, (J.H.D.); Crowle Waste, 1978, (R. Key); and Manton, 1979.
L.F.P. Sallows, Willow, Aspen.

792. *Rheumaptera hastata L.* **Large Argent and Sable**
(Argent and Sable)
A species which is found flying in the daytime in Birch Woods in May and June. Lep. Lincs. 3 described it as "local", but recorded it from a large number of areas. (e.g. Langworth, Legsby, Linwood, Saxilby, Hartsholme, Skellingthorpe etc.).
It was frequently seen in Birch Woods this century up to 1960.

For example it was at: Newball and Skellingthorpe, 25th May 1926, (A.E.M.); Legsby, 1950, (M.P.G.); Sweeting Thorns, June 1959, "common on Bramble flowers", (J.H.D.); Linwood, 26th May 1960, (J.H.D. and G.A.T.J.). Since 1960 it seems to have disappeared completely from the county.
L.F.P. Birch.

794. *Epirrhoe rivata Hubn.* **Wood Carpet**
The Wood Carpet is found in July and August in Woodland localities. It looks very much like the Common Carpet *(Epirrhoe alternata Mull.)* which is on the wing a little earlier than this, and again in August and September. L.N.U. records note that these species are frequently confused, but place the Wood Carpet at Linwood, 25th June 1959, (A. D. Torless); and Riseholme Wood, 18th August 1967, (A. D. Townsend – "requires confirmation"). etc.
L.F.P. Bedstraws.

795. *Epirrhoe alternata Mull.* **Common Bedstraw Carpet**
 (sociata Borkh.) **(Common Carpet)**
Very common, and widespread, in two generations, (see note on previous species). In the records for: Pickworth, Willingham, Humberstone, Stourton Park (Horncastle), Tumby Wood, Tealby, Denton Reservoir, Elsham Park, Stainton Wood etc.
L.F.P. Bedstraws.

796. *Epirrhoe tristata L.* **Small Argent-and-Sable**
A moorland or upper heathland species in the British Isles, flying in May/June, and again in some areas in August.
There is a very doubtful record for the 1800s for the Gainsborough district (F.M.B.), It cannot be substantiated.
L.F.P. Heath Bedstraw.

797. *Epirrhoe galiata Schiff.* **Galium Carpet**
Chiefly found in June on chalk/limestone areas. There is a single mention in L.N.U. records dating from the time of Lep. Lincs. 3, and reading "N. Gainsborough, F.M.B.".
L.F.P. Bedstraws.

800. *Chesias legatella Schiff.* **Streaked Carpet**
 (spartiata Fuessl.) **(The Streak)**
Can be common (in October generally) where a good growth of Broom occurs. It was considered rare early in the century, but recent records placed it at: Skellingthorpe, 1946, (C. Taylor); Woodhall Spa, 15th October 1954, (F.L.K.); Greetwell, 1965, "larvae plentiful", (J.H.D.); and Linwood, October 1972, (A.D.T.).
L.F.P. Broom.

802. *Odezia atrata* L. **Chimney-sweeper**
(chaerophyllata L.)
South (1961) described the species as "always very local, frequents moist fields, borders of woods and even waysides", It was "common and abundant in some localities" at the time of Lep. Lincs. 3, but today is found locally in suitable areas throughout the county in June and July.

Recorded from: Ropsley Rise wood, July 1928, (A.E.M.); Irby Dales, 1955, (L.N.U.); Holme Plantation, June 1959, (J.H.D.); New Denton Reservoir, 13th June 1964, (G.N.H.); Woolsthorpe by Belvoir, 2nd July 1966, (G.N.H.); Normanby le Wold, 22nd June 1974, "common each year", (J.H.D.).

L.F.P. Earthnut.

803. *Anaitis plagiata* L. **Slender Treble-bar**
(The Treble-bar)
Flies in two generations (May/June and August/September) in sandy areas. In 1895 it was found at Broughton (A. E. Hall), and it was found at Broughton again in May 1978, (L.N.U. Meeting – J.H.D.).

Other records had it at: Holywell Wood, 16th June 1951, (L.N.U.); and Castle Bytham, 3rd June 1972, (A.E.S.).

L.F.P. St. John's Wort.

805. *Carsia sororiata* Hubn. **Manchester Treble-bar**
(paludata Thunb.)
This species inhabits swampy moorland areas in the North of England and Scotland. There is mention in the records of "an odd one occurring near Lincoln" at the time of Lep. Lincs. 3 – in a letter from a Mr. Burton. Since, however, Mr. Burton admits in the letter to obtaining specimens from a number of correspondents (including one in Manchester), the record, perhaps, ought not to be taken seriously.

L.F.P. Cowberry, Bilberry, Cranberry.

807. *Horisme vitalbata* Schiff. **Umber Waved Carpet**
(Small Waved Umber)
This species is found in May/June and August along hedgerows in the southern counties of England where Traveller's Joy occurs. There is a single L.N.U. entry for Boston, where a fertile female specimen was found by R.E.M.P. in the 1950s.

L.F.P. Traveller's Joy.

RECORDS OF LINCOLNSHIRE SPECIES

809. *Horisme tersata Schiff.*   **Fern Carpet**
(The Fern)
Occurs along hedgerows, or in gardens where the foodplant flourishes, in July. There is one mention in Lep. Lincs. 3 for Hammeringham. The subsequent records are for: Sleaford, 17th July 1956, (G.N.H.); Linwood, 1972 and 1975, (M.T.); and S. Thoresby, 1978, (R.E.M.P.).
        L.F.P. Clematis and Ranunculus.

810. *Lobophora halterata Hufn.*   **Large Seraphim**
(hexapterata Schiff.)   **(The Seraphim)**
A very local species where the chief foodplant (Aspen) occurs – flying in May and June. J.H.D. discovered the larvae of the species and bred it from: Sweeting Thorns, May 1947; Laughton, 17th May 1972; and Holme Plantation, May 1977. Elsewhere in Lincolnshire it has been found at Moor Farm, Kirkby Moor, 1973, (R.E.M.P.)
        L.F.P. Aspen, Poplars, Sallows.

811. *Mysticoptera sexalata Retz.*   **Small Seraphim**
(sexalisata Hubn.)
Found locally in woodland in May and June. There are two sightings of it in L.N.U. records. Firstly, it was found at Skellingthorpe at the turn of the century (C.P.A. in Lep. Lincs 3), and interestingly enough, at Skellingthorpe again in 1950, (T. C. Taylor).
        L.F.P. Sallows and Willows.

812. *Acasis viretata Hubn.*   **Brindle-barred Yellow**
(Yellow-barred Brindle)
A local species, out early in the year (May/June) with a possible second generation in the early autumn, (August/September). It appears scarce in the county, as we have only three records for: North Somercotes, 24th May 1918, (Rev. S. Proudfoot); S. Thoresby, 1970, and Kirkby Moor, 1973, (Both R.E.M.P.).
        L.F.P. Flowers and leaves of
        Holly, Ivy, Privet etc.,

813. *Trichopteryx polycommata Schiff.* **Barred Tooth-striped**
This species is known to be found locally in March and April. It is certainly rare in Lincolnshire, as we only have a single record for Stainfield, near Wragby, (C.P.A. – Lep. Lincs. 3).
        L.F.P. Privet.

814. *Trichopteryx carpinata Borkh.*   **Early Tooth-striped**
   (lobulata Hubn.)
A woodland species, found flying locally in April and May. It was recorded from Linwood in 1907 and 1909, (S.B.S.), and it turned up there again 50 years later, when recorded 14th April 1959, (J.H.D.). Other localities for the species are: Roughton Moor, 11th April 1964, "to sallow blossom", and Kirkby Moor, 2nd May 1964, (Both G.N.H.); and New Park Wood, 14th May 1966, (A.D.T.).
L.F.P. Honeysuckle, Birch, Sallow.

815. *Orthonama lignata Hubn.*   **Oblique Carpet**
   (vittata Borkh.)
A rare species in the county – it flies in marshy districts in May and June, with a second generation in the autumn. It was recorded for Lep. Lincs 3 at Cowbit, Skellingthorpe and Wyberton. Later records had it at: Amcotts – "from a female specimen found June 1962, several larvae were reared to emerge in October that year", (J.H.D.); Theddlethorpe, 1973, and Saltfleetby, 1975, (both R.E.M.P.).
L.F.P. Bedstraws.

816. *Ortholitha mucronata Scop.*   **Common Lead-belle**
   (umbrifera Prout)
A local species, but recorded in reasonable numbers in June where Furze and Broom occur. It was found at Scotton on a number of occasions between 1951 and 1969. Other localities in the records are: Irby Dales, Laughton, Brocklesby, Manton and Epworth Turbary.
L.F.P. Furze and Broom.

818. *Ortholitha chenopodiata L.*   **Shaded Broad-bar**
   (limitata Scop.)
Found abundantly in grassy places – flying in the daytime in July and August. It has been widely recorded from e.g. Kirton Marsh, Tetney Haven, Byards Leap, Ashbyville, Freiston Shore, Ancaster, Benington Marsh, Crowle Waste etc.
L.F.P. Clover, Vetch, Grasses.

821. *Ortholitha bipunctaria Schiff.*   **Local Chalk Carpet**
                                        (Chalk Carpet)
A species of the chalk downs, or limestone hills, flying in July and August. It was first recorded in the county in 1966, when a small, but thriving colony was discovered in the restricted area of an old limestone quarry near Kirton Lindsey. It has been seen regularly, in fair numbers, up to the time of writing (1979). It has been

suggested that in years gone by it may have occurred more widely on the Wold areas of the county – but that the intensive cultivation and farming of the region has subsequently led to its decrease.
L.F.P. Clover and Trefoils.

822. *Larentia clavaria Haw.* **Mallow Carpet**
(cervinata Schiff.) **(The Mallow)**
A species which occurs late in the year – in September and October. Early lepidopterists seemed to find it fairly commonly, but since Lep. Lincs 3 there have been few records for it. Single specimens have been seen in the Scunthorpe area, (J.H.D.). Others have turned up at: Ruskington, October 1963, and Sleaford, 19th September 1966, (Both G.N.H.); and Boston, 5th October 1968, (A.E.S.).
L.F.P. Mallow.

823. *Pelurga comitata L.* **Dark Spinach**
Some years ago this species occurred commonly in gardens in the Scunthorpe area. South confirms that it is found around market gardens, and also mentions waste places and sandy coasts. It is seen in July and August. It was at Grimsby, 8th July 1950, (G.A.T.J.); Sleaford, August 1958, (G.N.H.); Scunthorpe, 19th August 1962, (J.H.D.); Gibraltar Point, 9th August 1969, (A.D.T.); Boston, 9th August 1974, (A.E.S.).
L.F.P. Goosefoot, Common Orache.

824. *Oporinia autumnata Borkh.* **Large Autumnal Carpet**
(The Autumnal Moth)
Out late in the year (September/October and later), and there are few L.N.U. records. The earlier come from N. Ashby (Brigg) and Louth, 1888-93 (R.E.C. and G.W.M.). Two later entries refer to Barton on Humber 1951, (C. G. Else), and Haverholme, 15th November 1964, (G.N.H.).
L.F.P. Birch, Alder, Larch etc.

825. *Oporinia filigrammaria H.-S.* **Small Autumnal Carpet**
Appears in August and September – found on moorlands in the more western and northern counties of England, and in other areas as a casual migrant. It was found at South Thoresby, September 1978, "a single specimen", (R.E.M.P.). On the same date, it was also recorded at Lissington (Mr. G. Haggett).
L.F.P. Bilberry.

224 THE BUTTERFLIES AND LARGER MOTHS

826. *Oporina dilutata Schiff.*   **November Carpet**
(nebulata Thunb. non Scop.)   **(November Moth)**
A common autumn species – found around woodland from late September through to November. Localities in the records include: Pelham's Pillar Woods, 28th September 1959, (G.A.T.J.); Heckington, 11th October 1959, (G.N.H.); Boston, 20th October 1968, (A.E.S.); and Scawby, 17th October 1979, (R.J.) etc.,
L.F.P. Foliage of many trees, e.g. Elm, Oak, Birch.

827. *Oporinia christyi Prout*   **Christy's Carpet**
(The Pale November Moth)
Another woodland species flying in October and November. It is very similar in appearance to the previous species *(dilutata Schiff.)* and was originally considered to be a colour form of it. Due to this confusion, there are L.N.U. records for the species but they certainly cannot be read confidently, as applying to *christyi*.
L.F.P. Foliage of Deciduous trees like Beech, Elm etc.

828. *Operophtera brumata L.*   **Common Winter**
(Winter Moth)
According to Lep. Lincs 3, this species was "far too common" and "very destructive in orchards" early this century. The same may be said today, as on occasion the larvae are capable of defoliating a group of trees in orchards or woodland. The moth appears between the end of autumn one year, and the early months of the next. It has been widely recorded from such places as: Donna Nook, Nocton Wood, Freshney Bog, Holland Division, Grimsby, Knaith, Scawby, Haverholme etc.
L.F.P. Trees such as Oak, Apple etc.

829. *Operophtera fagata Scharf.*   **Northern Winter**
(boreata Hubn.)
Often found commonly in areas of Birch woodland in October and November, and sometimes later. Recorded from: Wrawby Moor, Sweeting Thorns, Sleaford, Hendale Woods etc.
L.F.P. Birch.

830. *Asthena albulata Hufn.*   **White Waved Carpet**
(candidata Schiff.)   **(Small White Wave)**
Mainly found in woodland in May/June, with a second generation occasionally. It was thought "locally common" at the time of Lep. Lincs 3, and recorded from: Pelham's Pillar Woods, Hartsholme and Skellingthorpe. There are later records for: Skellingthorpe,

June 1976, (J.H.D.); Bourne Wood, 16th June 1956, (G.N.H.); and Austacre Wood, Bardney, 1972, (A.D.T.).
L.F.P. Birch, Hazel, etc.

831. *Minoa murinata* Scop. **Drab Carpet**
(euphorbiata Schiff.) **(Drab Looper)**
Found in woodland in the spring (May/June) and occasionally again in the autumn. The only L.N.U. record stems from R.E.M.P. who wrote that it was to be found in High Hall Wood, Woodhall, in the 1950s.
L.F.P. Wood Spurge.

832. *Hydrelia flammeolaria* Hufn. **Yellow Waved Carpet**
(luteata Schiff.) **(Small Yellow Wave)**
Flies locally in June and July where the larval foodplant occurs. There are recent records for: Scawby, 1954, (J.H.D.); Linwood, 1971, (A.D.T.); Swallow Vale, 23rd June 1979, (R.J.).
L.F.P. Maple and Alder.

833. *Hydrelia testaceata* Don. **Sylvan Waved Carpet**
(sylvata Schiff.) **(Waved Carpet)**
This species occurs in damp, wooded areas in June. There are no records since Lep. Lincs. 3, when it was at: Legsby, 1893, "10 specimens in June, but absent in the 3 following years", (G.H.R.); Pelham's Pillar Wood, 1909, (S.B.S.); Tothill, "a single specimen", 10th July 1909, (G.W.M.).
L.F.P. Alder, Birch, Sallow.

834. *Euchoeca nebulata* Scop. **Dingy Shell**
(obliterata Hufn.)
Appears in June/July in damp regions where Alder flourishes. It was considered "rare" early this century – and mentioned from Linwood, (1907/8), and Haverholme Priory, (June 1896).
It next turned up at Doddington, 1913, (Rev. Proudfoot and Rev. Blathwayt). In the mid 1960s, according to Geo. Hyde, it was to be found in numbers in Owlet Plantation, Laughton.
It was discovered in the Broughton Woodlands on 14th June 1974, (R.J. and J.H.D.). From this female, ova were laid which proved infertile. A subsequent search of the area revealed one larva which pupated.
L.F.P. Alder.

836. *Discoloxia blomeri Curt.* **Blomer's Ripplet**
This woodland species flies in June and July. It was first recorded for the L.N.U. at Hendale Wood, June 1955, (J.H.D.). It was seen at this locality again on 4th July 1957, (J.H.D.). The last mention of it is for Limber, when it was "attracted to car lights", 1st July 1965, (G.A.T.J. and J.H.D.).
L.F.P. Wych Elm.

837. *Anticollix sparsata Treits.* **Dentated Pug**
(sparsaria Hubn.)
A species found in Fens and in marshy woodlands in June and early July. Most L.N.U. records come from Mr. R. E. M. Pilcher. They are for: Crowle Waste, (no date); Woodhall, "Larvae", 1st October 1971, Moor Farm, Woodhall, "common", 1979. More recently (4.6.82) it was found at the Messingham B.I.S. reserve (J.H.D.).
L.F.P. Yellow Loosestrife.

838. *Eupithecia pini Retz.* **Cloaked Pug**
(togata Hubn.)
In South we can read that "this fine pug was first noted in England in 1845, in a plantation of Spruce Fir at Black Park, Buckinghamshire". It flies in May and June. The only L.N.U. record reads: "Larvae were taken in the Boston area in 1965" (R.E.M.P.).
L.F.P. Immature seeds in Spruce Cones.

839. *Eupithecia subumbrata Schiff.* **Shaded Pug**
(scabiosata Borkh.)
Found over rough areas "inland, or more frequently on the coast. It also occurs in fens, marshy places in woods etc." (South) in June/July. The only L.N.U. record is for South Thoresby, (R.E.M.P).
L.F.P. Flowers of Hawkbits, Hawk's-beard etc.

840. *Eupithecia subnotata Hubn.* **Plain Pug**
This species, which flies in July, was described as "scarce" in Lep. Lincs. 3 and given for: Ashby, Brigg, (R.T.C.); Cleethorpes, 14th July 1905, (F.W.S.); and Haverholme Priory, (J.D.C.). There are three later records for Boston 1965, (larvae); along the Lincolnshire coast, 1971; and S. Thoresby (1974) – (all R.E.M.P.).
L.F.P. Orache and Goosefoot.

RECORDS OF LINCOLNSHIRE SPECIES

842. *Eupithecia distinctaria* H.-S. **Thyme Pug**
As the name implies, this species inhabits areas where wild Thyme abounds (in June and July). The only L.N.U. record stems from Lep. Lincs 3 when it was reputedly found south of Haverholme Priory (J.D.C.).
L.F.P. Wild Thyme.

843. *Eupithecia tenuiata* Hubn. **Slender Pug**
A species flying in damp, marshy places in June and July. It was "scarce" in Lep. Lincs 3, but it was noted that "it is a species which requires working for in the larval state". It was reocrded from N. Linwood, 1907 and 1908, bred from sallow catkins, (S.B.S.).
Later records came from R.E.M.P. for: Boston, 1965; Moor Farm, 1974 and "Larvae – centre of Lincs", 1978.
L.F.P. Sallow Catkins.

844. *Eupithecia inturbata* Hubn. **Maple Pug**
 (subciliata Doubl.)
Can be found in July and August where the foodplant occurs. There is one early record for "Market Rasen, at heather bloom, J. A. Hardy, Nat. Chron. 1895, p.112". The only further record is for Bardney Forest, 1974, (R.E.M.P.).
L.F.P. Maple (preferably the flowers).

845. *Eupithecia haworthiata* Doubl. **Haworth's Pug**
 (isogrammaria H.-S.)
Most frequently found in the south of England, where it flies in the sunshine in June and July, where Old Man's Beard occurs.
Again we have a single early L.N.U. record – for Well, 18th July 1909, (F. J. Whittle), and a single later record for "Larvae at Ancaster", 1975, (R.E.M.P.).
L.F.P. Traveller's Joy.

846. *Eupithecia plumbeolata* Haw. **Lead-coloured Pug**
Out in May and June where the larval foodplant thrives. Lep. Lincs. 3 recorded it at Legsby, 12th June 1896, (J.H.R.). More recently it was seen at: Sleaford, 10th July 1956, "3 specimens", (G.N.H.); and Mablethorpe in the 1950s and 1960s, (T. R. New).
L.F.P. Yellow Rattle and Cow-wheat.

847. *Eupithecia linariata* Schiff. **Toadflax Pug**

This species is fairly widespread where Toadflax grows in plenty, with larvae feeding in the seedheads. Lep Lincs. 5 placed it at "Limber, bred from Toadflax, May and June, from seedheads gathered the previous summer" – 1914, (G.W.M.).

It was later found at: Barton on Humber, 1946, (C.G.E.); Scawby, "larvae" August 1957, (J.H.D.); Sleaford, 6th July 1958, "to light", (G.N.H.); Boston, 1965, and Bardney Forest 1974, (R.E.M.P.).

L.F.P. Toadflax.

848. *Eupithecia pulchellata* Steph. **Foxglove Pug**

Flies in May and June where the larval foodplant occurs. Its presence in the county has been confirmed by R. E. M. Pilcher, but data is not available at time of writing.

L.F.P. Foxglove.

849. *Eupithecia irriguata* Hubn. **Marbled Pug**

A species found early in the year (April to May) in Oak Woodlands. The only reference to it in L.N.U. records reads: "Division 2 (now S. Humberside region) 4 larvae taken by Rev. S. Proudfoot".

L.F.P. Oak.

850. *Eupithecia exiguata* Hubn. **Mottled Pug**

Found along hedgerows in May and June. It was considered "rather scarce" in Lep. Lincs. 3, at: Ashby, Brigg (R.T.C.); Binbrook, 31st May 1909, (G.W.M.); Great Carlton (C. D. Ash); Panton and Cowbit (Chas. M. Hufton).

The only later mention of it is for Boston, "larvae found in 1965", (R.E.M.P.).

L.F.P. Hawthorn, Sloe, Sallow etc.

851. *Eupithecia insigniata* Hubn. **Pinion-spotted Pug**
(consignata Borkh.)

This species is out in May and June mainly. All L.N.U. records stem from R. E. M. Pilcher, who first recorded it at Kirkby Moor, 17th May 1974. He later found it "widespread in central Lincolnshire in 1978" and at South Thoresby and in the Lincoln area in June 1979.

L.F.P. Apple, Hawthorn.

852. *Eupithecia valerianata* Hubn.　　　　　　　　**Valerian Pug**
Found in damp, marshy areas, flying in May and June. It was first recorded for the L.N.U. by R.E.M.P. at S. Thoresby in 1973. He also found it "widespread" in Bardney Forest in 1974.

L.F.P. Valerian.

853. *Eupithecia pygmaeata* Hubn.　　　　　　　　**Marsh Pug**
(palustraria Doubl.)
Flies in May and June, and sometimes a little later, with a possible second generation.
According to South (1961) "the species occurs in all Eastern counties of England". It was thought scarce early this century, but recorded from: Ashby, Legsby, Newball and Panton, (G.H.R.); and two had been taken at Market Rasen in 1893 (W.L.). It was found at Freshney Bog, 1910, as "a very local species, the best time to take it being late afternoon," (G.W.M., L.N.U. meeting). It was next taken at Theddlethorpe in 1973 and 1975 (R.E.M.P.).

L.F.P. Stitchwort and Chickweed.

854. *Eupithecia venosata* F.　　　　　　　　**Netted Pug**
Seen in June and July mainly, where larvae feed on the seed capsules of Campions. It was considered "scarce" in Lep. Lincs. 3, but there are later records for quite a number of areas. It was at: Scawby, May 1957, "emerged from larvae found the previous year", (J.H.D.); Boston area, "larvae 1965" (R.E.M.P.); Kirton Lindsey quarries 23rd July, 1967, (J.H.D.); S. Thoresby, 8th July 1979, (R.E.M.P.); and Scunthorpe, "in gardens", July 1979, (J.H.D.).

L.F.P. Campions.

855. *Eupithecia centaureata* Schiff.　　　　　　　　**Lime-speck Pug**
(oblongata Thunb.)
Occurs quite commonly across Lincolnshire between June and August. It has been typically recorded from: Barton on Humber, Boston, Grimsby, Sleaford, Scunthorpe, Frieston Shore, Theddlethorpe, Gibraltar Point etc.

L.F.P. Ragwort, Scabious etc.,

856. *Eupithecia trisignaria* H.-S.　　　　　　　　**Triple-Spotted Pug**
Flies in June and July, and according to South is "rarely met with in the open". It is a rare species in Lincolnshire, the sole record being for the centre of the county, (Woodhall area) June 1978, (R.E.M.P.).

L.F.P. Angelica, Cow-parsnip.

858. *Eupithecia satyrata* Hubn. **Satyr Pug**
A rare species in this county. It flies normally over heathland in May and June. There are L.N.U. records for: "Linwood, 9th June 1908, and 2 reared 1909 (S.B.S.)"; Bardney Forest 1974, and S. Thoresby 16th June 1979, (Both R.E.M.P.).
L.F.P. Knapweed, Heath, Sallow.

859. *Eupithecia tripunctaria* H.-S. **White-spotted Pug**
(albi-punctata Haw; non Hufn.)
Usually considered quite scarce in the county – flies in May and June. There are two early records for: Market Rasen, "a few larvae" 1895 (W.L.); and near Binbrook, 1907 (S.B.S.). There are later records for: Brocklesby, 26th June 1957, (Mr. P. Cue and G.A.T.J.); Normanby Park, 15th June 1971, (J.H.D.); Lincoln, 1976 and Gibraltar Point 1978, (both M.T.); and Holme Plantation, 19th June 1981, (J.H.D.). The most recent records to hand (end of May, early June 1982) are for Bagmoor and Hendale (R.J. and J.H.D.) when the species turned up in numbers.
L.F.P. Angelica and other Umbelliferae.

860. *Eupithecia absinthiata* Clerck. **Wormwood Pug**
(minutata Schiff.)
Fairly frequently found in June and July across the county. It has been typically recorded from: Belton, Gibraltar Point, Scunthorpe, Sleaford, Boston etc.
L.F.P. Wormwood, Ragwort, Mugwort etc.

861. *Eupithecia goossensiata* Mab. **Ling Pug**
(minutata Doubl. non Schiff.)
Occurs commonly on heathland in June, July and August. There are more recent records for: Scotton Common, 4th June 1955, (G.A.T.J.); Kirkby Moor, 12th August 1962, (G.N.H.); Boston area 1965, and South Thoresby, 16th June 1979, (R.E.M.P.).
L.F.P. Heath and Ling.

862. *Eupithecia expallidata* Doubl. **Bleached Pug**
A very local species. It is mainly found in July and August in counties more to the south of England. There was one early record for Binbrook, "taken in garden at privet blossom in July 1905 (S.B.S.)". Larvae were also found in the Boston area in 1965, (R.E.M.P.).
L.F.P. Golden Rod, Ragwort.

863. *Eupithecia assimilata* Doubl. **Currant Pug**
A species which frequents gardens in May and June, with a later generation in the autumn. It is quite rare in Lincolnshire – there are few early records, and later ones for: Sleaford, 9th August 1956, "from larvae on Blackcurrant", and taken again in 1957 (Both G.N.H.); and Mablethorpe in the 1950s (T. R. New).
L.F.P. Currant.

864. *Eupithecia vulgata* Haw. **Common Pug**
Commonly seen across the county from May to July. Places mentioned in the records include: Grimsby, Laughton, Limber, Sleaford, Syston Park, Scunthorpe, Mablethorpe etc.
L.F.P. A wide variety of plants such as Bramble, Sallow, Hawthorn, Ragwort etc.

865. *Eupithecia denotata* Hubn. **Bell-flower Pug**
(campanulata H.-S.) **(Campanula Pug)**
A rare species in Lincolnshire – emerging in May and June. It was first recorded in the county in 1974 by R. E. M. Pilcher, and larvae were discovered at Lissington in 1975 by G. Haggett.
L.F.P. Campanula.

866. *Eupithecia castigata* Hubn. **Grey Pug**
Fairly common in May, June and July. There are recent records for: Sleaford, 3rd July 1958, (G.N.H.); Mablethorpe, 1952-1965, (T. R. New); S. Thoresby, 16th June 1979, (R.E.M.P.).
L.F.P. Widely polyphagous – eats foliage of many plants.

867. *Eupithecia icterata* Vill. **Tawny Speckled Pug**
This species was considered scarce early in the century, but it is quite well distributed over the county today in July, August and September. It was recorded in Lincolnshire, 1921, (Rev. F. S. Alston, L.N.U. Trans. Vol.5, p.154). It was at: Grimsby, 1944 (A. Smith); Laughton, 4th August 1959, (J.H.D.); Linwood Warren, 15th September 1963, "to m.v. light", (A.D.T.); Boston, 1965, (R.E.M.P.); and Scunthorpe, 1975, (J.H.D.).
L.F.P. Yarrow, Mugwort, Tansy.

868. *Eupithecia succenturiata* L. **Bordered Pug**
Early this century this species was considered "rare", with a single record for Lincoln. It flies in July and early August.
However, it was found in Grimsby in July 1948, 1954, and 1956 (G.A.T.J.); Scunthorpe, 1956, 1957, and 1959 (Quite plentiful), 1963, ("Common at house lights, a good year for many pugs"), (All J.H.D.); Boston, 1964, 1965, and S. Thoresby, 5th July 1979, (All R.E.M.P.).
L.F.P. Mugwort, Yarrow.

869. *Eupithecia indigata* Hubn. **Ochreous Pug**
A species found flying in May and June in Pine Woodlands. There are a few L.N.U. records for the early part of the century mentioning Ashby, Linwood and Skellingthorpe. The only further reliable record is for S. Thoresby, 6th June 1979, (R.E.M.P.).
L.F.P. Needles of Pine and Larch.

870. *Eupithecia pimpinellata* Hubn. **Pimpinel Pug**
(denotata Guen.)
A scarce species in Lincolnshire (out in June and July). An early record placed it at Barton on Humber "larvae common in a disused chalk pit on Burnet Saxifrage", 15.9.1897, (J.W.M. – identified by Mr. B. R. Bankes – Lep. Lincs 3).
Recently, larvae were found in the Boston area in 1965, more inland in 1971, and at Claxby in 1974 (all R.E.M.P.).
L.F.P. Burnet-saxifrage (on the flowers).

871. *Eupithecia extensaria* Freyer **Scarce Pug**
A coastal species which flies in June and July. Larvae were first recorded for the L.N.U. at Gibraltar Point in 1952 on Sea Wormwood (A. E. Smith, L.N.U. Trans. Vol.13 p.234). The species was later found at: the Saltings at Boston in 1964 (R.E.M.P.); Fishtoft, 1967 (J.B.); Boston, 21st August 1971, "3 larvae on Saltings" (A.E.S.) and "Lincolnshire, along coast 1971" (R.E.M.P.) etc.
L.F.P. Sea Wormwood.

872. *Eupithecia nanata* Hubn. **Narrow-winged Pug**
Found on Heathland in May, June and July – as indicated by the Lep. Lincs. 3 records for: Linwood, Wrawby Moor, Scotton Common etc. Most recently recorded at: Kirkby Moor, 15th May 1960, (G.N.H.); Stapleford Moor, 23rd July 1964, (L.N.U.): Woodhall Spa, 16th July 1965, (J.B.); Lincs. inland, 1971, (R.E.M.P.).
L.F.P. Ling, Heather.

873. *Eupithecia innotata Hufn.* **Angle-barred Pug**
Found in coastal regions in June/July. Lep. Lincs. gave the status as "rare". The records read "N. Skegness, 2 examples 16.7.1897 (G.T.P.)" plus "In the year 1879, Mr. G. Porritt took upon the sandhills at Skegness, Lincolnshire, two specimens which on examination proved to be genuine *E.innotata*".
There are only further records for the Boston area in 1964 and 1965, (R.E.M.P.).
L.F.P. Ash, Wormwood, Mugwort.

874. *Eupithecia fraxinata Crewe* **Ash Pug**
Scarce in the county (June, July and August). The first L.N.U. record was for Binbrook, 8th August 1907/8, (S.B.S.). There are three later records for Scunthorpe, 1959, 1960 and 1962 (all J.H.D.), one record for South Thoresby, 5th July 1979, (R.E.M.P.), and a record for Bagmoor, 4th June 1982 (R.J.).
L.F.P. Ash.

876. *Eupithecia virgaureata Doubl.* **Goldenrod Pug**
A very scarce species in Lincs. with the moth out in May and June. The only reliable records are from R.E.M.P. who discovered larvae in "central Lincolnshire" in 1978, and in the "Lincoln area" in June 1979.
L.F.P. Golden-rod, Ragwort.

877. *Eupithecia abbreviata Steph.* **Brindled Pug**
Flies in Oak woods in April and May. It was thought "scarce" early in the century, but there are later L.N.U. records for: Division 2 (S. Humberside) 1917, (Rev. Proudfoot); Kirkby Moor, 2nd May 1964, (G.N.H.); Woodhall Spa, "common" 1st May 1965, (G.A.T.J. and A.D.T.); New Park Wood, 14th May 1966, (A.D.T. and G.N.H.); and Holme Plantation, Messingham, May 1972, (J.H.D.).
L.F.P. Oak.

878. *Eupithcia dodoneata Guen.* **Oak-tree Pug**
Another species of Oak woodland, out in May and June. There is one early L.N.U. record for: Ropesley, "rare", (W. A. Atmore – Lep. Lincs. 3); a record for Bardney Forest, "reared from larvae", June 1979, (R.E.M.P.); and a record for Twigmoor Woodlands, 31st May 1982, (R.J.).
L.F.P. Oak.

880. *Eupithecia sobrinata Hubn.* **Juniper Pug**
(pusillata auct.)
South indicated that "The moth is out from late July to early October and may be found in nearly all parts of the British Isles where the food plant occurs". Lep. Lincs. 3 noted it at Market Rasen and Wyberton, (Rare).
More recently it was recorded at South Thoresby, as "widespread since 1949", (R.E.M.P.), and it remains to date at this locality (e.g. 12th July 1979).
L.F.P. Juniper.

882. *Eupithecia lariciata Freyer* **Larch Pug**
This species flies in May and June. It was found quite widely according to the early records, wherever Pine woods occur in the county, (e.g. Holton Le Moor, Ashby, Elsham, Pelham's Pillar Woods, Welton, Roxton, Hartsholme, Wrawby Moor, Limber etc.). It has been seen in Pine woods in the 1970s by R.E.M.P. and J.H.D.
L.F.P. Larch, Spruce.

883. *Eupithecia tantillaria Boisd.* **Dwarf Pug**
(pusillata Schiff.)
Out in May and June in Pine woods. There are only 3 L.N.U. records for: Limber, May 1948, (G.A.T.J.); Scawby Woods, "several 1953", (J.H.D.); and S. Thoresby, 1973, (R.E.M.P.).
L.F.P. Spruce.

884. *Chloroclystis coronata Hubn.* **V Pug**
Can be found from May to October (in two generations) in wooded districts. It was only known from Hartsholme Priory early this century. There are, however, a good number of further records indicating an increase since Lep. Lincs. 3. Later records include: Skellingthorpe, 1915, (Revs. Blathwayt and Proudfoot); Grimsby, 19th May 1954, (G.A.T.J.); Near Kelby, 20th June 1961, "bred from larvae", (G.N.H.); Scunthorpe, 2nd September 1962, (J.H.D.); Boston, August 1970, (A.E.S.); Riseholme, 1971, (A.D.T.); New Park Wood, Bardney, 11th August 1979, (S. Van den Bos).
L.F.P. Blossoms of Sweet Chestnut, Hawthorn, Bramble etc.

885. *Chloroclystis debiliata Hubn.* **Bilberry Pug**
A species normally found in June and July. The only mention in the records stems from Lep. Lincs. 3 when it was supposedly taken near Gainsborough in 1860 (F.M.B.).
L.F.P. Bilberry.

886. *Chloroclystis rectangulata* L. **Green Pug**
Can occur commonly in orchards in June and July. It was at Grimsby, 28th June 1952, (G.A.T.J.); Scunthorpe, 1959, '60, '62, (J.H.D.); Sleaford, 29th June 1965, (G.N.H.); Boston, 1965, (R.E.M.P.); Boston, 26th June 1972, (A.E.S.); and Messingham, 11th June 1982, (R.J.); etc.
L.F.P. Flowers of Apple, Hawthorn etc.

887. *Gymnoscelis pumilata* Hubn. **Double-striped Pug**
Occurs in two generations between April and August. There was one locality known in Lep. Lincs 3 – Lincoln, (F.M.B.). It was found later at: Sleaford, May 1964, (G.N.H.); Kirkby Moor, 2nd May 1964, "common on Broom flowers", (G.N.H.); Scotton Common, "larvae plentiful – and bred – on Gorse flowers, in early 1970s" (J.H.D.).
L.F.P. Flowers of Furze, Broom, Hawthorn etc.,

887A. *Chlorocclystis chloerata* Mabille
L.N.U. records note that this species was new to the British List on 16th April 1971, (E. C. Pelham – Clinton). There is one county record for Wragby Road, Lincoln, 1972, (R.E.M.P.). According to the above Entomologist's Gazette (Vol.23, p.152) however, E. C. Pelham – Clinton notes that "During 1972 larvae have been found in many parts of Southern England from Essex to Gloucestershire, and specimens have been found in collections which show, that the species has been a British resident for many years". (written 31.V.1972). Later in the Gazette (p.220) B. Goater wrote: "It is a sobering thought that here is yet another species of macrolepidoptera which must have been with us from time immemorial; moreover, it is evidently widespread, at least in southern England, and by no means uncommon. Yet we have passed it by".
L.F.P. Blackthorn blossom and leaves.

DEILINIINAE

888. *Abraxas sylvata* Scop. **Clouded Magpie**
 (ulmata F.*)*
A woodland species found commonly where the larval foodplants thrive, (May and June). It has been recorded from, for example, Elsham, 24th May 1952, (M.P.G.); Irby Dales, 25th June 1955, (L.N.U.); Sturton, 23rd June 1959, (J.H.D.); Crowle Crossroads, July 1975, (P. W. R. Walter); and Hendale Wood, 14th July 1979, "swarming – extremely common", (R.J.); etc.
L.F.P. Wych Elm, Beech.

889. *Abraxas grossulariata* L. **Common Magpie**
Abundant all over the county in gardens, woodlands, lanes etc. in July and August, with larvae an occasional pest on Gooseberry, Currant etc.
It has been taken from Alford, Irby Dales, Humberstone, Scunthorpe, Sleaford, Welland Washes, Laughton, Grimsby, Ancaster, Boston, etc.
L.F.P. Sloe, Hawthorn etc.

891. *Lomaspilis marginata* L. **Clouded Border**
Found commonly among Sallows and Poplars in May, June and July.
There are numerous records for: Newball, Skellingthorpe, Wrawby Moor, Scawby, Tattershall, Bardney, Manton, Marton, Welland, Epworth Turbary, Muckton Wood, Elsham Park, Roughton Moor, Scotgrove Wood, etc.
L.F.P. Sallows, Aspen, Willows.

892. *Ligdia adustata* Schiff. **Scorched Silver**
(Scorched Carpet)
Out in two generations (April to June, and August/September), but an uncommon species in Lincolnshire.
There are early records for Wyberton, 1896, (J.C.); Panton, "one at light", 27.8.1895, (G.H.R.); and Hartsholme, 18th May 1903, (J.F.M.), (Lep. Lincs 3). The only more recent records are for Welton Wood near Alford, 22nd May 1971, "to light", (J.H.D.), and for Dunholme, 2nd June 1978, (K.S.).
L.F.P. Spindle.

894. *Bapta bimaculata* F. **White-pinion Spotted**
(taminata Schiff.)
This moth is out in May and June along hedgerows or in wooded areas. It is rare and local in Lincolnshire, with L.N.U. records since Lep. Lincs 3 for: Woodhall Spa, 26th May 1957, (F.L.K.); Boston, 8th June 1974," and Boston again, June 1979, and 1980 (all A.E.S.); and Washingborough, 19th May 1981, (Mrs. A. E. Binding).
L.F.P. Hawthorn, Wild Cherry, Plum.

RECORDS OF LINCOLNSHIRE SPECIES 237

895. *Bapta Temerata Schiff.* **Clouded Silver**
(punctata Hubn.)
Fairly common and widespread in woodlands in May and June. There are a large number of localities for it in the records, including Alford, Hendale Wood, Scawby, Aswardby Thorns, Manton, Bradley Woods, Limber, Laughton, etc.
L.F.P. Blackthorn, Bird Cherry, Plum.

896. *Deilinia pusaria L.* **White Waved Silver**
(common White Wave)
An abundant species among Birch scrub, woodlands, heathland etc. in two broods (May/June and August/September). Typically recorded from: Nocton Wood, Limber, Easton Hall, Grimsby, Twigmoor, Scotton, Tumby Woods, Epworth Turbary, Muckton Wood, Market Rasen, Woodhall, Lincoln, Twyford Forest etc.
L.F.P. Birch, Alder etc.

897. *Deilinia exanthemata Scop.* **Common Waved Silver**
(Common White Wave)
Common in May, June and July in the damper areas where Sallows abound. Recorded from: Stapleford, Pickworth, Tattershall, Scotton Common, Laughton, Crowle Waste, Scotgrove Wood, etc.
L.F.P. Sallow, Aspen.

898. *Ellopia fasciaria L.* **Barred Red**
(prosapiaria L.)
Not too common a species, but, as reported in Lep. Lincs. 3, "Generally to be found sparingly in woods where there are plenty of Scotch Fir trees". It is on the wing in June and July. There are L.N.U. records for: Linwood, 1923, (A. E. Musgrave); Grimsby, 12th July 1955, (G.A.T.J.); Swallow Vale, 2nd August 1958, (G.A.T.J.); Woodhall Spa, 17th July 1964, (G.A.T.J.); Laughton and Twigmoor, 20th July 1974, (J.H.D.), etc.
L.F.P. Scots Pine

899. *Campaea margaritata L.* **Barred Light-green**
(margaritaria L.) **(Light Emerald)**
Found plentifully in woodlands in June and July, and comes in numbers to light. Localities mentioned in L.N.U. records include: Barton-on-Humber, Grimsthorpe Park, Bottesford, Elsham, Easton Hall, Aswardby, Brocklesby, Linwood, Ancaster Valley, Woodhall, Riseholme Freshney Bog, Kingsthorpe, Boston, Bardney, etc.
L.F.P. Oak, Beech, Birch etc.

900. *Angerona prunaria* L. **Orange Thorn**
(The Orange Moth)
A very local species in Lincolnshire – restricted to a few wooded areas where it flies in June and July. There are early records for: Market Rasen and Newball Wood; Owersby, 1879, (F.A.L.); Tothill, June 1880, (C. D. Ash); Bourne Wood and Holbeach District, (L.M.C.); and Skellingthorpe in 1915, 1946, 1949, and 1951. It was at Stainton Wood in June 1919, (L.N.U. Trans. Vol.5, p.57); Doddington, Old Hay Wood, 15th July 1958, (L.N.U. – G.A.T.J.); and Bardney, 1975, (M.T.).
L.F.P. Hawthorn, Birch, Sloe, Honeysuckle etc.

901. *Semiothisa notata* L. **Blunt Peacock Angle**
(Peacock Moth)
A very local species in Birch wooded areas between May and August. The only L.N.U. record is for South Thoresby, "to light", 22nd August 1976, (R.E.M.P.).
L.F.P. Birch, Sallow.

902. *Semiothisa alternaria* Hubn. **Sharp Peacock Angle**
(alternata Schiff.) **(Sharp-angle Peacock)**
Flies in May/June and occasionally again in July/August, and is always very locally distributed. The only L.N.U. records are from Lep. Lincs 3 for near Gainsborough and Lincoln (F.M.B.).
L.F.P. Sallow, Alder, Sloe.

903. *Semiothisa liturata* Clerck **Tawny-barred Angle**
Frequently seen in Pine woodlands in June and July mainly, but with a possible second generation.
The more recent records include: Hendale Wood, 10th July 1955, (M.P.G.); Bourne Woods, 16th June 1956, (G.N.H.); Scawby Woods, June and July 1957 and 1958, (J.H.D.); Linwood, 1971, (A.D.T.); Boston, 3rd August 1972, (A.E.S.); and Laughton, 30th June 1973, *"ab. nigrofulvata"*, (R.J.).
L.F.P. Scots Pine, Larch.

904. *Theria rupicapraria* Schiff. **Early Umber**
(Early Moth)
Abundant all over the county in January and February. Found at Sleaford, Scunthorpe District, Kelby, Bardney, Barton etc.
L.F.P. Hawthorn, Sloe.

905. *Erannis leucophaearia Schiff.* **Spring Umber**
A locally common species (February/March) where Oak occurs, e.g. Scawby, Kexby, Glentworth, Bourne Woods, Messingham, all the Bardney Woodlands etc.
L.F.P. Oak.

906. *Erannis aurantiaria Hubn.* **Scarce Umber**
Emerges towards the end of the year (October/November), and larvae can often be found to be locally abundant in woodlands in the following spring.
Typical records include: Bottesford, 17th November 1955, (M.P.G.); Knaith, larvae on 25th May 1957, (L.N.U.); Sleaford, November 1959, (G.N.H.); Holme Plantation, October 1976, (J.H.D.) etc.
L.F.P. Oak, Birch, Hawthorn.

907. *Erannis marginaria F.* **Dotted Border**
(progemmaria Hubn.)
A common species all over the county, emerging during warm spells in the spring (February, March, April) among hedgerows and in woodlands.
Localities for it include: Howsham, Osgodby Moor, Grimsby, Bourne, Roughton Moor, Bardney Woodlands, Scunthorpe, Boston, etc.
L.F.P. Hawthorn, Birch, Sallow, Oak, etc.

907. *Erannis marginaria F.* **Dotted Border**
(progemmaria Hubn.)
A common species all over the county, emerging during warm spells in the spring (February, March, April) among hedgerows and in woodlands.
Localities for it include: Howsham, Osgodby Moor, Grimsby, Bourne, Roughton Moor, Bardney Woodlands, Scunthorpe, Boston, etc.
L.F.P. Hawthorn, Birch, Sallow, Oak, etc.

908. *Erannis defoliaria Clerck* **Mottled Umber**
Appears abundantly late in the year (October to December) and sometimes during the following spring (January to March). Some years the larvae are so numerous that extensive defoliation of trees can occur.
It has been found at: Nocton Wood, the Holland Division, Holton Le Moor, Skellingthorpe, Holme Plantation, Nettleton, Haverholme Priory, Sleaford, Bardney etc.
L.F.P. Foliage of trees such as Birch, Oak etc.

ENNOMINAE

909. *Anagoga pulveraria* L. **Barred Umber Thorn**
(Barred Umber)
A woodland species, on the wing in May and early June, but not too common in Lincolnshire. There are records for: Linwood, 5th June 1911 (L.N.U.); Newball Wood, 1926 (A.E.M.); New Park Wood, 14th May 1966, (A.D.T.); Skellingthorpe, 1952, 1955, 1976, (M.P.G. and J.H.D.). L.F.P. Sallow, Birch, Ash, etc.

910. *Ennomos autumnaria* Wernb. **Large Thorn**
(alniaria Schiff. non L.)
A migrant species, normally found coming to light in the autumn, around the coast to the south and east of the country.
It was first recorded for the L.N.U. at Tothill Wood, August 1976, (R.E.M.P.), and it turned up at Gibraltar Point, 30th August 1976, (M.T.). L.F.P. Birch, Hawthorn, Sloe etc.

911. *Ennomos quercinaria* Hufn. **August Thorn**
(angularia Hubn.)
An uncommon species in Lincolnshire – seen in August and September. Early records placed it at: Lincoln, Market Rasen, Pelham's Pillar Woods, Haverholme Priory and Holbeach. Subsequently it has been found at: Twyford Forest, 1932, (A.E.M.); Grimsby, "one only", 13th September 1957, (G.A.T.J.); Riseholme, 15th September 1968, (A.D.T.).
L.F.P. Birch, Hawthorn, Oak etc.

912. *Deuteronomos alniaria* L. **Canary-shouldered Thorn**
(tiliaria Borkh.)
Widely distributed in damp areas in late August and September. It was thought "not common" early this century, but it is certainly plentifully recorded in suitable localities today – e.g. Sutton on Sea, Barton, Candlesby, Laughton, Woodhall, Grimsby, Epworth, Saltfleetby, Boston, Riseholme, Burton Pits, Greetwell, etc.
L.F.P. Birch, Alder, Sallow etc.

913. *Deuteronomos fuscantaria* Steph. **Dusky Thorn**
Not plentiful, but occurs over a wide area in September mainly, in wooded habitats. Following Lep. Lincs. 3 it was next recorded in: Scunthorpe, 1953 (J.H.D.); Lincoln, 11th September 1955, (B.J.); Scotton Common, 10th September 1955, (M.P.G.); Laughton, "5 seen" 9th September 1959, (G.A.T.J. and J.H.D.); Sleaford, 10th September 1964, (G.N.H.); Boston, 22nd September 1968, (A.E.S.); Gibraltar Point, 4th October 1969, (A.D.T.); Alkborough, 1970, (M.P.G.). L.F.P. Ash.

RECORDS OF LINCOLNSHIRE SPECIES 241

914. *Deuteronomos erosaria Schiff.* **September Thorn**
This species was considered "very rare" at the time of Lep. Lincs. 3. It now turns up in a number of areas where Oak predominates, in August and September, though still not in large numbers.
The more recent records include: Barton on Humber, 1946, (C.G.E.); Scunthorpe 1955, Manton Common 1956, and Laughton, July 1958, (all J.H.D.); Skellingthorpe, 12th August 1962, (A.D.T.); Sutton on Sea, 18th September 1970, (G.H.); and New Park Wood, Bardney, 11th August 1979, "several to light", (J.H.D.).

L.F.P. Oak.

915. *Selenia bilunaria Esq.* **Early Thorn**
 (illunaria Hubn.)
Widespread across the county – occurs commonly in two generations (spring – March to May, Summer – July and August.)
Localities in the records include: Irby Dales, Grimsby, Barton, Sleaford, Swallow, Bardney, Ruskington, Roughton, Mablethorpe, Boston, Lincoln, etc.

L.F.P. Birch, Sallow, Hawthorn etc.

916. *Selenia lunaria Schiff.* **Lunar Thorn**
A local species, out between May and July. It was found early this century at: Market Rasen, Haverholme Priory, Holbeach (var. *delunaria Hubn.)* etc. (Lep. Lincs. 3). It was more recently at: S. Thoresby, 1970 and 1971, (R.E.M.P.); and Whitgift and Fockerby – during the "Humber Survey" between late May and July 1976, (Andrew Grieve).

L.F.P. Sloe, Oak, Birch.

917. *Selenia tetralunari Hufn.* **Purple Thorn**
 (illustraria Hubn.)
There are two generations of this species (April/May and July/August) in wooded areas. It was only once recorded early this century, but is now found more frequently. Records from the 1960s onwards include: Woodhall Spa, 1st May 1965, (G.A.T.J.); New Park Wood, 14th May 1966, (A.D.T.); S. Thoresby, 1970 and 1971, (R.E.M.P.); Boston, 9th May 1973, (A.E.S.); and Gibraltar Point, 5th August 1979 (R.E.M.P.).

L.F.P. Birch, Alder, Sallow, Oak etc.

918. *Apeira syringaria* L.   **Lilac Thorn**
(Lilac Beauty)

A woodland species found very locally in June and July, in small numbers, over a wide area.

Following Lep. Lincs. 3 it has been seen at: Denton, 1924, (H. L. Bond); Wrawby Moor, 26th May 1951, "larvae", (M.P.G.); Welton Le Marsh, 7th June 1952, (L.N.U.) Skellingthorpe Woods, 25th May 1955, (M.P.G.); Dunsby Wood, 5th June 1959, (J.H.D.); Irnham, June 1962, "larvae and pupae", (G.N.H.); Broughton, May 1978, (L.N.U.); Normanby le Wold, 23rd August 1978, (D.B.); etc.

                        L.F.P. Privet and Honeysuckle.

919. *Gonodontis bidentata* Clerck.   **Scalloped Hazel Thorn**

Very common in woodland (May and June) across the county, with larvae being found commonly on a number of plants in gardens. It is a very variable species, with some moths being extremely dark in shading.

Localities mentioned in the records include: Boston, Manton, Grimsby, Brocklesby, Scunthorpe, Bardney, Bourne, etc.

                       L.F.P. Foliage of many trees –
                       e.g. Birch, Oak, Hawthorn, Sallow
                                               etc.

920. *Colotois pennaria* L.   **Feathered Thorn**

Found plentifully in wooded districts in the autumn (October to December).

Recorded localities include: Skellingthorpe, Welton Le Marsh, Holme Plantation, Bardney, Knaith, Sleaford, Elsea Wood near Bourne, Grimsby, etc.

        L.F.P. Foliage of trees like Poplar, Birch, Hawthorn etc.

921. *Crocallis elinguaria* L.   **Scalloped Oak Thorn**

Emerges in wooded districts in July and August, and is quite common and widespread in the county.

It has been found at: Alford, 1950, (F.L.K.); Boston, 23rd July 1952, (Miss B. Hopkins); Grimsby, 27th July 1955, (G.A.T.J.); Sleaford, July 1957, (G.N.H.); Kelby, 30th July 1960, (G.N.H.); Chapel St. Leonards, 25th August 1960, (J.H.D.); and Lincoln, 25th August 1969, (A.D.T.); etc.

                       L.F.P. Foliage of most trees.

922. *Plagodis dolobraria* L. **Scorched-Wing**
Found in woodlands in May and June – the species is seen quite frequently, but not in large numbers.
The more recent records include: Wragby, 1940, (R. K. Cornwallis); Bottesford, 5th June 1955, (M.P.G.); Grimsby, 4th June 1958, (G.A.T.J.); Lincoln, 1971, (A.D.T.); Goxhill, 16th June 1973, (G. Potts); and Hendale Wood, 20th June 1975, (R.J.), etc.
L.F.P. Oak, Birch, Sallow.

923. *Opisthographtis luteolata* L. **Sulphur Thorn**
  (crataegata L.*)* **(The Brimstone)**
An abundant species almost everywhere in May, June and July. Of the many places mentioned in the records, a few examples are: Irby Dales, Brocklesby, Twigmoor, Riby, Sleaford, Scunthorpe, Woodhall, Skellingthorpe, Riseholme, Kelby, Boston, Barrow Haven, etc.
L.F.P. Hawthorn.

924. *Epione repandaria* Hufn. **Common Bordered-beauty**
  (apiciaria Schiff.*)* **(Bordered Beauty)**
On the wing in July, August and September, but not commonly recorded in Lincolnshire – as instanced by the comment in Lep. Lincs. 3: "As this species has the habit of flying late at night it may be commoner than the records appear to show".
It has been recorded at: Little Coates, 1914, (F. W. Sowerby); Lincoln, 1943, (T. C. Taylor); Scotter, 14th August 1954, (L.N.U.); Woodhall, 15th August 1966, (A.D.T.); Epworth Turbary, 1967, (L.N.U.); and Brumby Nature Reserve, September 1978, (S.B.).
L.F.P. Sallow, Willow, Alder.

927. *Pseudopanthera macularia* L. **Speckled Yellow**
A woodland species which flies in the sunshine during May and June. It is quite common in areas of the South of England, but not in Lincolnshire. There is a single mention of it in L.N.U. records, implying that it was once found in woods near Gainsborough, (F.M.B.). The record cannot be confirmed.
L.F.P. Wood Sage, Dead Nettle, etc.

OURAPTERYGINAE

928. *Ourapteryx sambucaria* L. **Swallow-tailed Elder**
(Swallow-tailed Moth)
Very common in all kinds of habitat in July. There are many records for such localities as: Grimsby, Woodhall, Barton, Sleaford, Saltfleetby, Twigmoor, Kirkby Underwood, Bardney, Mablethorpe, Boston, etc.
L.F.P. Hawthorn, Elder, etc.

244 THE BUTTERFLIES AND LARGER MOTHS

BISTONINAE

929. *Phigalia pedaria* F. **Pale Brindled-beauty**
(pilosaria Schiff.)
Commonly found in the early spring (March) in a forest areas. The species has been typically found at: Holton le Moor, Nocton Wood, Skellingthorpe, Glentworth, Knaith, Sleaford, Boston, Bardney, Twigmoor, etc.
A high proportion of the specimens seen are of the melanic form.
L.F.P. Most forest trees – Birch, Oak, Elm, Lime, etc.

930. *Apocheima hispidaria Schiff.* **Small Brindled-beauty**
This species emerges early in the year (February/March) in woodlands. In Lep. Lincs. 3. it was reported to be "scarce", and at Ashby, Brigg, (R.T.C.); and Gainsborough, (F.M.B.). The only further records are for Woodhall and Bardney, "early and abundant", in March and April, (R.E.M.P.).
L.F.P. Hawthorn, Birch, Elm.

933. *Lycia hirtaria* Clerck **London Brindled-beauty**
(The Brindled Beauty)
In Lincolnshire the range of this species covers mainly the southern half of the county. It flies in March, April and May. Lep. Lincs. 3 called it scarce, but since then it has been found at: Sleaford, 1946, (A. Pilkington); Woodhall, 28th April 1954, (F.L.K.); Grimsby, 30th April 1959, "to m.v. light", (G.A.T.J.); Boston, 1961/62, (R.E.M.P.); New Park Wood, 14th May 1966, (G.N.H.); Lincoln, Riseholme, Hatton Sykes and Southrey, 1972, (A.D.T.); Messingham Sand Quarries, "larvae July 1976", (J.H.D.).
L.F.P. Lime, Elm, Willow.

934. *Biston strataria* Hufn. **Oak Brindled-beauty**
(prodromaria Schiff.) **(Oak Beauty)**
A woodland species which emerges in March and April. Lep. Lincs. 3 noted its status as "scarce". It remains rare in the South Humberside region, but has been found more commonly to the south of the county (e.g. Boston).
It has been recorded more recently at: Alford, 24th March 1950, (F.L.K.); Manton Warren, 12th May 1956, (J.H.D.) and G.A.T.J.); Glentham, "larvae" 1958, (M.P.G.); Boston, 1961 and 1962, "early and abundant", (R.E.M.P.); Roughton Moor, 11th April 1965, (G.A.T.J. and G.N.H.); Boston, 18th March 1974, (A.E.S.).
L.F.P. Oak, Birch, Elm etc.

935. *Biston betularia* L. **Pepper-and-salt**
(Peppered Moth)
A species found commonly across the county in June and July. At the time of Lep. Lincs. 3 it was recorded that "the type form appears to be rapidly decreasing in numbers, being superseded by the melanic variety", (i.e. the black form *carbonaria*). The melanic form became dominant, and was virtually the only form seen up to the end of the 1950s. From this point, the numbers of "typical" specimens noted have seemed to increase, and the intermediate form *(insularia)* has put in an appearance in recent years.
There are many records, and they cover all areas of Lincolnshire.
L.F.P. Oak, Birch, Sallow etc –
a wide variety of trees and shrubs.

BOARMIINAE

936. *Menophra abruptaria* Thunb. **Waved Umber Beauty**
(Waved Umber)
Found rather locally in April and May. There are records for: Barton on Humber, 21st May 1951, (M.P.G.); Grimsby, May 1954, (G.A.T.J.); Sleaford, 7th May 1958, (G.N.H.); New Park Wood, 14th May 1966, (A.D.T. and G.N.H.); Mablethorpe, 1965, (T. R. New); Boston, 29th May 1969, (A.E.S.); Scawby, 17th April 1975, (R.J.).
L.F.P. Privet, Lilac.

938. *Cleora rhomboidaria* Schiff. **Willow Beauty**
(gemmaria Brahm*)*
Plentiful in July and August, with sometimes a partial second generation in September. There are a widespread variety of localities in the records, including: Little Coates, Grimsby, Scunthorpe, Sleaford, Syston Park, Boston, Bardney, Lincoln, etc.
L.F.P. Hawthorn, Birch, Ivy, Privet.

939. *Cleorodes lichenaria* Hufn. **Brussels Lace**
Very scarce in the county. There are no recent records, but entries in Lep. Lincs. 3 read:
N. Alford District, Mother Wood (Aby), 4.7.1891, (E.W.); Gainsborough and Lincoln, (F.M.B.); Theddlethorpe, 1904, (A. E. Gibbs – Ent. 1905 p.81.).
L.F.P. Lichen growing on tree branches.

940. *Deileptenia ribeata* Clerck. **Satin Beauty**
(abietaria Schiff.*)*
This moth is out between June and August. There is one mention of it in L.N.U. records, for: Tealby, 1877 to 1879, (F.A.L.).
L.F.P. Pine, Spruce, Oak etc.

941. *Alcis repandata* L. — **Mottled Beauty**

A very common moth all over the county in June and July, and a number of dark forms are found. There are records for: Limber, Scawby, Barton, Muckton Wood, Woodhall, Saltby, Boston, Newball, Stamford, etc.   L.F.P. Hawthorn, Birch, Hazel, etc.

944. *Boarmia roboraria* Schiff. — **Great Oak Beauty**

Emerges in June and July in Oak woodland. It was noted as "scarce and local" in Lep. Lincs. 3 and there are few subsequent records. These are for: Skellingthorpe, 1915, (Rev. Blathwayt and Rev. Proudfoot); Lincoln area, 1921 (H. C. Bee, L.N.U. Trans. Vol. 5., p.154); Sleaford, 1948, (H. Pilkington); Linwood, 25th June 1959, (A.D.T., M.P.G. and G.A.T.J.); Linwood, 22nd June 1960, (J.H.D.).   L.F.P. Oak.

945. *Pseudoboarmia punctinalis* Scop. — **Pale Oak Beauty**
(consortaria F.)

A species out in woodland in June and July. The only note in L.N.U. records mentions larvae taken by Rev. S. Proudfoot, 1917 in the area now known as S. Humberside.   L.F.P. Oak, Birch

946. *Ectropis biundularia* Borkh. — **Early Engrailed**
(bistortata auct.) **(The Engrailed)**

Appears in March/April and again in June/July, and has been widely recorded across the county.

There is some confusion in the records, however, as the following passage indicates: "There is so much difference of opinion as to whether this insect is the same species as *crepuscularia* Hubn., or whether they are to be regarded as 2 distinct species, that I have treated all records as referring to one species, *bistortata*, and I have accordingly set out in detail the dates of capture so far as possible", (Lepidoptera Records Secretary).

Localities mentioned include: Market Rasen, Coles Wood, Sweeting Thorns, Pillar Woods, Usselby, Owston Ferry, Nocton Wood, Haverholme Priory, Girsby Manor, Laughton, Kirkby Moor, Bardney, etc.   L.F.P. A wide range of trees.

947. *Ectropis crepuscularia* Schiff. — **Small Engrailed**
(biundularia Esp.)

Flies in May and June, and difficult to separate from the previous species. It has been reported from Newball, Skellingthorpe and Doddington, 1927, (A.E.M.); Linwood, 1961, (G.A.T.J.); New Park Wood, 14th May 1966, (A.D.T.); and Swallow Vale, 23rd June 1979, (J.H.D. and R.J.).

L.F.P. Foliage of deciduous trees.

950. *Aethalura punctulata* Schiff. **Grey Birch Beauty**
(punctularia Hubn.) **(The Grey Birch)**
Commonly found between late April and June, in woods and on heathland where Birch occurs.
Examples of localities are: Scawby, Girsby Manor, Manton Warren, Bourne Woods, Scunthorpe, Kirkby Moor, Bardney Woodlands, Scotton Common, etc.

L.F.P. Birch.

952. *Pachycnemia hippocastanaria* Hubn. **Horse-chestnut Longwing**
Flies in April and May, with a second generation in August, and is a heathland species. There is one record – one larva was swept from Heather at Scotton Common by George Hyde in June 1977. This larva was verified as the above species by Gerald Haggett. This species is known to be a rather local insect on heaths in the South of the country. Its occurrence in Lincolnshire was a cause of some surprise.

L.F.P. Heather, Ling.

958. *Ematurga atomaria* L. **Common Heath Beauty**
(Common Heath)
Plentiful in May and June in all Lincolnshire heathlands. Areas mentioned in L.N.U. records include: Woodhall Spa, Osgodby Moor, Scawby, Linwood, Manton Common, Epworth Turbary, Scotton Common, Kirkby Moor, Welton Wood, Roughton Moor, etc.

L.F.P. Ling, Heath.

959. *Bupalus piniaria* L. **Bordered White Beauty**
A species found in Pine woods in May and June. The males fly in the day time, and can sometimes be seen in good numbers. The females are more rarely seen on the wing, and need to be beaten from cover. The females are also much more variable in colour form, ranging from red forms to more dark colouration. Many woods are mentioned in the records, e.g. Twigmoor, Wrawby Moor, Woodhall, Willingham Forest, Skellingthorpe, Kirkby Moor, Linwood, Haverholme, etc.

L.F.P. Pine.

248   THE BUTTERFLIES AND LARGER MOTHS

961. *Itame wauaria* L.   **V looper**
                                                     (V-moth)
This species, which is out in July and August, was described as "common in fruit gardens" in Lep. Lincs. 3, and was recorded in many of the county divisions. Today it is certainly more scarce.
    All subsequent records are either for Grimsby (1948, '54, '55, '56, '57, '59, '62 G.A.T.J.) or Scunthorpe (1958, '59, '60 – J.H.D.).
                                        L.F.P. Gooseberry and Currant.

962. *Itame brunneata* Thunb.   **Rannoch Looper**
      (*fulvaria* Vill.)
This species, as its English name implies, is found mainly in Scotland, and is out in June and July.
    A "single specimen" appeared at Boston, 23rd June 1960, (R.E.M.P.), and as Mr. Pilcher noted for the records "This species does occur from time to time in East Anglia."
                                                  L.F.P. Bilberry.

963. *Lithina chlorosata* Scop.   **Brown Silver-lined**
      (*petraria* Hubn.)
Very common in woodlands among Bracken (May/June) all over Lincolnshire. It has been noted at: Limber, Woodhall, Wrawby Moor, Scawby, Pickworth, Tetford, Willingham Forest, Risby Warren, Broughton, Norton Disney, Newball Wood, etc.
                                                  L.F.P. Bracken.

964. *Chiasmia clathrata* L.   **Heath Lattice**
                                              (Latticed Heath)
Widespread in April and May, and again in July and August in limestone areas – including being found along road verges. It was thought "uncommon" in the late 1800s but there are recent records for many areas, e.g. Swallow Vale, 22nd June 1951, (G.A.T.J.); Ancaster, 12th July 1952, (M.P.G.); Pickworth, 19th June 1954, (L.N.U.); Legsby, 1954, (R. Cornwallis); Manton Common, 17th August 1957, (J.H.D.); Willingham Forest, 26th May 1956, (L.N.U.); Fishtoft and Gibraltar Point, 12th June 1971, (A.E.S.); etc.
                                        L.F.P. Trefoils, Clover.

965. *Dyscia fagaria* Thunb.   **Grey Scalloped Bar**
      (*belgaria* Hubn.)
Found in damp heathland in June and July. There is one entry in L.N.U. records for: "Woodhall, Scotton and Laughton, between 1930 and 1945, (R.E.M.P.)".
                                         L.F.P. Ling and Heath.

968. *Aspitates ochrearia* Rossi **Yellow Belle**
(citraria Hubn.)
There are two generations of this species (May/June and August/September) along the coast on dunes or rough land. Lep. Lincs. 3 gave it at Saltfleetby, "2 examples about 1890", (C.P.A.). The only other record is for Gibraltar Point, August 1958 and 1959, (J.H.D.).
L.F.P. Low plants like Plantain and Knotgrass.

969. *Perconia strigillaria* Hubn. **Grass-waved**
(Grass Wave)
A heathland species – out in June and July. Lep. Lincs. 3 described it as "rare" and recorded it at Roughton Moor. All further records are for: Kirkby Moor, June 1962, '65, '67, '68, (J.B., G.N.H., and A.D.T.); Roughton Moor, 18th June 1968, (G.N.H.); and Bardney, 1975, (M.T.).
L.F.P. Ling, Heath, Broom.

CHAPTER 4.
CONTRIBUTORS TO THE AUTHORS AND TO THE LINCOLNSHIRE NATURALISTS' UNION RECORDS

A.
Miss E. Maude Alderson
Mr. R.P. Alington
Mr. T.H. Allis
Rev. F.S. Alston
Mr. C.P. Arnold
Mr. C.D. Ash
Mr. E.A. Atmore
Dr. F.M. Aungier

B.
Mr. R.W. Bacon
Mr. F. Baines
Mr. F.T. Baker
Mr. R.R. Bankes
Mr.C.G. Barrett
Mr. J. Bebbington F.R.E.S.
Mr. H.C. Bee
Rev. W. Beecher
R. Benn
Mr. and Mrs. A. Binding
Dr. F.P.H. Birtwistle
Rev. F.L. Blathwayt
Mr. H.L. Bond
Mr. S. Van Den Bos
Mr. J.W. Boult
Mr. D. Brant
Mr. F. Braisier
Mr. R.F. Bretherton
Mr. W.H. Brooks
Mr. A. Bullock
Mr. D. Burton
Mr. F.M. Burton

C.
Mr. R. Carlton
Mr. C.S. Carter
Mr. J.W. Carr
Professor Carr
Mr. W.D. Carr
Dr. R.T. Cassal
Dr. Cassel
Mr. A.E. Chambers

Mr. J.C. Lane-Claypon
Mr. C. Clayton
Mr. F. Clayton
Mr. E.A. Cockayne
Rev. W.W. Cooper
Mr. H.H. Corbett
Mr. J. Cordeau
Mr. R.C. Cornwallis
Mr. T.H. Court
Mr. J. Coward
Mr. J.D. Coward
Mr. J.L. Cox
Mr. L.M. Curtiss
Captain W.A. Cragg
Mr. T.P.R. Crane
Mrs. Cross
Mr. B.T. Crow
Mr. V. Crow
Mr. P. Cue

D.
Rev. C.D. Dale
Mr. J.C. Dale
Annie Dows
Mr. J.H. Duddington

E.
Mr. C.G. Else
Mr. E. Exton

F.
Mr. B. Featherstone
Dr. E.H. Felton
Mr. P. Forrington
Canon Fowler
Mr. R.A. French

G.
Mr. R. Garfit
Mr. G. Gascoyne
Dr. George
Miss J. Gibbons

Mr. A.E. Gibbs
Mr. B. Goater
Mrs. A. Goodall
Mr. R. Goodall
Miss E.M. Goom
Miss N. Goom
Mr. M.P. Gooseman
Mr. R.W. Goulding
Mr. R. Goy
Mr. W. Gray
Mr. G.A. Grierson

H.
Mr. G. Haggett
Mr. A.E. Hall
Mr. G. Harmsworth
Mr. J.A. Hardy
Rev P.C. Hawker
Mr. I.R.P. Heslop
Mr. G.N. Holland
Miss B. Hopkins
Mr. J. Hossack
Canon G. Houlden
Mr. W. Houlden
Mr. S. Hudson
Mr. C. M. Hufton
Mr. George Hyde

J.
Mr. B. James
Mr. G.A.T. Jeffs, F.R.E.S.
Andrew Johnson
Mr. R. Johnson, F.R.E.S.
Mr. and Mrs. R. Johnson

K.
Miss E. Kay
Mr. H.W. Kew
Mr. R. Key
Mr. F.L. Kirk

L.
Mr. J. Larder
Dr. F. Arnold Lees
Mr. W. Levington
Lincolnshire Naturalists' Union
Lincolnshire Naturalists' Trust Records
Miss H. Lubin
Mr. T. Lamin

M.
Mr. E. Mason
Mr. J.E. Mason
Mr. G.W. Mason

Mr. R. May
Mr. E. Mead
Prof. R. Meldola. F.R.S.
Mr. H.W. Miles
Mr. E. Moody
Mr. Mossop
Mr. A.E. Musgrave
Mr. J.F. Musham

N.
Mr. A.H. Neale
Mr. T.R. New
Mr. L.H. Newman
Mr. F.J. North

O.
Miss J. Osgerby
Mr. C.L. Ottaway
Mr. J. Owen

P.
Sir W.P. Parker Bart
Rev. E.A. Woodruffe-Peacock
Mr. D.H. Pearson
Mr. W.M. Peet
Mr. E.C. Pelham-Clinton
Mr. A. Pilkington
Mr. R.E.M. Pilcher
Mr. G.T. Porritt
Mr. E. Porter
Mr. G. Posnett
Mr. C. Potts
Mr. R. Wood Powell
Rev. S. Proudfoot
Mr. L.B. Prout

R.
Rev. G.H. Raynor
Mr. E.J. Redshaw
Mr. James Rennie
Mr. A. Reynolds
Mr. Carey Rigall
Dr. P. Riley
Mr. Isaac Robinson
Mr. A. Roebuck

S.
Mr. E. Sanders
Mr. K. Saville
Mr. E. J. Scott
Mr. P.R. Seymour
Mr. M. Simpson
Mr. G. Skelton
Mr. H.M. Small

CONTRIBUTORS TO THE AUTHORS

Mr. A. Smith
Mr. A.E. Smith
Mr. A.E. Smith O.B.E.
Mr. H.M. Brice Smith
Mr. J. Snow
Mr. R. South
Mr. F.W. Sowerby
Mr. D.A.E. Spalding
Rev. H.E. Stancliffe
Dr. S.B. Stedman
Mr. P. Stevenson

T.
Mr. C. Taylor
Mr. T.C. Taylor
Mr. E. Tearle
Rev. A. Thornley
Rear Admiral A.D. Torlesse
Mr. A.D. Townsend
Mr. M.C. Townsend
Mr. J. Trinder

W.
Mr. E.R. Walker
Mr. R.W. Walker
Mr. J.C. Walter
Mr. P.W.R. Walter
Mr. A.H. Waters
Mr. J.H. White
Mr. J.W. White
Mr. F.G. Whittle
Mr. F.J. Whittle
Mrs. V. Wilkin
Mr. B. Wilkinson
Rev. C. Wilkinson
Mr. H.J.J. Winter
Mr. W.R. Withers
Mr. E.L. Wood
Mr. E. Woodthorpe
Miss F. Woodward
Mr. C.G. Wright
Mr. P. Wynne

CHAPTER 5.

LIST OF L.N.U. LEPIDOPTERA RECORDING SECRETARIES SINCE 1893

1893-94	(Rev. G.H. Raynor (R.W. Goulding	Panton Louth
1905-21	G.W. Mason	Barton-on-Humber
1922-27	(H.C. Bee (A.E. Musgrove	Lincoln Lincoln
1928-37	A.E. Musgrove	Lincoln
1938-43	F.L. Kirk B.Sc.	Spalding
1944	(F.L. Kirk B.Sc. (T.H. Court	Spalding Market Rasen
1945-46	T.H. Court	Market Rasen
1947-49	F.L. Kirk B.Sc.	Alford
1950-55	H.M. Small	Skellingthorpe
1956-67	G.A.T. Jeffs F.R.E.S.	Grimsby
1968	G.N. Holland B.Sc.	Sleaford
1969-73	Antony E.Smith B.Sc.	Boston
1973-	J.H. Duddington	Scunthorpe

CHAPTER 6.

REFERENCES CONSULTED

A.
ACKWORTH, B. A. — Butterfly Miracles and Mysteries, 1947
ALLAN, P. B. M. — A Moth Hunter's Gossip, 1947
ALLAN, P. B. M. — Moths and Memories, 1948
ALLAN, P. B. M. — Larval Foodplants, 1949
ALLAN, P. B. M. — Talking of Moths, 1975 edn.
ALLAN, P. B. M. — Leaves From a Moth-hunter's Notebook, 1980
AMATEUR ENTOMOLOGISTS' SOCIETY — Bulletins *(items noted for L.N.U. records)*.

B.
BARLEY, M. W. — Lincolnshire and the Fens, Ch.1. The Land, 1972
BARRETT, C. G. — Lepidoptera of the British Isles, 1893/6
BEAUFOY, S. — Butterfly Lives, 1947
BEIRNE, B. P. — British Pyralid and Plume Moths, 1954
BENINGFIELD, G., GOODDEN, R. — Butterflies, 1981
BLAMEY, M., FITTER, R., FITTER, A. — Wild Flowers of Britain and Northern Europe
BRADLEY, J. D., & FLETCHER, D. S. — Recorder's Log Book of British Butterflies and Moths, 1979
BRITISH BUTTERFLY CONSERVATION SOCIETY — B.B.C.S. News, 1977-1980
BULOW-OLSEN, A. — Plant Communities, 1978
BURTON, J. — Oxford Book of Insects, 1968

C.
CARTER, D. J. — Caterpillars, 1979
CHINERY, M. — Insects of Britain and Northern Europe, 1973

D.
DARLINGTON, A. — Natural History Atlas of Great Britain, 1969
DENNIS, R. L. H. — The British Butterflies, Their Origin and Establishment, 1977
DOWDESWELL, W. H. — The life of the Meadow Brown, 1981
DUDDINGTON, J. H. — Butterflies and Some Coastal Species of Moths from the Scunthorpe District, *Journal of the Scunthorpe Museum Society,* Vol.2, 1965

DUDDINGTON, J. H.　　　　The Ecological Diversity of Lincolnshire and South Humberside With Reference to the Macro-Lepidoptera, Presidential Address, *L.N.U. Transactions,* Vol.XIX, No.3, Parts 1 and 2, 1978

E.
ENTOMOLOGIST, THE　　　*(Quoted in L.N.U. records).*
ENTOMOLOGISTS'
MONTHLY MAGAZINE, THE *(Quoted in L.N.U. records).*
ENTOMOLOGIST, THE　　　Vol. 78 to Vol. 104, (1945 to 1971).
ENTOMOLOGISTS' GAZETTE Vol. 1 to Vol. 30, (1951 to 1980).

F.
FITTER, A.　　　　　　　　An Atlas of the Wild Flowers of Britain and Northern Europe.
FIRMIN, PYMAN & OTHERS　A Guide to the Butterflies and Larger Moths of Essex, 1975
FORD, E. B.　　　　　　　Butterflies, 1967
FORD, E. B.　　　　　　　Moths, 1967
FORD, R. L. E.　　　　　　Practical Entomology, 1963

G.
GIBBONS, E. J.　　　　　　The Flora of Lincolnshire, *Lincolnshire Naturalists' Union (Brochure No.6)* 1975
GOODDEN, R.　　　　　　Wonderful World of Butterflies and Moths, 1977.
GOODDEN, R.　　　　　　British Butterflies, 1978.
GRAY, D.　　　　　　　　Butterflies On My Mind, 1978

H.
HEATH, J.　　　　　　　　Lepidoptera Distribution Maps Scheme, Guide to the Critical Species, 1969-71
HEATH, J. (Editor)　　　　The Moths and Butterflies of Great Britain and Ireland, Vol.1. 1976
HEATH, J. (Editor)　　　　The Moths and Butterflies of Great Britain and Ireland, Vol. 9, 1979
HESLOP, I. R. P.　　　　　Indexed Check List of the British Lepidoptera, 1947
HESLOP, I. R. P., HYDE, G. E.,
STOCKLEY, R. E.　　　　　Notes and Views of the Purple Emperor, 1964
HESLOP, I. R. P.　　　　　Revised Indexed Check List of the British Lepidoptera, 1964 with the seven supplements published to 18th May 1970.
HIGGINS, L. G.　　　　　　Palearctic Melitaea, 1941
HIGGINS, L. G., RILEY N. D. Butterflies of Britain and Europe, 1970 and 1973 editions.
HIGGINS, L. G.　　　　　　The Classification of European Butterflies, 1975.
HODGSON, N. B.　　　　　Insects of the British Isles, 1946.
HOWARTH, T. G.　　　　　Colour Identification Guide to British Butterflies, 1973.

I.
IMMS, A. D.　　　　　　　Insect Natural History, 1947 and 1973.

REFERENCES CONSULTED

J.
JOHNSON, R.,
DUDDINGTON, J. H.,
WALTER, P. W. R. — A Checklist of Local Lepidoptera, *Scunthorpe Museum*, 1969.

K.
KOCH, M. — Wir Bestimmen Schmetterlinge, 3. Eulen, 1972.

L.
LEP. LINCS. 1. — Lepidoptera of Lincolnshire 1, *L.N.U. Transactions* Part 1, Vol. 1, 1905-1908, pp 174-191, G. W. Mason
LEP. LINCS. 2. — Lepidoptera of Lincolnshire 2, *L.N.U. Transactions* Part 2, Vol. 1, 1905-1908, pp 230-262, G. W. Mason
LEP. LINCS. 3. — Lepidoptera of Lincolnshire 3, *L.N.U. Transactions* Part 3, Vol. 2, 1909-1911, pp 73-103, G. W. Mason.
LEP. LINCS. 4. — Lepidoptera of Lincolnshire 4, *L.N.U. Transactions* Part 4, Vol. 2, 1909-1911, pp 177-219, G. W. Mason
LEP. LINCS. 5. — Lepidoptera of Lincolnshire 5, *L.N.U. Transactions* Part 5, Vol. 3, 1912-1915, p 170, G. W. Mason
LEP. LINCS. 6. — Lepidoptera of Lincolnshire 6, *L.N.U. Transactions* Part 6, Vol. 4, 1916-1918, p 16, G. W. Mason
LEWIS, H. L. — Butterflies of the World, 1973
LINCOLNSHIRE NATURALISTS' UNION TRANSACTIONS
LINCOLNSHIRE NATURALISTS' UNION — *(Volumes quoted in L.N.U. records)*
Lepidoptera Divisional Records in care of Lepidoptera Records Secretary – all records to date of publication.
LINCOLNSHIRE TRUST FOR NATURE CONSERVATION — Newsletters, 1972 and 1974
LINCS. BUTTS. — Lincolnshire Butterflies, *L.N.U. Transactions*, Vol. 1 1905-1908, pp 76-85, G. W. Mason

M.
MASON, G. W. — See *L.N.U. Transactions*, Lep. Lincs. 1 to Lep. Lincs 6, and *Lincs. Butts.* 1905-1918
MAY, J. — Prehistoric Lincolnshire, History of Lincolnshire 1, 1976
MAYS, R. — Henry Doubleday, 1978
MEASURES, D. G. — Bright Wings of Summer, 1976
MEYRICK, E. — A Revised Handbook of British Lepidoptera, 1970
MILES, P. M., MILES, H. B. — Seashore Ecology, 1966
MILES, P. M., MILES, H. B. — Town Ecology, 1967
MILES, P. M., MILES, H. B. — Chalk and Moorland Ecology, 1968
MILES, P. M., MILES, H. B. — Fresh Water Ecology, 1967
MILES, P. M., MILES, H. B. — Woodland Ecology, 1968
MILLER & SKETCHLY — Fenland, 1878, *(Quoted in L.N.U. records)*

260 THE BUTTERFLIES AND LARGER MOTHS

N.
NATURALIST, THE	*(Volumes quoted in L.N.U. records).*
NATURALISTS' WORLD, THE	May 1885, 1886 *(Quoted in L.N.U. records).*
NEWMAN, EDWARD	The Natural History of British Butterflies and Moths, 1869, 1st Edition
NEWMAN, L. HUGH.	British Moths and Their Haunts, undated.
NEWMAN, L. H.	Looking at Butterflies, 1959
NEWMAN, L. H.	Man and Insects, 1965
NEWMAN, L. H.	Hawkmoths of Britain and Europe, 1965
NEWMAN, L. H.	Create a Butterfly Garden, 1967
NOVAK, I.	Butterflies and Moths, 1980

O.
OLDROYD, H.	Insects and their World, 1966

R.
RUSSWURM, A. D. A.	Aberrations of British Butterflies, 1978
ROUGEOT, P. C., VIETTE, P.	Papillons Nocturnes d'Europe et D'Afrique-du Nord, 1978

S.
SANDERS, E.	A Butterfly Book, 1955
SCORER, ALFRED GEORGE.	The Entomologist's Log-Book, and Dictionary of the Life Histories and Food Plants of the British Macro-Lepidoptera, 1913
SEITZ, Prof. Dr. A.	Les Macrolépidoptéres du Globe, Vol. 2; Bombycides et Sphingides Palearctiques, 1913.
SEITZ. Prof. Dr. A.	Les Macrolépidoptéres du Globe, Vol. 3; Heteroceres Noctuiformes, 1914
SEITZ. Prof. Dr. A.	Les Macrolépidoptéres du Globe, Vol. 4; Géométrides, 1912
SMART, P.	Encyclopaedia of the Butterfly World, 1976
SMITH, A. E., CORNWALLIS, R. K.	The Birds of Lincolnshire, *Lincolnshire Natural History (Brochure 2,)* 1955
SMITH, A. E.	Nature Conservation in Lincolnshire, 1969
SOUTH, R.	The Moths of the British Isles, Series 1 and Series 2, 1961 and 1972 editions.
SOUTH, R.	The Butterflies of the British Isles, 1956
STOKOE, W. J.	Caterpillars of British Butterflies, 1950
STOKOE, W. J.	Caterpillars of British Moths, Series 1 and Series 2, 1958
STONE, J. L. S. & MIDWINTER, M. J.	Butterfly Culture, 1975

T.
TINBERGEN, N.	Curious Naturalists, 1974
TUTT, J. W.	British Lepidoptera, 1899

W.
WATSON, A.	Butterflies, 1981
WHALLEY, P.	Butterfly Watching, 1980
WHALLEY, P.	Butterflies, 1981
WILKINSON, J., & TWEEDIE, M.	Butterflies and Moths, 1980
WHITWELL, J. B.	Roman Lincolnshire, *History of Lincolnshire,* Vol. 2, Ch. 2, Geography of Lincolnshire, 1970.

CHAPTER 7.

SILHOUETTES TO ASSIST IN THE IDENTIFICATION OF A NUMBER OF THE LEPIDOPTERA FAMILIES

Drawn by J. G. A. Johnson.

A reader finding a butterfly or moth image, or larva, can get an idea of its family by comparing the insect's shape with the following silhouettes, and by them referring back in the text of Chapter 3 to the section indicated by the silhouette reference number.

7.

14.

22.

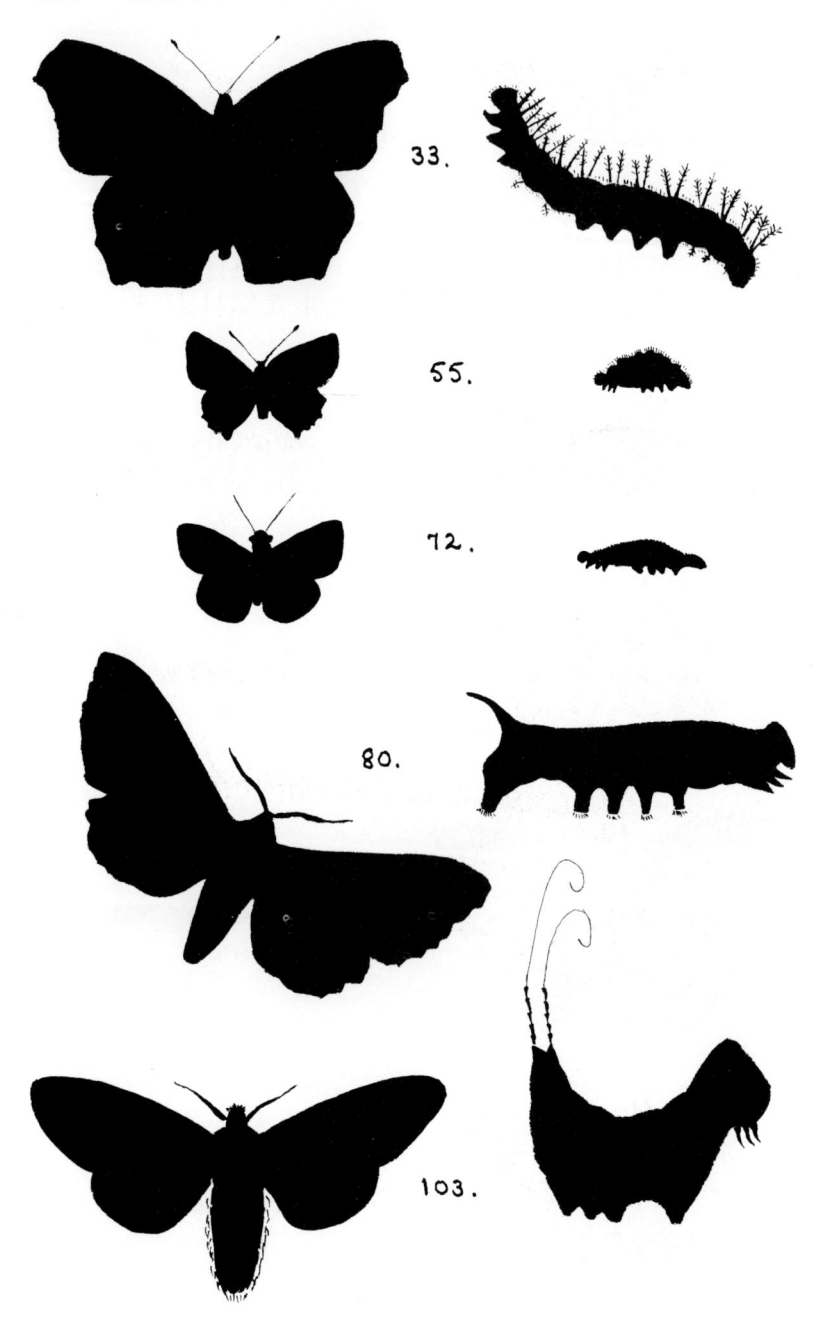

262   THE BUTTERFLIES AND LARGER MOTHS

33.

55.

72.

80.

103.

SILHOUETTES OF LEPIDOPTERA FAMILIES 263

111.

117.

135.

145.

264   THE BUTTERFLIES AND LARGER MOTHS

156.

165.

218.

331.

502.

SILHOUETTES OF LEPIDOPTERA FAMILIES 265

518.

610.

621.

672.

266   THE BUTTERFLIES AND LARGER MOTHS

830.

889.

928.

935.

CHAPTER 8.

SKETCH MAPS OF THE COUNTY

Drawn by Mr. E. J. Redshaw.

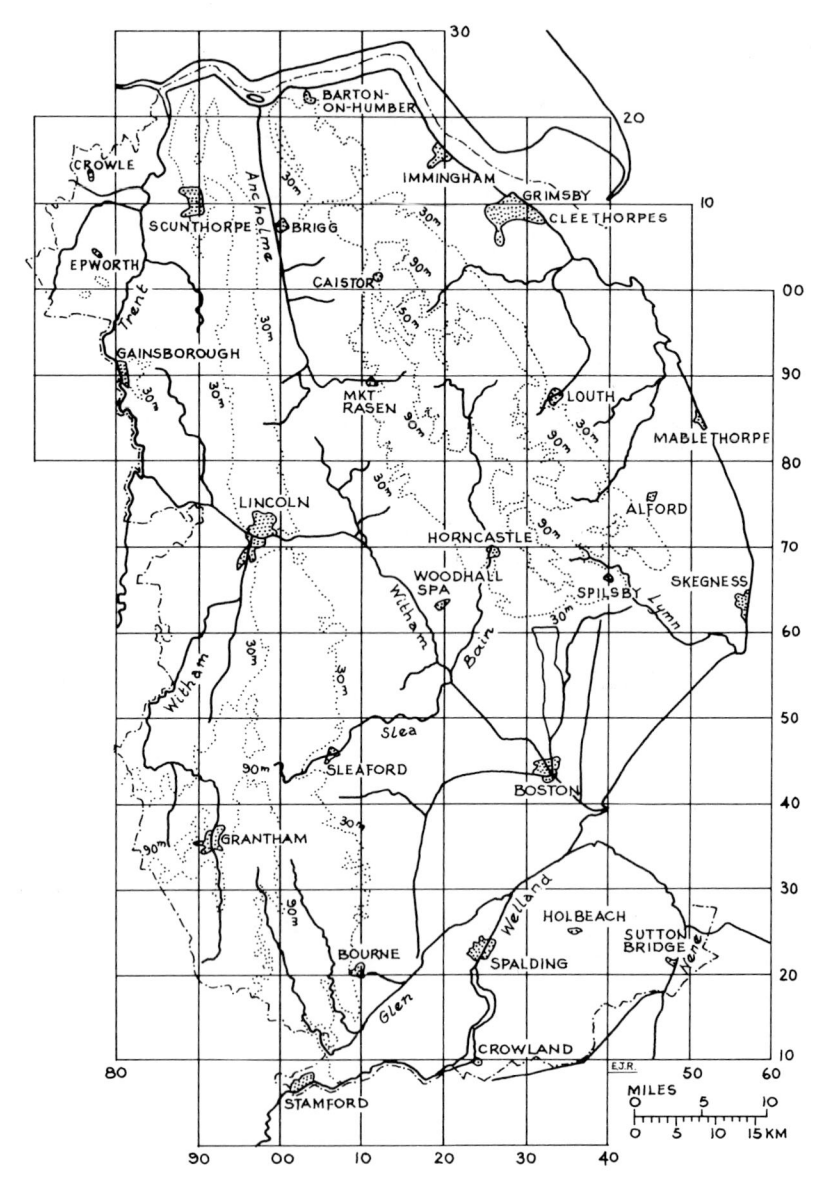

Fig. 1. – Topography of Lincolnshire and South Humberside. *(Based on the Ordnance Survey map with the sanction of the controller of H.M. Stationery Office Crown Copyright Reserved.)*

SKETCH MAPS OF THE COUNTY

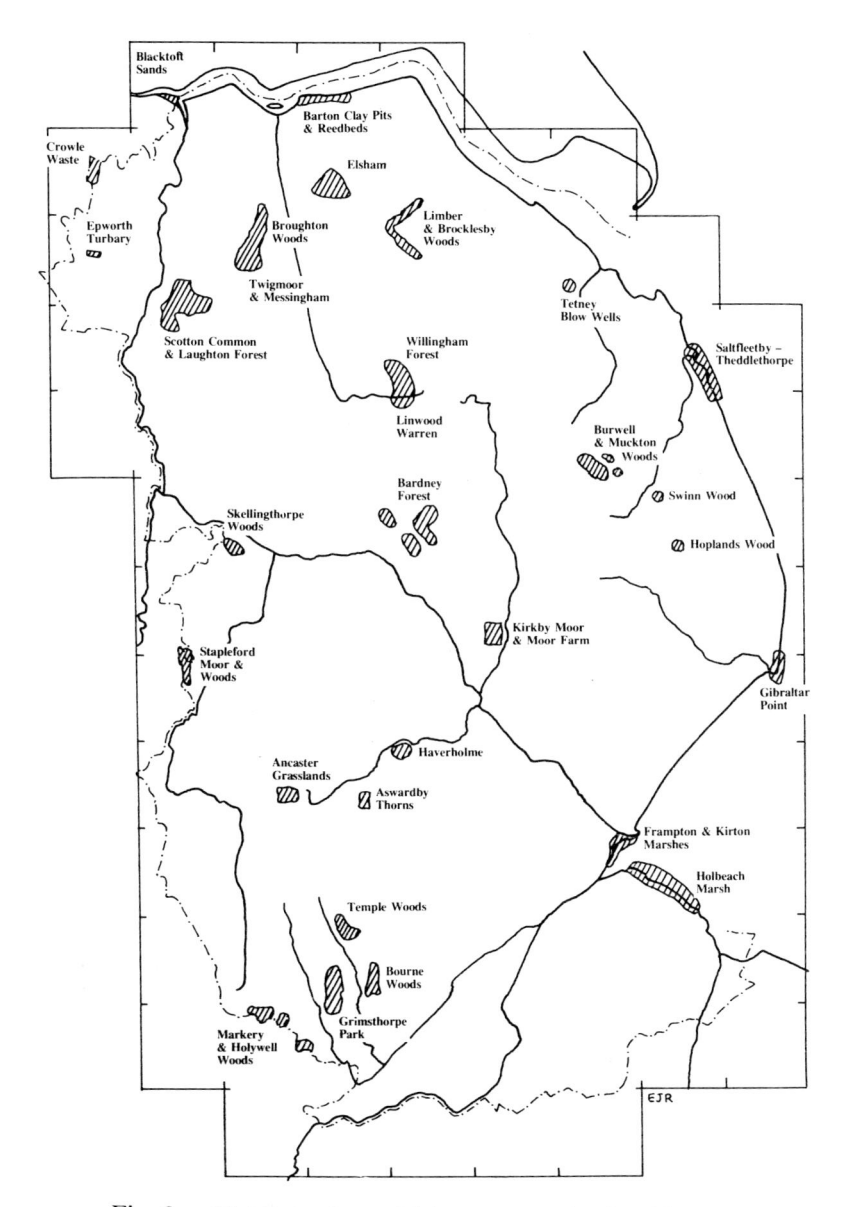

Fig. 2. – Habitats of special importance for Lepidoptera
(Based on the Ordnance Survey map with the sanction of the Controller of H.M. Stationery Office Crown Copyright Reserved).

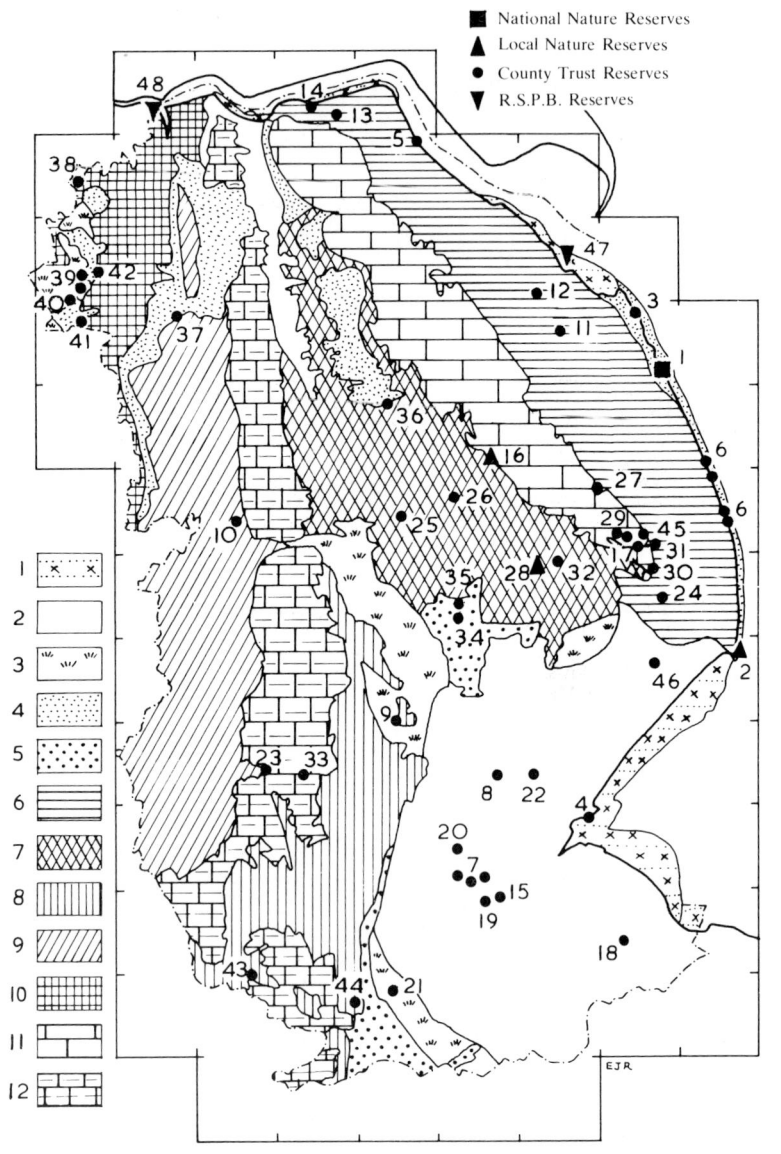

Fig. 3. – Soil Regions and Location of Nature Reserves.
(Based on maps by J. W. Blackwood and D. N Robinson with the sanction of the Controller of H.M. Stationery Office Crown Copyright Reserved.)

SKETCH MAPS OF THE COUNTY

LIST OF NATURE RESERVES

1. Saltfleetby – Theddlethorpe NNR
2. Gibraltar Point LNR
3. Donna Nook
4. Frampton Marsh
5. Killingholme Wader Pits
6. Sea Bank Clay Pits
7. Gosberton Clay Pits
8. Hubbert's Bridge Clay Pit
9. Ewerby Pond
10. Burton Old Gravel Pits
11. Covenham Reservoir
12. Tetney Blow-wells
13. Barrow Blow-wells
14. Barton Reedbeds
15. Surfleet Reedbed
16. Red Hill, Goulceby LNR
17. Dawber Lane Quarry
18. The Shrubberies, Long Sutton
19. The Yews Farm, Donington
20.
21. Baston Fen
22. Baptist Cemetery, Boston
23. Moor Closes, Ancaster
24. Heath's Meadows, Bratoft
25. Little Scrubbs Meadow
26. Sotby Meadows
27. Swaby Valley
28. Snipe Dales LNR
29. Mill Hill Claxby, and Skendleby Psalter Bank
30. Candlesby Hill
31. Willoughby Meadow
32. Furse Hill, Hagworthingham
33. Rauceby Warren
34. Kirkby Moor
35. Moor Farm
36. Linwood
37. Scotton Common
38. Crowle Waste
39. Epworth & Haxey Turbaries
40. Birds Wood
41. Langholme Wood
42. Axholme Light Railway
43. Tortoiseshell Wood
44. Dole Wood
45. Hoplands Wood
46. Friskney Decoy Wood
47. RSPB Tetney
48. RSPB Blacktoft

KEY TO SOIL REGIONS

1. Saltmarsh
2. Alluvial soils
3. Peats
4. Coastal dunes and inland cover sands
5. Gravels
6. Middle Marsh and coastal clays and silts
7. Central Lincs clays
8. South Kesteven boulder clays
9. Trent Vale clays and gravels
10. Isle of Axholme clays
11. Chalk soils
12. Limestone soils

CHAPTER 9.

INDEX OF SCIENTIFIC NAMES OF RECORDED SPECIES

Numbers refer to Species Number in Systematic List
(Chapter 2)

A
abbreviata Steph. 877
abietaria Schiff. 940
abjecta Hübn. 451
Abraxas 888
abruptaria Thunb. 936
absinthiata Clerck 860
absinthii L. 531
Acasis 812
aceris L. 513
Acherontia 82
Achlya 132
acis Schiff. 67
Acontia 641
Actebia 290
adippe L. 40
adusta Esp. 562
adustata Schiff. 892
advena Schiff. 389
advena Schiff. 348
Aegeria 226
aegeria L. 17
AEGERIINAE 225
aegon Schiff. 61
aenea Hübn. 649
aescularia Schiff. 669
aestivaria Hübn. 674
Aethalura 950
aethiops Esp. 20
affinis L. 494
affinitata Steph. 746
agathina Dup. 305
agestis Schiff. 63
aglaia L. 42
Aglais 37
Agrochola 575
AGROTINAE 271

Agrotis 277
albicillata L. 740
albicolon Hübn. 374
albipunctata Haw. 859
albipunctata Hufn. 682
albovenosa auct. 525
albula Schiff. 168
albulalis Hübn. 168
albulata Hufn. 830
albulata Schiff. 749
alchemillata L. 747
Alcis 941
algae Esp. 421
Allophyes 557
alni L. 515
alniaria L. 912
alniaria Schiff. 910
alpium Osbeck 511
alsines Brahm 431
Alsophila 669
alternaria Hübn. 902
alternata Müll 795
alternata Schiff. 902
amata L. 681
Amathes 305
Amphipyra 502
AMPHIPYRINAE 477
Anagoga 909
Anaitis 803
Anaplectoides 320
Anarta 342
ANARTINAE 342
Anchoscelis 579
Anepia 373
Angerona 900
angularia Hübn. 911
annulata Schulze 684

Anthocharis 9
Anticlea 739
Anticollix 837
antiopa L. 34
antiqua L. 135
Antitype 568
Apamea 441
APAMEINAE 437
Apatele 512
APATELINAE 506
Apatura 28
APATURINAE 28
Apeira 918
Aphantopus 26
apiciaria Schiff. 924
apiformis Clerck 223
Apocheima 930
Apoda 208
Aporia 4
Aporophyla 553
aprilina L. 559
arbuti F. 501
ARCHIERINAE 667
Archiearis 667
Arctia 200
ARCTIIDAE 171
ARCTIINAE 187
arcuosa Haw. 478
Arenostola 413
areola Esp. 546
argiolus L. 68
argus L. 61
ARGYNNINAE 39
Argynnis 39
Aricia 62
armigera Hübn. 341
artemis Schiff. 49

274 THE BUTTERFLIES AND LARGER MOTHS

arundinis F. 423
Asphalia 131
assimilata Doubl. 863
associata Borkh. 770
asteris Schiff. 528
Asthena 830
astrarche Bergst. 63
atalanta L. 30
Atethmia 581
Atolmis 171
atomaria L. 958
atrata L. 802
atropos L. 82
augur F. 297
aurago Schiff. 583
aurantiaria Hübn. 906
aurinia Rott. 49
autumnaria Wernb. 910
autumnata Borkh. 824
avellana L. 208
aversata L. 717
Axylia 319

B

badiata Schiff. 738
baja Schiff. 311
bajularia Schiff. 673
barbalis Clerck 665
basilinea Schiff. 448
batis L. 126
belgaria Hübn. 965
bella Borkh. 302
bembeciformis Hübn. 224
Bena 592
berbera Rungs. 502A
berberata Schiff. 737
betulae L. 51
betularia L. 935
bicolorana Fuessl. 593
bicolorata Hufn. 363
bicolorata Hufn. 773
bicoloria Vill. 467
bicruris Hufn. 368
bidentata Clerck 919
bifaciata Haw. 750
bifida Brahm 101
bilineata L. 758
bilunaria Esp. 915
bimaculata F. 894
binaria Hufn. 161
bipunctaria Schiff. 821
biselata Hufn. 719
bisetata Rott. 719
Biston 934
BISTONINAE 929
bistortata auct. 946
biundularia Borkh. 946
biundularia Esp. 947
blanda Schiff. 432

blandina F. 20
blomeri Curt. 836
Boarmia 944
BOARMIINAE 936
Bombycia 552
BOMBYCIDAE 157
bombycina Hufn. 361
BOMBYCINAE 157
BOMBYCOIDEA 100
bombyliformis Esp. 99
Bombyx 158
Bomolocha 652
boreata Hübn. 829
Brachionycha 550
bractea Schiff. 626
brassicae L. 5
brassicae L. 345
brassicae Ril. 632
brumata L. 828
brunnea Schiff. 298
brunneata Thunb. 962
bucephala L. 121
Bupalus 959

C

caenosa Hubn. 140
Caeruleocephala L. 619
caja L. 200
c-album L. 38
Callimorpha 191
Callophrys 55
Calothysanis 681
camelina L. 117
camilla L. 29
Campaea 899
campanulata H.-S. 865
candidata Schiff. 830
cannae Ochs. 421
capsincola Hübn. 368
captiuncula Treits. 468
capucina L. 117
Caradrina 430
Caradrininae 429
cardamines L. 9
cardui L. 31
carmelita Esp. 118
carpinata Borkh. 814
carpini Schiff. 159
carpophaga Borkh. 371
Carsia 805
carterocephalus 78
cassinia Schiff. 550
castanea Esp. 310
castigata Hübn. 866
castrensis L. 146
Catocala 608
Catocalinae 608
Celaena 481
Celama 169

Celastrina 68
Celerio 88
celerio L. 93
centaureata Schiff. 855
Ceramica 353
Cerapteryx 378
Cerastis 323
certata Hübn. 790
Cerura 103
CERURINAE 100
cervinalis Scop. 790
cervinata Schiff. 822
cespitis Schiff. 377
chaerophyllata L. 802
chamomillae Schiff. 529
Chaonia 107
chaonia Hübn. 107
chenopodiata L. 818
chenopodii Schiff. 355
Chesias 800
chi L. 569
Chiasmia 964
chilodes 428
chloerata 887A
Chlorissa 675
Chloroclysta 774
chloroclystis 884
chlorosata Scop. 963
christyi Prout 827
chrysitis L. 623
chrysoprasaria Esp. 679
chrysorrhoea L. 138
Cidaria 772
CILICINAE 165
Cilix 165
cinerea auct. 281
cinxia L. 48
circellaris Hufn. 577
Cirrhia 585
citrago L. 582
citraria Hübn. 968
citrata L. 778
Citria 584
clathrata L. 964
clavaria Haw. 822
clavipalpis Scop. 435
clavis Hufn. 280
Cleorodes 939
clorana L. 594
Clossiana 44
Clostera 122
Clytie 613A
c-nigrum L. 313
Coenobia 427
Coenonympha 24
coerulata F. 785
cognata Thunb. 781
Colias 11
Colocasia 617
Colotois 920

INDEX OF SCIENTIFIC NAMES

Comacla 173
comes Hübn. 327
Comibaena 673
comitata L. 823
comma L. 77
comma L. 400
complana L. 181
compta Schiff. 367
concolor Guen. 414
confusalis H.-S. 169
conigera Schiff. 408
connexa Borkh. 450
consignata Borkh. 851
consortaria F. 945
conspersa Schiff. 366
conspicillaris L. 380
convolvuli L. 83
coridon Poda 65
coronata Hübn. 884
corticea Schiff. 280
corylata Thunb. 764
coryli L. 617
Cosmia 493
COSSIDAE 263
COSSINAE 265
Cossus 265
cossus L. 265
costaestrigalis Steph. 658
Cosymbia 682
crabroniformis Lew. 224
Craniophora 524
crassalis Treits. 652
crassicornis Haw. 411
crataegata L. 923
crataegi L. 4
crataegi L. 147
crenata Hufn. 447
crepuscularia Schiff. 947
cribralis Hübn. 663
cribrumalis Hübn. 663
Crocallis 921
croceago Schiff. 572
croceus Fourc. 13
cruda Schiff. 384
Cryphia 506
cucubali Schiff. 370
cucullatella L. 166
Cucullia 526
CUCULLIINAE 526
culiciformis L. 234
cultraria F. 162
Cupido 69
cursoria Hufn. 272
curtula L. 122
Cyaniris 67
Cybosia 176
Cycnia 195
cydippe L. 40
cynipiformis Esp. 232

cytherea F. 476
cytisaria Schiff. 671

D

dahlii Hübn. 301
Daphnis 94
daplidice L. 8
Dasychira 136
Dasypolia 567
DASYPOLIINAE 547
debiliata Hübn. 885
decolorata Hübn. 748
defoliaria Clerck 908
Deilephila 95
DEILEPHILINAE 88
Deileptenia 940
Deilinia 896
DEILINIINAE 888
denotata Guen. 870
denotata Hübn. 865
denticulatus Haw. 281
dentina Esp. 354
deplana Esp. 178
depressa Esp. 178
derasa L. 125
derivata Schiff. 739
designata Hufn. 728
despecta Treits. 427
Deuteronomos 912
Diacrisia 196
Diarsia 298
Diataraxia 351
didyma Esp. 456
didymata L. 736
diffinis L. 495
diluta Schiff. 131
dilutaria Hübn. 704
dilutata Schiff. 826
dimidiata Hufn. 707
dipsacea L. 335
Discoloxia 836
dispar Haw. 56
dissimilis Knoch 358
dissoluta Treits. 425
distinctaria H-S. 842
dodonaea Schiff. 106
dodoneata Guen. 878
dolabraria L. 922
dominula L. 203
dotata L. 771
Drepana 160
DREPANDIDAE 160
Drepaninae 160
dromedarius L. 111
Drymonia 106
Dryobotodes 565
dubitata L 789

duplaris L. 129
Dypterygia 438
Dyscia 965
Dysstroma 776

E

Earias 594
Earophila 738
echii Borkh. 373
Ecliptopera 765
Ectropis 946
Ectypa 616
edusa F. 13
Eilema 178
Electrophaes 764
elinguaria L. 921
Ellopia 898
elpenor L. 96
elymi Treits. 417
emarginata L. 720
Ematurga 958
emutaria Hübn. 695
Enargia 497
ENNOMINAE 909
Ennomos 910
Epione 924
Epirrhoe 794
Episema 619
epomidion Haw. 446
Erannis 905
Erebia 19
eremita F. 565
Eremobia 461
Eriogaster 149
erosaria Schiff. 914
Erynnis 72
Euchoeca 834
Euclidimera 615
Eumenis 18
Eumichtis 561
euphrosyne L. 44
Euphydryas 49
Eupithecia 838
Euplexia 472
Euproctis 138
Eupsilia 571
Eurois 321
EUSTROTIINAE 597
Euxoa 272
exanthemata Scop. 897
exclamationis L. 285
exigua Hübn. 436
exiguata Hübn. 850
expallidata Doubl. 862
expolita Staint. 468
exsoleta L. 544
extensaria Freyer 871
extrema Hübn. 414

276 THE BUTTERFLIES AND LARGER MOTHS

F
fagana F. 592
fagaria Thunb. 965
fagata Scharf. 829
fagi L. 104
falcataria L. 163
fascelina L. 136
fasciana L. 603
fasciaria L. 898
fasciuncula Haw. 465
ferrugata Clerck 725
ferrugata Staud. 726
ferruginea Esp. 577
festiva Schiff. 299
festucae L. 627
fibrosa Hübn. 482
filigrammaria H. H.S. 825
filipendulae L. 218
fimbria L. 332
fimbriata Schreber 332
firmata Hübm. 782
fissipuncta Haw. 458
flammea Curt. 392
flammea Schiff. 391
flammeolaria Hufn. 832
flavago F. 584
flavago Schiff. 490
flavicincta Schiff. 568
flavicornis L. 132
flavofasciata Thunb. 748
flexula Schiff. 666
fluctuata L. 730
fluviata Hübn. 731
fluxa Hübn. 415
fontis Thunb. 652
formicaeformis Esp. 235
fraxinata Crewe 874
fraxini L. 608
fuciformis L. 98
fuliginosa L. 197
fulva Hübn. 413
fulvago Hübn. 497
fulvago L. 585
fulvaria Vill. 962
fulvata Forst. 772
fumata Steph. 689
furcata Thunb. 784
furcula Clerck 102
furuncula Schiff. 467
furva Schiff. 453
fuscantaria Steph. 913
fusconebulosa Deg. 268
fuscovenosa auct. 702

G
GALATHEA L. 27
galiata Schiff. 797
galii Rott. 91

gamma L. 635
Gastropacha 156
GASTROPACHINAE 155
geminipuncta Haw. 424
gemmaria Brahm 938
genistae Borkh. 357
Geometra 672
GEOMETRIDAE 667
Geometrinae 670
GEOMETROIDEA 667
gilvago Schiff. 586
glareosa Esp. 309
glauca Hübn. 361
glaucata Scop. 165
glyphica L. 616
gnoma F. 109
Gonepteryx 14
Gonodontis 919
GOPNOPTERINAE 651
gonostigma auct. 134
gopossensiata Mab. 861
Gportyna 488
gothica L. 382
gracilis Schiff. 390
graminis L. 378
Graphiphora 297
Griposia 559
grisealis Schiff. 662
griseola Hübn. 179
grossulariata L. 889
Gymnoscelis 887
Gypsitae 322

H
Habrosyne 125
Hada 354
HADEDNINAE 345
halterata Hufn. 810
Hamearis 50
Harpyia 100
hastata L. 792
haworthiata Doubl. 845
haworthii Curt. 481
hecta L. 270
Heliophobus 374
HELIOTHINAE 333
Heliothis 335
hellmanni Ev. 415
helvola L. 579
Hemaris 98
Hemistola 679
Hemithea 674
hepatica Clerck 347
hepatica L. 446
HEPIALIDAE 266
Hepialinae 266
HEPIALOIDEA 266
Hepialus 266
herbida Hübn. 320

hermelina auct. 101
Herminia 665
Herse 83
HESPERIIDAE 71
Hesperiinae 73
HETEROGENEINAE 208
hextapterata Schiff. 810
hippocastanaria Hübn. 952
Hippotion 93
hirtaria Clerck 933
hispidaria Schiff. 930
holosericata Dup. 704
Horisme 807
humuli L. 266
hyale F. 11
Hydraecia 484
Hydrelia 832
Hydrillula 480
Hydriomena 784
Hyloicus 87
HYLOPHILIDAE 592
Hypena 653
HYPENINAE 652
hyperantus L. 26
HYPSINAE 202

I
icarus Rott. 64
icterata Vill. 867
icteritia Hufn. 585
illunaria Hübn. 915
illunaris Hübn. 613A
illustraria Hübn. 917
imitaria Hübn. 694
immaculata Thunb. 679
immanata Haw. 778
immutata L. 698
impluviata Schiff. 785
impudens Hübn. 397
impura Hübn. 395
incerta Hufn. 387
indigata Hübn. 869
infesta Ochs. 452
innotata Hufn. 873
inornata Haw. 716
insigniata Hübn. 851
instabilis Schiff. 387
interjecta Hübn. 330
interjectaria Boisd. 702
interrogationis L. 636
inturbata Hübn. 844
io L. 33
ipsilon Hufn. 286
iris L. 28
irregularis Hufn. 373
irriguata Hübn. 849

INDEX OF SCIENTIFIC NAMES 277

isogrammaria H.-S. 845
Itame 961

J

jacobaeae L. 191
janira L. 22
janthina Schiff. 329
Jodis 680
jota L. 630
jurtina L. 22

L

lacertinaria L. 164
lactata Haw. 699
lactearia L. 680
Laelia 140
Lampra 332
Lampropteryx 762
lanestris L. 149
Laothoe 80
Laphygma 436
Larentia 822
LARENTIINAE 722
lariciata Freyer 882
Lasiocampa 150
LASIOCAMPIDAE 145
LASIOCAMPINAE 145
Laspeyria 666
latruncula Schiff. 463
legatella Schiff. 800
lepida Esp. 371
leporina L. 512
Leptidea 10
Leucania 393
LEUCANIINAE 392
leucographa Schiff. 322
Leucoma 142
leucophaea View. 350
leucophaearia Schiff. 905
leucostigma Hübn. 482
libatrix L. 651
lichenaria Hufn. 939
lichenea Hübn. 563
Ligdia 892
lignata Hübn. 815
ligniperda F. 265
ligula Esp. 591
ligustri L. 86
ligustri Schiff. 524
LIMACODIDAE 208
LIMENITINAE 29
Limenitis 29
limitata Scop. 818
linariata Schiff. 847
linearia Hübn. 688
lineata F. 92
lineola Ochs. 74

lineolata Schiff. 760
literosa Haw. 466
Lithacodia 603
Lithina 963
Lithomoia 537
Lithophane 538
lithorhiza Borkh. 546
Lithosia 177
LITHOSIINAE 171
lithoxylaea Schiff. 441
litoralis Curt. 399
litura L. 580
liturata Clerck. 903
livornica Esp. 92
Lobophora 810
lobulata Hübn. 814
Lomaspilis 891
lonicerae Scheven 217
Lophopteryx 116
lota Clerck. 575
lubricipeda auct. 194
lubricipeda L. 192
lucens Freyer 486
lucina L. 50
lucipara L. 472
luctuosa Schiff. 641
lunaria Schiff. 916
lunosa Haw. 574
lunula Stroem 555
Luperina 469
lupulina L. 269
lurideola Zinck. 180
lutea Hufn. 194
lutea Stroem 584
luteata Schiff. 832
luteolata L. 923
lutosa Hübn. 411
lutulenta Schiff. 553
Lycaena 56
LYCAENIDAE 51
LYCAENINAE 56
lychnidis Schiff. 578
Lycia 933
Lycophotia 289
Lygephila 644
Lygris 767
Lymantria 143
LYMANTRIIDAE 134
LYMANTRIINAE 134
Lyncometra 761
Lysandra 65
lythargyria Esp. 407
Lythria 722

M

machaon L 1
macilenta Hübn. 576
MACROGLOSSINAE 97

Macroglossum 97
Macrothylacia 152
macularia L. 927
Malacosoma 145
malvae L. 71
Mamestra 346
Maniola 22
margaritaria L. 899
margaritata L. 899
marginaria F. 907
marginata F. 334
marginata L. 891
marginepunctata auct. 692
maritima Tausch. 428
matura Hufn. 476
maura L. 505
megacephala Schiff. 514
megera L. 16
Melanargia 27
Melanchra 346
Melanthia 759
Meliana 392
Melitaea 47
mellinata F. 770
mendica Clerck 195
mendica F. 299
Menophra 936
menthastri Esp. 192
menyanthidis View. 520
Meristis 429
Mesoleuca 740
mesomella L. 176
Mesotype 760
meticulosa L. 473
mi Clerck 615
miata L. 775
micacea Esp. 488
Milochrista 174
Mimas 79
miniata Forst. 174
minima Haw. 478
minimus Fuessl. 69
miniosa Schiff. 383
minorata Treits. 751
minutata Doubl. 861
minutata Schiff. 860
Moma 511
monacha L. 144
moneta F. 621.
monoglypha Hufn. 444
montanata Schiff. 729
Mormo 505
morpheus Hufn. 430
mucronata Scop. 816
multistrigaria Haw. 735
munda Schiff. 388
mundana L. 172
muscerda 186
myopaeformis Borkh. 233

myrtilli L. 342
Mysticoptera 811

N

Naenia 324
nana Hufn. 354
nana Rott. 366
nanata Hübn. 872
napi L. 7
nebulata Scop. 834
nebulata Thunb. 826
nebulosa Hufn. 349
neglecta Hübn. 310
NEMEOBIIDAE 50
NEMEOBIINAE 50
nemoralis F. 662
nerii L. 94
neustria L. 145
ni Hübn. 632
nictitans Borkh. 484
nigra Haw. 555
nigricans L. 273
nigrofasciaria auct. 739
nitens Haw. 348
Noctua 331
NOCTUIDAE 271
NOCTUINAE 325
NOCTUOIDEA 271
Nola 166
NOLIDAE 166
NOLINAE 166
Nonagria 421
NONAGRIINAE 410
notata L. 901
notha Hübn. 668
Notodonta 110
NOTODONTIDAE 100
NOTODONTINAE 104
Nudaria 172
nupta L. 610
Nycteola 595
NYCTEOLINAE 595
Nycterosea 731
NYMPHALIDAE 28
NYMPHALINAE 30
Nymphalis 33

O

obelisca Schiff. 276
obeliscata Hübn. 779
obliquaria Schiff. 801
obliterata Hufn. 834
oblonga Haw. 451
oblongata Thunb. 855
obscura Brahm. 295
obscura Haw. 454
obsoleta Hübn. 398

obstipata F. 731
occulta L 321
ocellata L. 81
ocellata L. 761
Ochlodes 76
ochracea Hübn. 490
ochrearia Rossi 968
ochroleuca Schiff. 461
Ochropleura 304
octogesima Hübn. 127
ocularis L. 127
oculea L. 484
Odezia 802
Odontosia 118
OENOCHROMINAE 669
oleracea L. 351
omicronaria Schif. 684
Omphaloscelis 574
Operophtera 828
OPHIDERINAE 640
ophiogramma Esp. 457
opima Hübn. 389
Opisthograptis 923
Oporinia 824
or Schiff. 128
orbona F. 327
orbona Hufn. 328
Orgyia 134
orion Esp. 511
ornitopus Hufn. 543
Ortholitha 816
Orthonama 815
Orthosia 382
ORTHOSIINAE 382
OURAPTERYGINAE 928
Ourapteryx 928
oxyacanthae L. 557

P

pabulatricula Brahm. 450
Pachycnemia 952
palaemon Pall. 78
paleacea Esp. 497
pallens L. 393
PALPINA Clerck 120
paludata Thunb. 805
paludis Tutt. 485
palustraria Doubl. 853
palustris Hübn. 480
pamphilus L. 24
Panaxia 203
Panemeria 501
Panemeria 501
paniscus F. 78
Panolis 391
PANTHEINAE 617

paphia L. 39
Papilio 1
papilionaria L. 672
PAPILIONIDAE 1
PAPILIONINAE 1
PAPILIONOIDEA 1
Pararge 16
Parasemia 199
Parastichtis 564
Pareulype 737
parthenias L. 667
pastinum Treits. 644
pavonia L. 159
pectinataria Knoch 733
pedaria F. 929
Peliosa 186
peltigera Schiff. 340
Pelurga 823
pendularia auct. 682
pennaria L. 920
Perconia 969
Peridroma 292
Perizoma 743
perla Schiff. 506
persicariae L. 346
petasitis Doubl. 489
Petilampa 478
petraria Hübn. 963
phaeorrhoea Don. 138
Phalera 121
Pheosia 108
Phigalia 929
Philereme 787
Philudoria 154
phlaeas L. 58
Phlogophora 473
Photedes 468
Phragmatobia 197
phragmitidis Hübn. 419
Phytometra 649
PIERIDAE 4
PIERINAE 4
Pieris 5
pigra Hufn. 124
pilosaria Schiff. 929
pimpinellata Hübn. 870
pinastri L. 87
pinastri L. 438
pini Retz. 838
piniaria L. 959
piniperda Panz. 391
pisi L. 353
pistacina F. 578
plagiata L. 803
Plagodis 922
plantaginis L. 199
Plebejus 61
plecta L. 304
Plemyria 773
plumbeolata Haw. 846

INDEX OF SCIENTIFIC NAMES

PLUSIIDAE 597
PLUSIINAE 620
Poecilocampa 148
Polia 347
polychloros L. 36
Polychrisia 621
polycommata Schiff. 813
Polygonia 38
polyodon L. 444
POLYOMMATIINAE 59
Polyommatus 64
Polyploca 133
Pontia 8
popularis F. 376
populata L. 769
populeti F. 386
populi L. 80
populi L. 148
porata L. 686
porcellus L. 95
porphyrea Esp. 561
porphyrea Schiff. 292
potatoria L. 154
praecox L. 290
prasina Schiff. 320
prasinana L. 592
proboscidalis L. 653
procellata Schiff. 759
PROCRINAE 220
Procris 220
Procus 462
prodromaria Schiff. 934
progemmaria Hübn. 907
promutata Guen. 692
pronuba L. 331
propugnata Schiff. 728
prosapiaria L. 898
protea Schiff. 565
pruinata Hufn. 671
prunaria L. 900
prunata L. 767
Pseudoboarmia 945
Pseudoips 593
Pseudopanthera 927
Pseudoterpna 671
psi L. 518
psittacata Hübn. 774
PSYCHOIDEA 208
Pterostoma 120
pudibunda L. 137
pudorina Schiff. 397
pulchella L. 189
pulchellata Steph. 848
pulchrina Haw. 631
pulveraria L. 909
pulverulenta Esp. 384
pumilata Hübn. 887
punctaria L. 687
punctata Hübn. 895

punctinalis Scop. 945
punctularia Hübn. 950
punctulata Schiff. 950
purpuraria L. 722
pusaria L. 896
pusillata auct. 880
pusillata Schiff. 883
pustulata Hufn. 673
puta Hübn. 282
putris L. 319
pygmaeata Hübn. 853
pygmina Haw. 413
pyraliata Schiff. 771
pyralina Schiff. 493
pyramidea L. 502
Pyrginae 71
Pyrgus 71
pyrina L. 264
pyritoides Hufn. 125
pyrophila Schiff. 294
Pyrrhia 334

Q

quadra L. 177
quadrifascaria L. 723
quadrifasiata Clerck 723
quadripunctata F. 435
quercana Schiff. 593
quercifolia L. 156
quercinaria Hufn. 911
quercus L. 52
quercus L. 150

R

radius Haw. 282
rapae L. 6
ravida Schiff. 295
recens Hübn. 134
reclusa F. 124
RECTANGULATA L. 886
remissa Hübn. 454
remutaria Hübn. 699
repandaria Hufn. 924
repandata L. 941
reticulata Vill. 375
retusa L. 499
revayana Scop. 595
rhamnata Schiff. 788
rhamni L. 14
Rheumaptera 790
Rhizedra 411
rhizolitha F. 543
RHODOCERINAE 11
rhomboidaria Schiff. 938
Rhyacia 294
ribeata Clerck 940
ridens F. 133

ripae Hübn. 287
rivata Hübn. 794
Rivula 648
rivularis F. 370
rivulata Schiff. 747
roboraria Schiff. 944
rostralis L. 656
ruberata Freyer 786
rubi L. 55
rubi L. 152
rubi View. 302
rubiginata Schiff. 773
rubricollis L. 171
rubricosa Schiff. 323
rufa Haw. 427
rufata F. 801
ruficornis Hufn. 107
rufina L. 579
rumicis L. 523
rupicapraria Schiff. 904
rurea F. 447
Rusina 504
russata Borkh. 776
russula L. 196

S

sacraria L. 721
sagittata F. 743
salicata Hübn. 734
salicis L. 142
sambucaria L. 928
sannio L. 196
saponariae Borkh. 375
satellitia L. 571
satura Schiff. 561
Saturnia 159
SATURNIIDAE 159
SATURNIINAE 159
satyrata Hübn. 858
SATYRIDAE 16
SATYRINAE 16
saucia Hübn. 292
scabiosata Borkh. 839
scabriuscula L. 438
Scoliopteryx 651
scolopacina Esp. 455
Scopula 689
Scotogramma 355
scutulata Schiff. 707
secalis L. 456
segetum Schiff. 277
selene Schiff. 45
Selenia 915
semele L. 18
semiargus Rott. 67
semibrunnea Haw. 538
Semiothisa 901
senex Hübn. 173
serena Schiff. 363

seriata Schrank 710
sericealis Scop. 648
Sesia 223
SESIIDAE 223
SESIINAE 223
sexalata Retz. 811
sexalisata Hübn. 811
sexstrigata Haw. 317
sibylla L. 29
silaceata Schiff. 765
similis Fuessl. 139
simulans Hufn. 294
simulata Hübn. 781
Simyra 525
sinapis L. 10
siterata Hufn. 774
SMERINTHINAE 79
Smerinthus 81
sobrinata Hübn. 880
sociata Borkh. 795
solidaginis Hübn. 537
sordens Hufn. 448
sordida Borkh. 452
sordidata F. 784
sororcula Hufn. 185
sororiata Hübn. 805
spadicea Staint. 591
spadicearia Schiff. 726
Spaelotis 295
sparganii Esp. 422
sparsata Treits. 837
spartiata Fuessl. 800
Sphecia 224
SPHINGIDAE 79
SPHINGINAE 82
SPHINGOIDEA 79
Sphinx 86
Sphinx Hufn. 550
Spilosoma 192
stabilis Schiff. 385
statices L. 221
Stauropus 104
stellatarum L. 97
STERRHINAE 681
stigmatica Hübn. 316
straminata Borkh. 716
straminata Treits. 712
straminae Treits. 396
stramineola Doubl. 179
strataria Hufn. 934
strigata Müll. 674
strigilis Clerck 462
strigillaria Hübn. 969
strigula Thunb. 289
Strymonidia 53
suasa Schiff. 358
subciliata Doubl. 844
sublustris Esp. 442
subnotata Hübn. 840
subsequa Hübn. 328

subsericeata Haw. 711
subtusa Schiff. 500
subumbrata Schiff. 839
succenturiata L. 868
suffumata Schiff. 762
suffusa Schiff. 286
suspecta Hübn. 564
sylvanus Esp. 76
sylvata Schiff. 833
sylvata Scop. 888
sylvestraria Hübn. 712
sylvestris Poda 73
sylvina L. 267
SYNTOMIDAE 204
SYNTOMINAE 204
Syntomis 204
syringaria L. 918

T

tages L. 72
taminata Schiff. 894
tantillaria Boisd. 883
taraxaci Hübn. 432
tarsipennalis Treits. 661
temerata Schiff. 895
templi Thunb. 567
tenebrata Scop. 501
tenebrosa Hübn. 504
tenuiata Hübn. 843
ternata Schrank 689
tersata Schiff. 809
testacea Schiff. 469
testaceata Don. 833
testata L. 768
Tethea 127
tetralunaria Hufn. 917
thalassina Hufn. 359
Thalpophila 476
thaumas Hufn. 73
Thecla 51
THECLINAE 51
Thera 779
Theria 904
thetis Rott. 66
Tholera 376
Thyatira 136
THYATIRIDAE 125
THYATIRINAE 125
Thymelicus 73
Tiliacea 582
tiliae L. 79
tiliaria Borkh. 912
tincta Brahm 347
tiphon Rott. 25
tipuliformis Clerck 229
tithonus L. 23
tityus L. 99
togata Hübn. 838
tragopoginis Clerck 503

transversa Hufn. 571
transversata Hufn. 788
trapezina L. 496
tremula Clerck 108
trepida Esp. 114
triangulum Hufn. 315
Trichiura 147
trichopteryx 813
tridens Schiff. 517
trifolii Esp. 215
trifolii Hufn. 355
trigrammica Hufn. 429
trilinea Schiff. 429
trilinearia Borkh. 688
trimacula Esp. 106
tripartita Hufn. 639
Triphosa 789
triplasia L. 638
tripunctaria H. S. 859
trisignaria H.-S. 856
tristata L. 796
tritici L. 274
truncata Hufn. 776
tullia Müll. 25
typhae Thunb. 423
typica L. 324

U

ulmata F. 888
ulvae Hübn. 428
umbra Hufn. 334
umbratica auct. 504
umbratica L. 527
umbrifera Prout 816
umbrosa Hübn. 317
unanimis Hübn. 449
Unca 638
uncana L. 606
uncula Clerck 606
undulana Hübn. 595
undulata L. 791
unidentaria Haw. 725
unifasciata Haw. 750
urticae Esp. 193
urticae Hübn. 639
urticae L. 37
Utetheisa 189

V

vaccinii L. 590
valerianata Hübn. 852
valligera Schiff. 278
Vanessa 30
varia Vill. 289
velleda Hübn. 268
venata Br. & Grey 76
venosa Borkh. 525

INDEX OF SCIENTIFIC NAMES 281

venosata F. 854
verbascia L. 533
versicolor Borkh. 464
vespiformis L. 232
vestigialis Hufn. 278
vetulata Schiff. 787
vetusta Hübn. 545
villica L. 201
viminalis F. 552
vinula L. 103
viretata Hübn. 812
virgata Hufn. 760
virgaureata Doubl. 876
virgularia Hübn. 710
viridaria Clerck 649
viridaria F. 733
viridata L. 675
viriplaca Hufn. 335
vitalbata Schif. 807
vitellina Hübn. 404
vittata Borkh. 815
vulgata Haw. 864

W

w-album Knock 54
wauaria L. 961
WESTERMANNIINAE 592
w-latinum Hufn. 357

X

xanthographa Schiff. 318
Xanthorhoe 723
Xerampelina Esp. 581
Xylena 544
XYLENINAE 537
Xylocampa 546
Xylomyges 380

Y

ypsillon Schiff. 458

Z

Zanclognatha 661
Zenobia 499
Zeuzera 264
ZEUZERINAE 263
ziczac L. 110
Zygaena 211
ZYGAENIDAE 210
ZYGAENINAE 211

CHAPTER 10.

INDEX TO ENGLISH NAMES OF RECORDED SPECIES

Numbers refer to Species Number in Systematic List (Chapter 2).

A.

Admiral Red	30
Admiral White	29
Alder Dagger	515
Angle-barred Pug	873
Angle, Blunt Peacock	901
Angle-shades, Large	473
Angle-shades, Small	472
Angle, Sharp Peacock	902
Angle, Striped	497
Angle, Tawny-barred	903
Antler	378
Archer Dart	278
Arches, Beautiful	561
Arches, Buff	125
Arches, Common Light	441
Arches, Dark	444
Arches, Green	320
Arches, Grey	349
Arches, Kent Black	168
Arches, Least Black	169
Arches, Pale Shining	348
Arches, Reddish light	442
Arches, Short-cloaked Black	166
Arches, Silvery	347
Argent-and-sable Large	792
Argent-and-sable Small	796
Argus, Brown	63
Ash Pug	874
August Thorn	911
Autumnal Carpet, Large	824
Autumnal Carpet, Small	825
Autumnal Rustic	309
Autumn Green Carpet	775

B.

Barberry Carpet	737
Bar Carpet, Purple	761
Bar, Grey Scalloped	965
Barred Chestnut Clay	301
Barred Hook-tip	162
Barred Light-green	899
Barred Red	898
Barred Rivulet	750
Barred Sallow	583
Barred Straw Chevron	771
Barred Tooth-striped	813
Barred Umber Thorn	909
Barred Yellow	772
Bars, Small Purple	649
Bath White	8
Beaded Chestnut	578
Beau, Marbled	506
Beau, Pine	391
Beautiful Arches	561
Beautiful Carpet	740
Beautiful Golden Y	631
Beautiful Hook-wing	666
Beautiful Snout	652
Beautiful Yellow Underwing	342
Beauty, Bordered White	959
Beauty, Camberwell	34
Beauty, Common Heath	958
Beauty, Great Oak	944
Beauty, Grey Birch	950
Beauty, Mottled	941
Beauty, Pale Oak	945
Beauty, Satin	940
Beauty, Waved Umber	936
Beauty, Willow	938
Bedstraw Carpet, Common	795
Bedstraw Hawk	91
Bee Hawk, Broad-bordered	98
Bee Hawk, Narrow-bordered	99
Belle, Yellow	968
Bell-flower Pug	865
Bilberry Brind	537
Bilberry Pug	885

284 THE BUTTERFLIES AND LARGER MOTHS

Birch Beauty, Grey	950	Broad-bordered Yellow	
Birch Mocha	682	Underwing	332
Bird's wing	438	Brocade, Broom	353
Black-arched Tussock	144	Brocade, Dark	562
Black Arches, Kent	168	Brocaded Rustic, Great	321
Black Arches, Least	169	Brocade, Dusky	454
Black Arches, Short-cloaked	166	Brocade, Light	357
Blackneck, Plain	644	Brocade, Pale-shouldered	359
Black Rustic	555	Broken-barred Carpet	764
Black-veined White	4	Broom Brocade	353
Bleached Pug	862	Brown Argus Blue	63
Blomer's Ripplet	836	Brown Crescent	482
Blood-veined Wave, Small	694	Brown-eye, Bright-line	351
Blood-veined, Large	681	Brown Fanfoot	661
Blossom, Peach	126	Brown Fritillary, High	40
Blossom Underwing	383	Brown Hairstreak	51
Blotched Emerald	673	Brown, Light Marbled	106
Blue-bordered Carpet	773	Brown-line Wainscot	408
Blue, Brown Argus	63	Brown, Lunar Marbled	107
Blue, Chalk-hill	65	Brown, Meadow	22
Blue, Common	64	Brown, Northern	20
Blue, Holly	68	Brown Rustic, Deep	553
Blue, Mazarine	67	Brown Scallop	787
Blue, Silver-studded	61	Brown Silver-lined	963
Blue, Small	69	Brown-spot Chestnut	580
Blunt Peacock Angle	901	Brown-tail	138
Blush, Maiden's	687	Brown-veined Wainscot	425
Border, Clouded	891	Brown, Wall	16
Border, Dotted	907	Brussels Lace	939
Bordered-Beauty, Common	924	Buff Arches	125
Bordered Gothic	375	Buff Ermine	194
Bordered Orange	334	Buff Footman	178
Bordered Pug	868	Buff, Marsh	480
Bordered Straw, Dark	340	Buff, Small Dotted	478
Bordered Straw, Scarce	341	Buff-tip	121
Bordered White Beauty	959	Bulrush Wainscot	423
Brass, Common Burnished	623	Burgundy, Duke of	50
Brick	577	Burnet, Broad-bordered Five-spot	215
Bright-line Brown-eye	351	Burnet, Companion	616
Brimstone	14	Burnet, Narrow-bordered	
Brind, Bilberry	537	Five-spot	217
Brindle-barred Yellow	812	Burnet, Narrow-bordered	
Brindle, Cloud-bordered	447	Six-spot	218
Brindle, Confused	453	Burnished Brass, Common	623
Brindled-Beauty, London	933	Butterbur Ear	489
Brindled-beauty, Oak	934	Buttoned Snout	656
Brindled-beauty, Pale	929		
Brindled-beauty, Small	930	**C.**	
Brindled Green Mottle	565	Cabbage Dot	345
Brindle, Dismal	458	Camberwell Beauty	34
Brindled Ochre	567	Campion Coronet	370
Brindled Pug	877	Canary-shouldered Thorn	912
Brindle, Large Clouded	446	Carpet, Autumn Green	775
Brindle, Slender	455	Carpet, Barberry	737
Brindle, Small Clouded	449	Carpet, Beautiful	740
Broad-barred White Gothic	363	Carpet, Blue-bordered	773
Broad-bar, Shaded	818	Carpet, Broken-barred	764
Broad-bordered Bee Hawk	98	Carpet, Chestnut-coloured	781
Broad-bordered Five-spot Burnet	215	Carpet, Christy's	827

INDEX OF ENGLISH NAMES

Carpet, Common Bedstraw	795	Chi, Grey	569
Carpet, Common Marbled	776	Chinese Character	165
Carpet, Dark Marbled	778	Chocolate Tip, Large	122
Carpet, Dark Twin-spot	725	Chocolate-tip, Small	124
Carpet, Fern	809	Christy's Carpet	827
Carpet, Flame	728	Cinnabar	191
Carpet, Galium	797	Clay, Barred Chestnut	301
Carpet, Garden	730	Clay, Common Ingrailed	299
Carpet, Grey Mottled	735	Clay, Dotted	311
Carpet, Grey Pine	779	Clay, Plain	312
Carpet Juniper	783	Clay, Purple	298
Carpet, Large Autumnal	824	Clay, Square-spotted	316
Carpet, Large Twin-spot	723	Clay Triple-lines	688
Carpet, Local Chalk	821	Clay Wainscot	407
Carpet, Mallow	822	Clearwing, Currant	229
Carpet, Marsh	743	Clearwing, Large Red-belted	234
Carpet, Narrow-barred	731	Clearwing, Osier Hornet	224
Carpet, November	826	Clearwing, Poplar Hornet	223
Carpet, Oblique	815	Clearwing, Red-tipped	235
Carpet, Pretty Chalk	759	Clearwing, Small Red-belted	233
Carpet, Purple Bar	761	Clearwing, Yellow-legged	232
Carpet, Reddish Pine	782	Clifden Nonpareil	608
Carpet, Red-green	774	Cloaked Minor	467
Carpet, Red Twin-spot	726	Cloaked Pug	838
Carpet, Sandy	748	Cloud-bordered Brindle	447
Carpet, Shoulder-striped	738	Clouded Border	891
Carpet, Silver-ground	729	Clouded Brindle, Large	446
Carpet, Small Twin-spot	736	Clouded Brindle, Small	449
Carpet, Spring Green	733	Clouded Drab	387
Carpet, Streaked	800	Clouded Ermine	196
Carpet, Streamer	739	Clouded Magpie	888
Carpet, Striped Twin-spot	734	Clouded Silver	895
Carpet, Sylvan Waved	833	Clouded Yellow, Common	13
Carpet, Umber Waved	807	Clouded Yellow, Pale	11
Carpet, Water	762	Cloud, Silver	380
Carpet, White Waved	830	Cloudy Sword-grass	544
Carpet, Wood	794	Clover, Marbled	335
Carpet, Yellow Waved	832	Club, Heart and	280
Centre-barred Sallow	581	Coast Dart	272
Chalk Carpet, Local	821	Colon, White	374
Chalk Carpet Pretty	759	Comma	38
Chalk-hill Blue	65	Common Bedstraw Carpet	795
Chamomile Shark	529	Common Blue	64
Character, Chinese	165	Common Bordered-beauty	924
Character, Common Hebrew	382	Common Burnished Brass	623
Character, Setaceous Hebrew	313	Common Chestnut	590
Chequered Skipper	78	Common Chevron	768
Chestnut, Beaded	578	Common Clouded Yellow	13
Chestnut, Brown-spot	580	Common Ear	484
Chestnut, Clay Barred	301	Common Emerald	674
Chestnut-coloured Carpet	781	Common Fanfoot	665
Chestnut, Common	590	Common Footman	180
Chestnut, Dark	591	Common Forester	221
Chestnut, Flounced	579	Common Heath Beauty	958
Chestnut Rustic, Red	323	Common Hebrew Character	382
Chevron, Barred Straw	771	Common Ingrailed Clay	299
Chevron, Common	768	Common Lackey	145
Chimney-sweeper	802	Common Lappet	156

286 THE BUTTERFLIES AND LARGER MOTHS

Common Lead-belle	816	Dagger, Poplar	514
Common Light Arches	441	Dagger, Powdered	525
Common Magpie	889	Dagger, Sycamore	513
Common Marbled Carpet	776	Dark Arches	444
Common Marbled Coronet	366	Dark Bordered Straw	340
Common Merveille-du-jour	559	Dark Brocade	562
Common Orange-underwing	667	Dark Chestnut	591
Common Pug	864	Dark Dagger	517
Common Quaker	385	Dark Dart	286
Common Ringlet	26	Dark Green Fritillary	42
Common Rustic	456	Dark Marbled Carpet	778
Common Sallow	585	Dark Scallop	788
Common Shark	527	Dark Spectacle	638
Common Silver Y	635	Dark Spinach	823
Common Small Skipper	73	Dark Tussock	136
Common Snout	653	Dark Twin-spot Carpet	725
Common Sprawler	550	Dart, Archer	278
Common Swallowtail	1	Dart, Coast	272
Common Swift	269	Dart, Dark	286
Common Tissue	789	Dart, Double	297
Common Vapourer	135	Dart, Garden	273
Common Wainscot	393	Dart, Heart and	285
Common Waved Silver	897	Dart, Portland	290
Common Winter	828	Dart, Sand	287
Common Yellow-underwing	331	Dart, Shuttle-shaped	282
Companion, Burnet	616	Dart, Square-spot	276
Concolorous Wainscot	414	Dart, Stout	295
Confused Brindle	453	Dart Turnip	277
Convolvulus Hawk	83	Dart, White-line	274
Copper, Large	56	Death's-head Hawk	82
Copper, Small	58	December Eggar	148
Copper Underwing	502	Deep Brown Rustic	553
Coronet, Campion	370	Delicate Wainscot	404
Coronet, Common Marbled	366	Dentated Pug	837
Coronet, Lychnis	368	Dingy Footman	179
Coronet, Varied	367	Dingy Shell	834
Coxcomb Prominent	117	Dingy Skipper	72
Cream-bordered Green	594	Dismal Brindle	458
Cream-spot Tiger	201	Dog's-tooth	358
Cream Wave, Dwarf	702	Dot, Cabbage	345
Cream Wave, Greater	699	Dotted Border	907
Cream Wave, Lesser	698	Dotted Buff, Small	478
Crescent, Brown	482	Dotted Chestnut	573
Crescent, Green-brindled	557	Dotted Clay	311
Crescent, Haworth's	481	Dotted Fanfoot	663
Crescent Striped	451	Dotted Footman	186
Crimson-speckled Flunkey	189	Dotted Rustic	294
Crown	524	Dot, White	346
Currant Clearwing	229	Double Dart	297
Currant Pug	863	Double Kidney	499
Currant Spinach	770	Double-lobed	457
		Double Square-spot	315
		Double-striped Pug	887
D.		Drab, Clouded	387
		Drab, Lead-coloured	386
Dagger, Alder	515	Drab, Northern	389
Dagger, Dark	517	Drinker	154
Dagger, Dusky Knot-grass	523	Duke of Burgundy	50
Dagger, Grey	518	Dun-bar	496
Dagger, Light Knot-grass	520	Dusky Brocade	454

INDEX OF ENGLISH NAMES

Dusky Knot-grass Dagger	523
Dusky-lemon Sallow	586
Dusky Sallow Rustic	461
Dusky Thorn	913
Dusky Wave, Small	710
Dwarf Cream Wave	702
Dwarf Pug	883

E.

Ear, Butterbur	489
Ear, Common	484
Ear, Large	486
Early Engrailed	946
Early, Grey	546
Early Thorn	915
Early Tooth-striped	814
Early Umber	904
Ear, Orange	490
Ear, Rosy	488
Ear, Saltern	485
Eggar, December	148
Eggar, Oak	150
Eggar, Pale	147
Eggar, Small	149
Eight, Figure of	619
Eight, Silver	621
Eighty, Figure of	127
Elder, Swallow-tailed	928
Elephant Hawk, Large	96
Elephant Hawk, Small	95
Emerald, Blotched	673
Emerland, Common	674
Emerald, Greater Grass	671
Emerald, Large	672
Emerald, Lesser	679
Emerald, Little	680
Emerald, Small Grass	675
Emperor, Purple	28
Empress	159
Engrailed, Early	946
Engrailed, Small	947
Ermine, Buff	194
Ermine, Clouded	196
Ermine, Water	193
Ermine, White	192
Eyed Hawk	81

F.

Fanfoot, Brown	661
Fanfoot, Common	665
Fanfoot, Dotted	663
Fan-footed Wave, Small	719
Fanfoot, Small	662
Feathered, Brown	504
Feathered Gothic	376
Feathered Ranuncule	563
Feathered Rustic, Light	281

Feathered Thorn	920
Fen Wainscot	419
Fern Carpet	809
Festoon	208
Figure of Eight	619
Figure of Eighty	127
Five-spot Burnet, Broad-bordered	215
Five-spot Burnet, Narrow-bordered	217
Flame Carpet	728
Flame Rustic	319
Flame Shoulder	304
Flame Wainscot	392
Flounced Chestnut	579
Flounced Rustic	469
Flunkey, Crimson-speckled	189
Footman, Buff	178
Footman, Common	180
Footman, Dingy	179
Footman, Four-dotted	176
Footman, Large	177
Footman, Muslin	172
Footman, Orange	185
Footman, Red-necked	171
Footman, Rosy	174
Footman, Round-winged	173
Footman, Scarce	181
Forester, Common	221
Four-dotted Footman	176
Four-spot	641
Fox	152
Foxglove Pug	848
Fritillary, Dark Green	42
Fritillary, Glanville	48
Fritillary, High Brown	40
Fritillary, Large Pearl-bordered	44
Fritillary, Marsh	49
Fritillary, Silver-washed	39
Fritillary, Small Pearl-bordered	45
Frosted Green Lutestring	133

G.

Galium Carpet	797
Garden Carpet	730
Garden Dart	273
Garden Tiger	200
Garden White, Large	5
Garden White, Small	6
Gatekeeper	23
Ghost Swift	266
Glanville Fritillary	48
Glaucous Shears	361
Goat	265
Golden Swift	270
Golden Y, Beautiful	631
Golden Y, Plain	630
Gold Spangle	626
Gold Spot	627

288 THE BUTTERFLIES AND LARGER MOTHS

Gold-tail	139
Gothic, Bordered	375
Gothic, Broad-barred White	363
Gothic, Feathered	376
Gothic, Hedge	377
Gothic Type	324
Gothic, Viper's-bugloss	373
Grass Emerald, Greater	671
Grass Emerald, Small	675
Grass Rivulet	749
Grass-waved	969
Grayling	18
Great Brocaded Rustic	321
Greater Cream Wave	699
Greater Grass Emerald	671
Greater Swallow Prominent	108
Great Oak Beauty	944
Great Prominent	114
Green Arches	320
Green-brindled Crescent	557
Green Carpet, Autumn	775
Green Carpet, Spring	733
Green, Cream-bordered	594
Green Fritillary, Dark	42
Green Hairstreak	55
Green Lutestring, Frosted	133
Green Mottle, Brindled	565
Green Pug	886
Green Silver-lines	592
Green-veined White	7
Grey Arches	349
Grey Birch Beauty	950
Grey Chi	569
Grey Dagger	518
Grey Early	546
Grey Mottled Carpet	735
Grey Pine Carpet	779
Grey Pug	866
Grey Rustic	310
Grey Scalloped Bar	965
Grey Shoulder-knot	543
Grizzled Skipper	71
Ground Lackey	146

H.

Hairstreak, Brown	51
Hairstreak, Green	55
Hairstreak, Purple	52
Hairstreak, White-letter	54
Hawk, Bedstraw	91
Hawk, Broad-bordered Bee	98
Hawk, Convolvulus	83
Hawk, Death's-head	82
Hawk, Eyed	81
Hawk, Humming-bird	97
Hawk, Large Elephant	96
Hawk, Lime	79
Hawk, Narrow-bordered Bee	99
Hawk, Oleander	94

Hawk, Pine	87
Hawk, Poplar	80
Hawk, Privet	86
Hawk, Silver-striped	93
Hawk, Small Elephant	95
Hawk, Striped	92
Haworth's Crescent	481
Hazel Thorn, Scalloped	919
Heart and Club	280
Heart and Dart	285
Heath Beauty, Common	958
Heath, Large	25
Heath Lattice	964
Heath Rivulet	751
Heath Rustic	305
Heath, Small	24
Hebrew Character, Common	382
Hebrew Character, Setaceous	313
Hedge Gothic	377
Herald	651
High Brown Fritillary	40
Highflyer, July	784
Highflyer, May	785
Highflyer, Ruddy	786
Holly Blue	68
Hook, Silver	606
Hook-tip Barred	162
Hook-tip, Oak	161
Hook-tip, Pebble	163
Hook-tip, Scalloped	164
Hook-wing, Beautiful	666
Hornet Clearwing, Osier	224
Hornet Clearwing, Poplar	223
Horse-chestnut Longwing	952
Humming-bird Hawk	97

I.

Ingrailed Clay, Common	299
Iron Prominent	111

J.

July Highflyer	784
Juniper Carpet	783
Juniper Pug	880

K.

Kent Black Arches	168
Kidney, Double	499
Kidney, Olive	500
Kitten, Poplar	101
Kitten, Sallow	102
Knot-grass Dagger Dusky	523
Knot-grass Dagger, Light	520
Knot, True Lovers'	289

INDEX OF ENGLISH NAMES

L

Lace, Brussels	939
Lackey, Common	145
Lackey, Ground	146
Lady, Painted	31
Lappet, Common	156
Larch Pug	882
Large Angle-shades	473
Large Argent-and-sable	792
Large Autumnal Carpet	824
Large Chocolate-tip	122
Large Clouded Brindle	446
Large Copper	56
Large Ear	486
Large Elephant Hawk	96
Large Emerald	672
Large Footman	177
Large Garden White	5
Large Heath	25
Large Marbled Tort	595
Large Nutmeg	452
Large Pearl-bordered Fritillary	44
Large Phoenix	767
Large Ranuncule	568
Large Red-belted Clearwing	234
Large Rivulet	746
Large Seraphim	810
Large Skipper	76
Large Thorn	910
Large Tortoiseshell	36
Large Twin-spot Carpet	723
Large Wainscot	411
Lattice, Heath	964
Lead-belle, Common	816
Lead-coloured Drab	386
Lead-coloured Pug	846
Least Black Arches	169
Least Minor	468
Least Satin Lutestring	129
Least Yellow-underwing	330
Leopard, Wood	264
Lesser-bordered Yellow-underwing	329
Lesser Cream Wave	698
Lesser Emerald	679
Lesser Lutestring	131
Lesser-spotted Pinion	494
Lesser Swallow Prominent	109
Lesser Yellow-underwing	327
Light Arches, Common	441
Light Arches, Reddish	442
Light Brocade	357
Light Feathered Rustic	281
Light-green, Barred	899
Light Knot-grass Dagger	520
Light Marbled Brown	106
Light Orange-underwing	668
Light Shears	354
Light Spectacle	639
Lilac Thorn	918
Lime Hawk	79
Lime-speck Pug	855
Ling Pug	861
Little Emerald	680
Local Chalk Carpet	821
London Brindled-beauty	933
Longwing, Horse-chestnut	952
Looper, Rannoch	962
Looper, V	961
Lovers' Knot, True	289
Lunar Marbled Brown	107
Lunar-spotted Pinion	493
Lunar Thorn	916
Lunar Underwing	574
Lunar Yellow-underwing	328
Lutestring, Frosted Green	133
Lutestring, Least Satin	129
Lutestring, Lesser	131
Lutestring, Poplar	128
Lutestring, Yellow-horned	132
Lychnis Coronet	368
Lyme-grass Wainscot	417

M.

Magpie, Clouded	888
Magpie, Common	889
Maiden's Blush	687
Mallow Carpet	822
Manchester Treble-bar	805
Maple Mocha	684
Maple Pug	844
Map-winged Swift	268
Marbled Beau	506
Marbled Brown, Light	106
Marbled Brown, Lunar	107
Marbled Carpet, Common	776
Marbled Carpet, Dark	778
Marbled Clover	335
Marbled Coronet, Common	366
Marbled Minor	462
Marbled Pug	849
Marbled Tort, Large	595
Marbled White	27
Marbled, White-spot	603
March Usher	669
Marsh Buff	480
Marsh Carpet	743
Marsh Fritillary	49
Marsh Pug	853
May Highflyer	785
Mazarine Blue	67
Meadow Brown	22
Mere Wainscot	415
Merveille-du-jour, Common	559
Merveille-du-jour, Scarce	511
Middle-barred Minor	465
Miller	512
Minor, Cloaked	467

Minor, Least	468		Olive Kidney	500
Minor, Marbled	462		Orange, Bordered	334
Minor, Middle-barred	465		Orange Ear	490
Minor, Rosy	466		Orange Footman	185
Minor, Rufous	464		Orange Sallow	582
Minor Shoulder-knot	552		Orange Thorn	900
Minor, Tawny	463		Orange-tip White	9
Mocha, Birch	682		Orange-underwing, Common	667
Mocha, Maple	684		Orange-underwing, Light	668
Mother Shipton	615		Orange Upperwing	572
Mottle, Brindled Green	565		Osier Hornet Clearwing	224
Mottled Beauty	941			
Mottled Carpet, Grey	735		**P.**	
Mottled Pug	850			
Mottled Rustic	430		Painted Lady	31
Mottled Umber	908		Pale Brindled-beauty	929
Mottled Willow, Pale	435		Pale Clouded Yellow	11
Mottled Willow, Small	436		Pale Eggar	147
Mouse	503		Pale Mottled Willow	435
Mullein Shark	533		Pale Oak Beauty	945
Mullein Wave	692		Pale Prominent	120
Muslin Ermine	195		Pale Shining Arches	348
Muslin Footman	172		Pale-shouldered Brocade	359
			Pale Tussock	137
N.			Pale Wormwood Shark	531
			Peach Blossom	126
Narrow-barred Carpet	731		Peacock	33
Narrow-bordered Bee Hawk	99		Peacock Angle, Blunt	901
Narrow-bordered five-spot Burnet	217		Peacock Angle, Sharp	902
Narrow-bordered six-spot Burnet	218		Pearl-bordered Fritillary, Large	44
Narrow-winged Pug	872		Pearl-bordered Fritillary, Small	45
Netted Pug	854		Pearly Underwing	292
New Small Skipper	74		Pebble Hook-tip	163
Nonpareil, Clifden	608		Pebble Prominent	110
Northern Brown	20		Pepper-and-salt	935
Northern Drab	389		Phoenix, Large	767
Northern Spinach	769		Phoenix, Small	765
Northern Winter	829		Pimpinel Pug	870
November Carpet	826		Pine Beau	391
Nutmeg, Large	452		Pine Carpet, Grey	779
Nutmeg, Small	355		Pine Carpet, Reddish	782
Nut-tree Tuffet	617		Pine Hawk	87
			Pinion, Lesser-spotted	494
O.			Pinion, Lunar-spotted	493
			Pinion-Spotted Pug	851
Oak Beauty, Great	944		Pinion-streaked Snout	658
Oak Beauty, Pale	945		Pinion, Tawny	538
Oak Brindled-beauty	934		Pinion, White-spotted	495
Oak Eggar	150		Pink-barred Sallow	584
Oak Hook-tip	161		Plain Blackneck	644
Oak Thorn, Scalloped	921		Plain Golden Y	630
Oak-tree Pug	878		Plain Pug	840
Oblique Carpet	815		Plain Wave	716
Oblique-striped	760		Point, Straw	648
Obscure Wainscot	398		Poplar Dagger	514
Ochre, Brindled	567		Poplar Hawk	80
Ochreous Pug	869		Poplar Hornet Clearwing	223
Old-lady	505		Poplar Kitten	101
Oleander Hawk	94		Poplar Lutestring	128
			Portland Dart	290

INDEX OF ENGLISH NAMES

Powdered Dagger	525	Purple Clay	298
Powdered Quaker	390	Purple Emperor	28
Pretty Chalk Carpet	759	Purple Hairstreak	52
Privet Hawk	86	Purple Thorn	917
Prominent, Coxcomb	117	Puss	103
Prominent, Great	114		
Prominent, Greater Swallow	108	**Q.**	
Prominent, Iron	111		
Prominent, Lesser Swallow	109	Quaker, Common	385
Prominent, Lobster	104	Quaker, Powdered	390
Prominent, Pebble	110	Quaker, Red-line	575
Prominent, Scarce	118	Quaker, Small	384
Pug, Angle-barred	873	Quaker, Twin-spot	388
Pug, Ash	874	Quaker, Yellow-line	576
Pug, Bell-flower	865		
Pug, Bilberry	885		
Pug, Bleached	862	**R.**	
Pug, Bordered	868		
Pug, Brindled	877	Rannoch Looper	962
Pug, Cloaked	838	Ranuncule, Feathered	563
Pug, Common	864	Ranuncule, Large	568
Pug, Currant	863	Red Admiral	30
Pug, Dentated	837	Red, Barred	898
Pug, Double-striped	887	Red-belted Clearwing, Large	234
Pug, Dwarf	883	Red-belted Clearwing, Small	233
Pug, Foxglove	848	Red Chestnut Rustic	323
Pug, Goldenrod	876	Reddish Light Arches	442
Pug, Green	886	Reddish Pine Carpet	782
Pug, Grey	866	Red-green Carpet	774
Pug, Haworth's	845	Red-line Quaker	575
Pug, Juniper	880	Red-necked Footman	171
Pug, Larch	882	Red Sword-grass	545
Pug, Lead-coloured	846	Red-tipped Clearwing	235
Pug, Lime-speck	855	Red Twin-spot Carpet	726
Pug, Ling	861	Red Underwing	610
Pug, Maple	844	Reed Tussock	140
Pug, Marbled	849	Reed Wainscot	421
Pug, Marsh	853	Riband Wave	717
Pug, Mottled	850	Ringed Wave	712
Pug, Narrow-winged	872	Ringlet, Common	26
Pug, Netted	854	Ripplet, Blomer's	836
Pug, Oak-tree	878	Rivulet, Barred	750
Pug, Ochreous	869	Rivulet, Grass	749
Pug, Pimpinel	870	Rivulet, Heath	751
Pug, Pinion-spotted	851	Rivulet, Large	746
Pug, Plain	840	Rivulet, Small	747
Pug, Satyr	858	Rosy Ear	488
Pug, Scarce	871	Rosy Footman	174
Pug, Shaded	839	Rosy Minor	466
Pug, Slender	843	Rosy Wave	695
Pug, Tawny Speckled	867	Round-winged Footman	173
Pug, Thyme	842	Ruby Tiger	197
Pug, Toadflax	847	Ruddy Highflyer	786
Pug, Triple-spotted	856	Rufous Minor	464
Pug, V	884	Rufous Wainscot	427
Pug, Valerian	852	Rustic, Autumnal	309
Pug, White-spotted	859	Rustic, Black	555
Pug, Wormwood	860	Rustic, Common	456
Purple Bar Carpet	761	Rustic, Deep Brown	553
Purple Bars, Small	649	Rustic, Dotted	294

Rustic, Dusky Sallow	461	Shaded Broad-bar	818
Rustic, Flame	319	Shaded Pug	839
Rustic, Flounced	469	Shark, Chamomile	529
Rustic, Great Brocaded	321	Shark, Common	527
Rustic, Grey	310	Shark, Mullein	533
Rustic, Heath	305	Shark, Pale Wormwood	531
Rustic, Light Feathered	281	Shark, Starwort	528
Rustic, Mottled	430	Sharp Peacock Angle	902
Rustic, Red Chestnut	323	Shears, Glaucous	361
Rustic, Shoulder-knot	448	Shears, Light	354
Rustic, Six-striped	317	Shears, Tawny	371
Rustic, Smooth	432	Shell, Dingy	834
Rustic, Square-spot	318	Shell Scallop	791
Rustic, Union	450	Shell, Yellow	758
		Shining Arches, Pale	348
		Shipton, Mother	615
S.		Shore Wainscot	399
		Short-cloaked Black Arches	166
Sallow, Barred	583	Shoulder, Flame	304
Sallow, Centre-barred	581	Shoulder-knot, Grey	543
Sallow, Common	585	Shoulder-knot, Minor	552
Sallow, Dusky-lemon	586	Shoulder-knot, Rustic	448
Sallow Kitten	102	Shoulder-striped Carpet	738
Sallow, Orange	582	Shoulder-striped Wainscot	400
Sallow, Pink-barred	584	Shuttle-shaped Dart	282
Sallow Rustic, Dusky	461	Silky Wainscot	428
Saltern Ear	485	Silky Wave	704
Sand Dart	287	Silver Cloud	380
Sandy Carpet	748	Silver, Clouded	895
Satellite	571	Silver, Common Waved	897
Satin Beauty	940	Silver Eight	621
Satin Lutestring, Least	129	Silver-ground Carpet	729
Satin Wave	711	Silver Hook	606
Satin, White	142	Silver-lined Brown	963
Satyr Pug	858	Silver-lines, Green	592
Scallop, Brown	787	Silver-lines, Scarce	593
Scallop, Dark	788	Silver, Scorched	892
Scalloped Bar, Grey	965	Silver-striped Hawk	93
Scalloped Hazel Thorn	919	Silver-studded Blue	61
Scalloped Hook-tip	164	Silver V	632
Scalloped Oak Thorn	921	Silver-washed Fritillary	39
Scallop, Shell	791	Silver, White Waved	896
Scallop Wave, Small	720	Silvery Arches	347
Scarce Bordered Straw	341	Silver Y, Common	635
Scarce Footman	181	Silver Y, Scarce	636
Scarce Merveille-du-jour	511	Single-dotted Wave	707
Scarce Prominent	118	Six-spot Burnet, Narrow-bordered	218
Scarce Pug	871	Six-striped Rustic	317
Scarce Silver-lines	593	Skipper, Chequered	78
Scarce Silver Y	636	Skipper, Common Small	73
Scarce Tissue	790	Skipper, Dingy	72
Scarce Umber	906	Skipper, Grizzled	71
Scarce Vapourer	134	Skipper, Large	76
Scarlet Tiger	203	Skipper, New Small	74
Scorched Silver	892	Slender Brindle	455
Scorched-wing	922	Slender Pug	843
September Thorn	914	Slender Treble-bar	803
Seraphim, Large	810	Small Angle-shades	472
Seraphim, Small	811	Small Argent-and-sable	796
Setaceous Hebrew Character	313	Small Autumnal Carpet	825

INDEX OF ENGLISH NAMES

Small Blood-veined Wave	694	Straw Chevron, Barred	771
Small Blue	69	Straw, Dark Bordered	340
Small Brindled-beauty	930	Straw Point	648
Small Chocolate-tip	124	Straw Underwing	476
Small Clouded Brindle	449	Streaked Carpet	800
Small Copper	58	Streamer Carpet	739
Small Dotted Buff	478	Striped, Crescent	451
Small Dusty Wave	710	Striped Hawk	92
Small Eggar	149	Striped Twin-spot Carpet	734
Small Elephant Hawk	95	Striped Wainscot	397
Small Engrailed	947	Sulphur Thorn	923
Small Fanfoot	662	Suspected	564
Small Fan-footed Wave	719	Swallow Prominent, Greater	108
Small Garden White	6	Swallow Prominent, Lesser	109
Small Grass Emerald	675	Swallowtail, Common	1
Small Heath	24	Swallow-tailed Elder	928
Small Mottled Willow	436	Swift, Common	269
Small Nutmeg	355	Swift, Ghost	266
Small Pearl-bordered Fritillary	45	Swift, Golden	270
Small Phoenix	765	Swift, Map-winged	268
Small Purple Bars	649	Swift, Wood	267
Small Quaker	384	Sword-grass, Cloudy	544
Small Red-belted Clearwing	233	Sword-grass, Red	545
Small Rivulet	747	Sycamore Dagger	513
Small Scallop Wave	720	Sylvan Waved Carpet	833
Small Seraphim	811		
Small Skipper, Common	73		
Small Skipper, New	74	**T.**	
Small Square-spot	302		
Small Tortoiseshell	37	Tawny-barred Angle	903
Small Twin-spot Carpet	736	Tawny Minor	463
Small Wainscot	413	Tawny Pinion	538
Small Yellow Underwing	501	Tawny Shears	371
Smoky Wainscot	395	Tawny Speckled Pug	867
Smoky Wave	689	Thorn, August	911
Smooth Rustic	432	Thorn, Barred Umber	909
Snout, Beautiful	652	Thorn, Canary-shouldered	912
Snout, Common	653	Thorn, Dusky	913
Snout, Pinion-streaked	658	Thorn, Early	915
Southern Wainscot	396	Thorn, Feathered	920
Spangle, Gold	626	Thorn, Large	910
Speckled Pug, Tawny	867	Thorn, Lilac	918
Speckled Wood	17	Thorn, Lunar	916
Speckled Yellow	927	Thorn, Orange	900
Spectacle, Dark	638	Thorn, Purple	917
Spectacle, Light	639	Thorn, Scalloped Hazel	919
Spinach, Currant	770	Thorn, Scalloped Oak	921
Spinach, Dark	823	Thorn, September	914
Spot, Gold	627	Thorn, Sulphur	923
Spotted, White-pinion	894	Thyme Pug	842
Sprawler, Common	550	Tiger, Cream-spot	201
Spring Green Carpet	733	Tiger, Garden	200
Spring Umber	905	Tiger, Ruby	197
Square-spot Dart	276	Tiger, Scarlet	203
Square-spot, Double	315	Tiger, Wood	199
Square-spot Rustic	318	Tissue, Common	789
Square-spot, Small	302	Tissue, Scarce	790
Square-spotted Clay	316	Toadflax Pug	847
Starwort Shark	528	Tooth-striped, Barred	813
Stout Dart	295	Tooth-striped, Early	814

294 THE BUTTERFLIES AND LARGER MOTHS

Tort, Large Marbled	595
Tortoiseshell, Large	36
Tortoiseshell, Small	37
Treble-bar, Manchester	805
Treble-bar, Slender	803
Treble-line	429
Triple-lines, Clay	688
Triple-spotted Pug	856
True Lovers' Knot	289
Tuffet, Nut-tree	617
Turnip Dart	277
Tussock, Black-arched	144
Tussock, Dark	136
Tussock, Pale	137
Tussock, Reed	140
Twin-spot Carpet, Dark	725
Twin-spot Carpet, Large	723
Twin-spot Carpet, Red	726
Twin-spot Carpet, Small	736
Twin-spot Carpet, Striped	734
Twin-spot Quaker	388
Twin-spot Wainscot	424
Type, Gothic	324

U.

Umber Beauty, Waved	936
Umber, Early	904
Umber, Mottled	908
Umber, Scarce	906
Umber, Spring	905
Umber Thorn, Barred	909
Umber Waved Carpet	807
Uncertain	431
Underwing, Beautiful Yellow	342
Underwing, Blossom	383
Underwing, Copper	502
Underwing, Lunar	574
Underwing, Pearly	292
Underwing, Red	610
Underwing, Small Yellow	501
Underwing, Straw	476
Union Rustic	450
Upperwing, Orange	572
Usher, March	669

V.

Valerian Pug	852
Vapourer, Common	135
Vapourer, Scarce	134
Varied Coronet	367
Vestal	721
Viper's-bugloss Gothic	373
V Looper	961
V Pug	884
V Silver	632

W.

Wainscot, Brown-line	408
Wainscot, Brown-veined	425
Wainscot, Bulrush	423
Wainscot, Clay	407
Wainscot, Common	393
Wainscot, Concolorous	414
Wainscot, Delicate	404
Wainscot, Fen	419
Wainscot, Flame	392
Wainscot, Large	411
Wainscot, Lyme-grass	417
Wainscot, Mere	415
Wainscot, Obscure	398
Wainscot, Reed	421
Wainscot, Rufous	427
Wainscot, Shore	399
Wainscot, Shoulder-striped	400
Wainscot, Silky	428
Wainscot, Small	413
Wainscot, Smoky	395
Wainscot, Southern	396
Wainscot, Striped	397
Wainscot, Twin-spot	424
Wainscot, Webb's	422
Wall Brown	16
Water Carpet	762
Water Ermine	193
Waved Carpet, Sylvan	833
Waved Carpet, Umber	807
Waved Carpet, White	830
Waved Carpet, Yellow	832
Waved Silver, Common	897
Waved Silver, White	896
Waved Umber Beauty	936
Wave, Dwarf Cream	702
Wave, Greater Cream	699
Wave, Lesser Cream	698
Wave, Plain	716
Wave, Riband	717
Wave, Ringed	712
Wave, Rosy	695
Wave, Satin	711
Wave, Silky	704
Wave, Single-dotted	707
Wave, Small Blood-veined	694
Wave, Small Dusty	710
Wave, Small Fan-footed	719
Wave, Small Scallop	720
Wave, Smoky	689
Webb's Wainscot	422
White Admiral	29
White, Bath	8
White Beauty, Bordered	959
White, Black-veined	4
White Colon	374
White Dot	346
White Ermine	192
White, Green-veined	7

INDEX OF ENGLISH NAMES

		Y.	
White, Large Garden	5		
White letter Hairstreak	54		
White-line Dart	274	Y, Beautiful Golden	631
White, Marbled	27	Y, Common Silver	635
White-marked	322	Yellow, Barred	772
White, Orange-tip	9	Yellow Belle	968
White-pinion Spotted	894	Yellow, Brindle-barred	812
White Satin	142	Yellow, Common Clouded	13
White, Small Garden	6	Yellow-horned Lutestring	132
White-spot Marbled	603	Yellow-legged Clearwing	232
White-spotted Pinion	495	Yellow-line Quaker	576
White-spotted Pug	859	Yellow, Pale Clouded	11
White Waved Carpet	830	Yellow Shell	758
White Waved Silver	896	Yellow, Speckled	927
White, Wood	10	Yellow Underwing, Beautiful	342
Willow Beauty	938	Yellow-underwing,	
Willow, Pale Mottled	435	Broad Bordered	332
Willow, Small Mottled	436	Yellow-underwing, Common	331
Winter, Common	828	Yellow-underwing, Least	330
Winter, Northern	829	Yellow-underwing, Lesser	327
Wood Carpet	794	Yellow-underwing,	
Wood, Speckled	17	Lesser-bordered	329
Wood Swift	267	Yellow-underwing, Lunar	328
Wood Tiger	199	Yellow Underwing, Small	501
Wood White	10	Yellow Waved Carpet	832
Wormwood Pug	860	Y, Plain Golden	630
Wormwood Shark, Pale	531	Y, Scarce Silver	636

CHAPTER 11.

A CODE FOR INSECT COLLECTING

JOINT COMMITTEE FOR THE CONSERVATION OF BRITISH INSECTS

This Committee believes that with the ever-increasing loss of habitats resulting from forestry, agriculture, and industrial, urban and recreational development, the point has been reached where a code for collecting should be considered in the interests of conservation of the British insect fauna, particularly macrolepidoptera. The Committee considers that in many areas this loss has gone so far that collecting, which at one time would have had a trivial effect, could now affect the survival in them of one or more species if continued without restraint.

The Committee also believes that by subscribing to a code of collecting, entomologists will show themselves to be a concerned and responsible body of naturalists who have a positive contribution to make to the cause of conservation. It asks all entomologists to accept the following Code in principle and to try to observe it in practice.

1. COLLECTING – GENERAL

1.1 No more specimens than are strictly required for any purpose should be killed.

1.2 Readily identified insects should not be killed if the object is to 'look them over' for aberrations or other purposes: insects should be examined while alive and then released where they were captured.

1.3 The same species should not be taken in numbers year after year from the same locality.

1.4 Supposed or actual predators and parasites of insects should not be destroyed.

1.5 When collecting leaf-mines, galls and seed heads never collect all that can be found; leave as many as possible to allow the population to recover.

THE BUTTERFLIES AND LARGER MOTHS

1.6 Consideration should be given to photography as an alternative to collecting, particularly in the case of butterflies.

1.7 Specimens for exchange, or disposal to other collectors, should be taken sparingly or not at all.

1.8 For commercial purposes insects should be either bred or obtained from old collections. Insect specimens should not be used for the manufacture of 'jewellery'.

2. COLLECTING – RARE AND ENDANGERED SPECIES

2.1 Specimens of macrolepidoptera listed by this Committee (and published in the entomological journals) should be collected with the greatest restraint. As a guide, the Committee suggests that a pair of specimens is sufficient, but that those species in the greatest danger should not be collected at all. The list may be amended from time to time if this proves to be necessary.

2.2 Specimens of distinct local forms of Macrolepidoptera, particularly butterflies, should likewise be collected with restraint.

2.3 Collectors should attempt to break new ground rather than collect a local or rare species from a well-known and perhaps over-worked locality.

2.4 Previously unknown localities for rare species should be brought to the attention of this Committee, which undertakes to inform other organisations as appropriate and only in the interests of conservation.

3. COLLECTING – LIGHTS AND LIGHT-TRAPS

3.1 The 'catch' at light, particularly in a trap, should not be killed casually for subsequent examination.

3.2 Live trapping, for instance in traps filled with egg-tray material, is the preferred method of collecting. Anaesthetics are harmful and should not be used.

3.3 After examination of the catch the insects should be kept in cool, shady conditions and released away from the trap site at dusk. If this is not possible the insects should be released in long grass or other cover and not on lawns or bare surfaces.

3.4 Unwanted insects should not be fed to fish or insectivorous birds and mammals.

3.5 If a trap used for scientific purposes is found to be catching rare or local species unnecessarily it should be re-sited.

3.6 Traps and lights should be sited with care so as not to annoy neighbours or cause confusion.

4. COLLECTING – PERMISSION AND CONDITIONS

4.1 Always seek permission from landowner or occupier when collecting on private land.

CODE OF INSECT COLLECTING

4.2 Always comply with any conditions laid down by the granting of permission to collect.
4.2 Always comply with any conditions laid down by the granting of permission to collect.
4.3 When collecting on nature reserves, or sites of known interest to conservationists, supply a list of species collected to the appropriate authority.
4.4 When collecting on nature reserves it is particularly important to observe the code suggested in section 5.

5. COLLECTING – DAMAGE TO THE ENVIRONMENT

5.1 Do as little damage to the environment as possible. Remember the interests of other naturalists; be careful of nesting birds and vegetation, particularly rare plants.
5.2 When 'beating' for lepidopterous larvae or other insects never thrash trees and bushes so that foliage and twigs are removed. A sharp jarring of branches is both less damaging and more effective.
5.3 Coleopterists and others working dead timber should replace removed bark and worked material to the best of their ability. Not all the dead wood in a locality should be worked.
5.4 Overturned stones and logs should be replaced in their original positions.
5.5 Water weed and moss which has been worked for insects should be replaced in its appropriate habitat. Plant material in litter heaps should be replaced and not scattered about.
5.6 Twigs, small branches and foliage required as foodplants or because they are galled, e.g. by clearwings, should be removed neatly with secateurs or scissors and not broken off.
5.7 'Sugar' should not be applied so that it renders tree-trunks and other vegetation unnecessarily unsightly.
5.8 Exercise particular care when working for rare species, e.g. by searching for larvae rather than beating for them.
5.9 Remember the Country Code!

6. BREEDING

6.1 Breeding from a fertilised female or pairing in captivity is preferable to taking a series of specimens in the field.
6.2 Never collect more larvae or other livestock than can be supported by the available supply of foodplant.
6.3 Unwanted insects that have been reared should be released in the original locality, not just anywhere.
6.4 Before attempting to establish new populations or 'reinforce' existing ones please consult this Committee.

NOTES

NOTES

NOTES

NOTES

NOTES